in SEARCH of FALLING STARS

in SEARCH of FALLING STARS

H.H. Nininger's Classic *Find a Falling Star* Revisited & Expanded

Harvey H. Nininger
Edited by Jim Banks

BookPress®
publishing

Copyright © 2024 by Jim Banks. All rights reserved.

No portion of this book may be reproduced, stored in a retrieval system, or transmitted in any form or by any means—electronic, mechanical, photocopy, recording, scanning, or other—except for brief quotations in critical reviews or articles, without prior written permission of the publisher. Requests to the publisher for permission or information should be submitted via email at info@bookpresspublishing.com.

Any requests or questions for the editor should be submitted to him directly at jim@insearchoffallingstars.com.

Published in Des Moines, Iowa, by:

Bookpress Publishing
P.O. Box 71532
Des Moines, IA 50325
www.BookpressPublishing.com

Publisher's Cataloging-in-Publication Data

Names: Nininger, Harvey Harlow, 1887-1986, author. | Banks, Jim, 1957-, editor.
Title: In search of falling stars : H.H. Nininger's classic find a falling star , revisited & expanded / H. H. Nininger; edited by Jim Banks.
Description: Includes bibliographical references and index. | Des Moines, IA: Bookpress Publishing, 2024.
Identifiers: LCCN: 2024915177 | ISBN: 978-1-960259-01-1
Subjects: LCSH Nininger, Harvey Harlow, 1887-1986. | Meteorites. | BISAC BIOGRAPHY & AUTOBIOGRAPHY / Science & Technology | SCIENCE / Space Science / Astronomy | SCIENCE / Research & Methodology | SCIENCE / History | SCIENCE / Earth Sciences / Mineralogy
Classification: LCC QB755 .N49 2024 | DDC 523.5/1--dc23

First Edition

Printed in the United States of America

10 9 8 7 6 5 4 3 2 1

CONTENTS

About This Book vii

Foreword xi

1   Open Fields 1
2   Paths 24
3   Stopping Places 46
4   The Challenge 80
5   Ways and Half-Ways 98
6   Finding a Way 152
7   The Path Behind 187
8   Peaks and Valleys 198
9   On Various Trails 229
10 Not All Knowledge is in Books 259
11 Not All the Big Ones Get Away 304
12 Prospects and Prospectors 338
13 Plateau 365
14 A New Lease 379
15 Discovery 429
16 Home 450
17 Slowing Down 483
18 New Paths and Byways 533

Glossary 553

List of Photographs 556

Original Dedication: *Find a Falling Star* 562

Original Preface: *Find a Falling Star* 563

Original Introduction: *Find a Falling Star* 566

Index 570

About This Book
A Note From the Editor

Find a Falling Star, my grandfather H. H. Nininger's book about his life working with meteorites, is well-known among both meteorite scientists and aficionados. What is not so well-known, however, is that the book as published in 1972 is only part of the story of his life as he wished it to be told.

The first draft of his autobiography, totaling more than 1,500 double-spaced, typewritten pages, was completed in 1965. He continued to work on subsequent drafts for years, through the late 1960s and into the early 1970s. All three Nininger offspring, son Robert Nininger and daughters Doris Nininger Banks and Margaret Nininger Huss, were involved in the project. My mother, Doris Nininger Banks, served as the editor. After years of revisions and editing, the manuscript had been reduced to just under 900 pages.

Various versions of the manuscript were shopped around to various publishers under the working title, *Follow Thy Star*. The consensus of most was that the book was just "too long" to be feasible.

In 1972, in a deal facilitated by Herbert Fales, H. H. Nininger's long-time friend and supporter, a small, independent publisher was selected to publish the book. An "outside" editor was hired by the publisher, and the manuscript was cut quickly, and in some cases a little haphazardly, by nearly half. The revised book was to focus only on meteorites, leaving out my grandfather's backstory and early adventures. The cuts also eliminated some of his earliest experiences trying to make a career out of studying meteorites. A new title was chosen by the family: *Find a Falling Star*.

It was my grandfather's hope to publish the "rest of the story" as a separate volume, but this never materialized. In 1982, my aunt, Margaret Nininger Huss, put together many of the excerpts in a separate, unpublished volume that she titled, *It Wasn't Always Meteorites*, and distributed it to the family. The fact that the entire book was never published seemed to be somewhat of a thorn in the family's side, but it was considered a "done deal," and that was that. Now, for the first time, the complete autobiography of H. H. Nininger has been re-assembled.

In Search of Falling Stars is my grandfather's story as it was originally intended to be told. It has been compiled from multiple original manuscripts (including handwritten comments by my grandfather, my mother, and others) and boxes of notes and correspondence. I cut some non-critical material from the unpublished chapters, added back some salient parts omitted from the published chapters, and did my best to make sure to include all the portions I believed my grandfather considered most important.

This book includes all of the material published in the original version of *Find a Falling Star*, with some minor revisions and editing, plus several substantive additions to the previously published chapters. The book also includes four completely new chapters. Many new and seldom-seen photographs have been added to round out the story.

My grandfather often talked about living a life that quite literally reached from the the time of the horse-and-buggy to the space age. His life story covers this same span of time. Despite the fact that his autobiography was written more than a half century ago, it holds up remarkably well and illustrates his forward-thinking approach to life.

Jim Banks
Sedona, Arizona

Foreword

Fifty-two years ago, the autobiography of my grandfather, Dr. Harvey H. Nininger, was published. *Find a Falling Star* was a significantly shortened version of the autobiography that Dr. Nininger had hoped to publish. It left out the description of his early life and the experiences that formed his character and drove his interest in science. It was not a given that a man who had grown up as Harvey Nininger had would end up as a scientist. But even in his early life, he exhibited the inquisitiveness, persistence, and strength of character that led him to become "the father of modern meteoritics."

Nininger was always interested in the natural world. Even though he grew up in a household with only two books, the Montgomery Ward Catalog and the Bible, where science was considered the work of the devil, Nininger taught himself about nature by observation and experimentation. His early school years were spent in primitive country schools in the farming communities where he lived. He "graduated" from common school—equivalent to passing the eighth grade—at age 19. His thirst for knowledge led him to find a way to go to the state Normal School (part high school and part college) and then to earn a college degree in biology. After a variety of teaching and work experiences, Nininger became a biology professor at McPherson College in Kansas in the fall of 1920. His enthusiasm for learning made a tremendous impression on his students, many of whom became life-long friends.

In August of 1923, Nininger was introduced to meteorites through an article in *The Scientific Monthly*. The article caught his imagination. The old saying, "When the student is ready, the teacher appears," proved true. Three months later, a large fireball passed over McPherson, Kansas, while Nininger was walking home. He resolved to find the meteorite that had made the fireball. Although he was unsuccessful, he did recover two other meteorites, and this set into motion the work that would occupy the rest of his life.

Nininger was the first person to make a career out of finding meteorites. When he started working on them, few thought that they were more than a curiosity, and no one believed that one could deliberately go out and find them. Despite this negative culture, Nininger believed that meteorites were important and could be found. He created a methodology that produced specimens from 226 meteorite falls (including eight witnessed falls) previously unknown to science. By the 1940s, he was finding half of the meteorites recovered worldwide. The key to this success was enlisting the aid of the people who lived on the land. He purchased the meteorites they found, and he honored the property rights of the finders and the landowners.

Nininger's enthusiasm for meteorites was contagious. His school lectures about meteorites held children spellbound, and sometimes a student would even help him locate one. He would walk into a bank or another place where people gathered and set a meteorite on a table or counter where people could see it. When someone asked what it was, he would tell them about meteorites, drawing everyone within earshot to hear the story.

Some of my earliest memories are of the Nininger Meteorite Museum in Sedona, Arizona. My parents, Margaret Nininger Huss and Glenn Huss, moved to Sedona shortly after I was born to help Nininger run the museum. From those early days through the early years of my graduate-school career, my grandfather was there to discuss meteorites and the natural world in general with me. This immersive atmosphere infected me with the "meteorite bug," and I never outgrew it.

Nininger was not just interested in collecting meteorites; he was interested in everything about them. He studied the whole meteorite—its size, shape, surface features, internal structure, mineralogy, etc. He could recognize essentially all of the known meteorites by sight. Although Nininger did not have the means to do mineral or isotope chemistry, he would often send out new meteorites for chemical analysis.

Throughout Nininger's career, more scientists became interested in meteorites, but the questions that occupied them were not the ones Nininger thought most interesting. Nininger wondered: How much of an impact record survives on Earth's surface? What is the role of meteorite impact in producing the Earth that we live on? And what was the role of meteorite impact on the evolution of life? How effectively do falling meteorites sample the asteroid belt? How many kinds of meteorites do we fail to recognize because they do not match our preconceived ideas of what meteorites should look like? How long does it take a meteorite to be altered beyond recognition on earth? Are there any fossil meteorites, and if so, what would they look like? I remember distinctly at the 1988 Meteoritical Society Meeting in Fayetteville, Arkansas, sitting in the audience of a presentation and suddenly realizing that all of the questions Nininger had been discussing when I was growing up were now part of the scientific program of the Meteoritical Society.

In Search of Falling Stars is the autobiography that Dr. Nininger hoped to publish. It describes in first-person detail the events and influences that shaped his life, how he got his education and became a scientist, and how he learned about meteorites. Dr. Nininger became the "father of modern meteoritics" because of a unique combination of his upbringing, the people who influenced his early life, the choices he made, his thirst for knowledge, an adventurous spirit, persistence, and a rare ability to synthesize information obtained from personal observations. Early in his career, the head curator of geology at the Smithsonian Institution, Dr. George Merrill, told him during a visit to the Smithsonian, "Young man, if we gave you all the money your program required and you spent the rest of your life doing what you propose, you might find one meteorite." Dr. Nininger lived 99 years and one month, found multitudes more than the one meteorite predicted, and established the foundation from which the modern science of meteoritics developed. This autobiography describes how it all happened.

Gary R. Huss
Research Professor and Director
W. M. Keck Cosmochemistry Laboratory
University of Hawai'i at Mānoa

1

Open Fields

*If thou follow thy star,
thou canst not fail
of glorious haven.*

—Dante, *Inferno*

My brothers and father nicknamed me "the squirrel dog." When bad weather prevented work in the fields, we often went squirrel hunting. Our old dog was a good squirrel hunter, but many times I saw the squirrels before he smelled them.

My eyes were always busy. I knew where all the birds were nesting, where the cottontail went to nurse her babies, which trees the 'possums were holed up in, where a skunk family dwelled under a big rock overhanging the creek. In the spring I would spy the first hummingbird nest and report the first covey of newly hatched quail. I watched the footprints of all wild things in our forest and knew at once when the ducks visited our tiny pond.

There were no books on natural history in our home, our neighbors' homes, nor in our school. Whatever we learned of nature, we learned by using our eyes.

I was born January 17, 1887, the third of four sons, in a one-room house four-and-a-half miles southwest of Conway Springs, Kansas. Our house was steep-roofed, weather-boarded, middle-aged, and unpainted, with a single door. One of my early memories is the enlargement of our living quarters in preparation for the fourth child, my brother Roy, who was born in September 1889.

A cookstove near the middle of the original one room served also as heater. The enlargement provided an adjoining kitchen, the walls formed of one-by-twelve-inch boards set vertically. A second addition was made by bringing over the two-room Landis house from its hill a quarter mile to the north of us.

These building activities loom in my memory. Fascinated, I watched Father as he framed and enclosed the new kitchen. The addition of the Landis house was even more exciting. I recall standing at the north window, my nose pressed to the glass, watching the amazing sight of a house slowly creeping toward us between rows of catalpa trees. The moving must have taken more than a day.

I remember the day of Roy's birth very clearly. When I returned from wherever I had been sent, Mother invited me to view my little brother. She laid back the bed covers and smiled. I do not recall the sight of him impressing me greatly, but the picture of Mother is very clear in my memory.

My other brothers were Jake and John, five and three years older than I. Our only sister, Naomi, arrived some years later.

We had a little vineyard and a peach orchard. A row of cottonwoods marked the west edge of our small pasture. Our fields were planted to corn.

I was always curious, but sometimes there were painful consequences to my curiosity. I couldn't resist trapping a bumble bee by closing a squash blossom over him as he sipped nectar at the bottom of the large funnel-shaped corolla. I had been warned—but how could the bee know where to sting me if my fingers were outside the closed blossom? Apparently he knew where. Still worse pain came when I chased and caught a big red velvet ant. But how else was I to learn that "ants" could sting? The biological fact that this furry creature was not an ant but a mutillid would not have lessened the hurt of her sting even if I had known it.

Our days were occupied by the ordinary childhood activities of small boys on the frontier farms of the early 1890s. There was a small pond a short distance east of the house where I whiled away many hours catching frogs, crawdads, tadpoles, turtles, and other crawling or swimming creatures of the water's edge while my older brothers fished or swam. The fish in our pond were merely carp and catfish, but they were abundant and frequently graced our table.

The four of us boys, tempted to fun in any interlude of our work/church/school days, were the torment of all living things from the kitchen to the barn. Our faces bore grime along with freckles, our trousers would bag and tear, our pockets carried bulging loads of spools, nuts, bugs, string—anything we could gather about the place. We quarreled, emptied the cookie jar, raided the tree of green apples. We'd sneak off to the creek after church, snagging our Sunday best on barbed wire, then pay for our transgressions.

We made our own entertainment and most of our toys. When my two older brothers began to help in the fields at about the age of nine, and while Roy was too young, I often was out in the fields alone. Two activities bring vivid memories: playing with tumbleweeds and catching grasshoppers. As a youngster of five or seven years, I had great fun chasing tumbleweeds on windy days, tying several together with three or four feet of binder twine between each, then watching and following as they raced before the wind.

I often caught grasshoppers and would come back with a pocketful. To me, that aggregate of grasshoppers was a treasure. They were not mere grasshoppers. There were the red wings, the blue wings with yellow borders, the clear wings, the yellow wings with gray borders and the black wings with yellow borders. The last-named was a species which twenty years later would precipitate an argument between myself and a professor.

***

Our farm was about twenty-five miles north of the Oklahoma line. The day before the government opened the Cherokee Strip, several men came into our yard for water and feed for their horses on their way to the boundary to take part in the land rush. Others passed by in one-horse carts and buggies, in spring wagons, or even on great, ungainly high-wheeled bicycles. On the day of the opening, September 16, 1893, as we sat down to our noon meal, Father reminded us that at twelve o'clock the cannon signal would be fired; we all listened, though I don't recall that we actually heard it.

Evidently our Kansas years were not propitious. I have a very clear memory of the three-day dust and sand storm that proved to be the last straw and caused Father to decide to leave the state in 1895. For three days and nights, we could see neither sun, moon nor stars and could scarcely see to move between house and barn. We hung quilts and blankets over the windows and kept a lamp burning in the house during the day. When the storm was over, much of our corn had disappeared completely, and much of the soil as well.

Somehow Father made a sale or trade, and that fall, we moved to a farm near Warrensburg, Missouri. More memories cluster around the four years we lived in Missouri than surround any equal time of my boyhood. From the plains where scarcely a rock lay and where the only trees had been planted by man, we moved into the edge of a forest near a creek in the foothills of the Ozarks, where coal was mined from shale between beds of limestone in one-man operations within a few hundred yards of our house.

There is something about a forest that stimulates a boy's imagination, and a creek, however small, is a never-ending source of interest. The house we moved into had been built of logs before the Civil War. A few years before we bought it, the logs, many of them black walnut, had been covered with siding. There were eight rooms—no plumbing of course—and to us, it seemed a veritable mansion. Church was eight miles away and town, nine miles. School was a mile and three-quarters. Apart from footwork, our mode of transportation was by lumber wagon. A trip to town or church was a whole day's project. Roads were terrible. The unpaved streets of Warrensburg were a place to stay away from in wet weather. The roads would be churned into a perfect loblolly by the hauling of coal from the little one-man mines at Bristle Ridge. Wagons occasionally were abandoned hub-deep right in Main Street.

In those last years of the century, Missouri was still on the frontier. Each man owned a piece of land or hoped to own one. He raised the food his family needed with very few exceptions. The other things he required, he bought from the man who made them. Each town had its harness-maker and a shoemaker. There was a grist mill where grain each farmer grew was ground for toll—no need for cash. Each farm was a self-sufficient unit except for a few pieces of simple machinery—and many of those were made in the local blacksmith shop. Axe handles were hewn by us from young ash saplings.

I have no recollection of ever seeing my mother open tin cans from the store during my boyhood. The vegetable cellar held a winter's supply of potatoes, squash, cabbage, turnips, carrots, apples, sauer-kraut, and so on. Fruit cellar shelves were lined with fruits in glass jars and vegetables of almost every kind. Hams, shoulders, side bacon and dried beef hung in the smoke house. Mincemeat was preserved in glass jars. Sausage was "fried down," packed in lard or stuffed in entrail casings which we cleaned at butchering time. Souse and head cheese were preserved in glass jars or packed into flour-sack casings. Molasses was made from sorghum grown on our farm.

We produced our own butter, milk, eggs, and feed for the livestock. Mother dried fruit and sweet corn for winter. She ground and sometimes roasted our coffee. She baked our bread. I never saw bakery pies, cakes, or cookies until I was grown. Soda crackers were the only baked item carried by the general store. I tasted my first soda water about the time I got my first store clothes at the age of ten or eleven.

A boy was expected to work on the farm. My ninth birthday came in the January soon after our move. Hay was our principal crop, and the next summer, I was made a hand in the hayfield. I could run the rake and load. Because I was small and could work in close quarters, it fell to me to arrange the hay in the mow of the barn after it was so nearly filled that a larger person could not work. It seemed an oven as I sweated and struggled with heavy forkfuls right next to the roof.

We were watched over by strict parents, a sober church, and a stern conscience. We belonged to the German Baptist Brethren, at that time and place a fundamentalist sect, a relic of Calvinism, better known as the Dunkards. I was taught from childhood that anyone who did not live according to the teachings of the Dunkard church would suffer in hell through all eternity. Our acceptance of this dogma was literal and genuine.

In actuality, our little church was not a fair example of the denominational thinking. We were on the frontier, remote from any enlightening influence. Even my parents managed a considerable mental metamorphosis during their later years. Probably the church of my childhood was narrow and primitive in the same way that our schools, mired in the backwash of isolation, failed to reflect the modernization and broadening of outlook taking place elsewhere in the nation.

There can be little quarrel with faith in temperance, the simple life, brotherhood, and obedience to Christ, yet the church of my younger days stressed the outward manifestations of these virtues beyond the point of moderation. The laudable democratic features of the Church of the Brethren—local autonomy with actions and decisions governed by elected representatives and guided by New Testament teachings—became in that climate the very source of iron-clad rule. During my childhood and young adult days, the Church was the ultimate arbiter of thought and behavior.

The Dunkards were marked by certain peculiarities of dress which were important in the life of the sect. I am very sure that until I had grown up and left home, the question of dress was considered by me to be the most important single item in the living of a Christian life. The men wore collarless coats and no neckties. The women wore prayer coverings and bonnets instead of hats. All jewelry was forbidden (my wife wore no wedding ring until we moved to Denver, Colorado, in the mid-1930s).

Musical instruments were not allowed in the church building. There was some disagreement as to whether they should be allowed in the home. I was about fourteen years old before my father finally consented to bringing an organ into our house. It was stipulated at first that nothing but hymns should be played on it, though gradually folk songs were permitted. The violin was positively forbidden as "an instrument of the devil."

I once heard my mother severely rebuked for asking a question which seemed to indicate doubt as to the authority of the Apostle Paul.

Father made his answer very positive and frightening: "You must never raise such a question. You are supposed to accept it on faith."

I remember a discussion among my father and other local church leaders in which all agreed that every word of the Bible was inspired and infallible. The question was then raised as to whether there were other writings which also were inspired, and if not, why not? The answer reached was that the minutes of the Church's "Annual Meeting" were inspired and were also infallible, so long as they did not conflict with the Bible.

***

Our school in Missouri was abominable. Classes were ungraded. Discipline was nonexistent. Grown young men attended with no other objective than to demonstrate their power to thwart the teacher. They boasted that no woman could manage the school and that the present incumbent would be forced out as had others before her. Pupils could enroll and attend when and as they chose, and they might repeat the same level year after year.

The small one-room schoolhouse held twenty-two desks, each intended to accommodate two children. It was forced to admit sixty-three pupils ranging from six to twenty-one years of age. Water supply? A pump. Plumbing? Two filthy outhouses, subject to frequent upsetting by pranksters. Heat? A pot-bellied stove. Ventilation? Cracks in walls, floors, under the door, and broken windows. Boys and girls never graduated from that school. They just quit after reaching the age when Father or Mother could no longer afford to do without their help at home.

A minister who preached in the schoolhouse one Sunday asked me the grade I was in. I was eleven years old, and my mother had to come to my rescue to explain that I was in the "Fourth Reader." Thus catching on to his meaning, I added that we were studying long division and spelling.

It would be my guess that we brothers were no more advanced in school learning when we left Missouri after four years than we had been when we left our school in Kansas.

2

In December, 1899, we moved to Oklahoma. My two older brothers drove a covered wagon, one of a train of nine moving west. Mother, my younger brother, my baby sister, and I went by train. Father accompanied our household goods and livestock in a railroad car. Life in Oklahoma marked changes in our lives. The country was new—Oklahoma wouldn't achieve statehood for seven more years. The part of the Territory we moved into had been Indian land until just a few years before, and wagon caravans of Indians still passed within a half-mile of our house on their treks between reservations.

Our land was mostly in timber, much of which we cleared during our first few years on the place. The wood was saleable either as fuel or logs for building. All the work had to be done by hand. Stumps had to be grubbed out. The ground was plowed as soon as the trees were gone.

Our principal crop was cotton, entirely new to us. Cotton farming was hard work, but we boys loved it because Father announced that he would pay us for picking it. Pay for work was new to us, also.

Previously the only money I ever had was a few dollars that passed through my hands on its way to various church missions. In Missouri, our parents would give each of us a little plat on which to raise onions, potatoes, or some other garden vegetable—but all the money earned therefrom was to go to missions. Once I was given a setting of eggs and was lent a hen to hatch and raise the brood. I have never forgotten the grand price I received for the fryers—seven cents a pound—and this, too, went to missions. Father had offered each of us a dollar if and when we would read the Bible through cover to cover. I earned that dollar and was induced to spend it on a Bible of my own. We would gather wild nuts—walnuts, hickory nuts, and hazel nuts—some of which we sold for fifty cents a bushel. Again, for missions.

Now we were to have money of our own. The regular wage paid for cotton picking was seventy-five cents per a hundred pounds. Father said he would pay us fifty cents a hundred. We were to buy our own clothes, but whatever was left, we could save or spend, though the latter, of course, was not advised. At fifty cents a hundred, one could earn seventy-five cents a day. If he were an exceptional picker and the crop was good, he could earn a dollar in a day. Not many men nor boys of the neighborhood could average 200 pounds even in good cotton, but my brother John and I could do it, and it was not long before I opened a bank account with $25. I've never felt as rich since.

In those times, a man's wages might be seventy-five cents a day. A hired hand worked on the farm for $15 or $20 a month. A neighbor woman who went about the various homes to wash and iron received fifty cents for her day's work.

Mother had always made most of my clothes, but in Oklahoma, I could go and buy myself a suit or pair of shoes, with my parents along, of course. When I was about twelve years old, my first suit of store-bought clothes cost $4. Whatever the restrictions, that feeling of independent ownership did wonders for all of us boys.

We raised other crops—corn, wheat, oats, hay. We also cleared off timber land, sawed and chopped wood, built fences, milked cows, fed hogs, plowed, harrowed, and performed scores of other chores. All work other than picking cotton was without pay. Our work day was supposed to be from sunup to sundown, but there were always a couple of hours of chores to be done after the day's work. If we were through by nine o'clock, that was considered good. We were ready for bed. I have no recollection of sitting around evenings after supper, except for evening worship while one member of the family read a Bible chapter and we all knelt for prayer. Sometimes this ceremony was observed in the morning.

At times when work was not too heavy, we would go on a night hunt for 'possums and 'coons. After rains, when the ground was too wet to work in the fields, we might—unless we had wood to chop, fence to build, or stables to clean—go squirrel hunting. Once or twice a year, we would take a half-day off and fish in the Cimarron River, three miles away.

In spite of the almost constant grind of hard work, I remember it as a good life. Except for hoeing and picking cotton, we did our field work with horses, and I liked horses. I liked all kinds of livestock and I loved nature. Birds, insects, and animals of all kinds interested me. It is difficult now to reconcile our killing of game with the liking I had for living things, but hunting was an esteemed pastime in the community. It was a favorite subject for conversation among the grownups, and their tales of exploits stimulated us young chaps.

***

A frontier-type rural life measures its days and years by events directly tied to nature. One year's crops are good or they are bad, the rains come too heavily or they fail to come, the muds were deepest or the dust was darkest in such-and-such a season, one night's hunting or an afternoon's fishing would be marked for a generation by a particular memory of "the time when."

Our food came from the soil; our well-being was dependent upon it. Our time was swallowed up by it. Our games were rooted in the things of nature. The wild creatures that lived near our farm were directly involved in our lives. They destroyed our chickens or they provided quarry for our hunt.

On a morning frosty with autumn, for some forgotten reason, I was school bound alone, dinner pail in hand. My path lay through an unspoiled stand of oak, hickory, and ash, where golden sprays of sunlight patterned the leafy carpet and no breath of wind stirred the tree tops. A woodpecker flew, chattering to join another hammering a resonant rhythm on a weather-beaten trunk. A squirrel rippled along the forest floor with an acorn for his cache. I stopped in my tracks, caught in enchantment. Only an occasional leaf fell with a subdued poetic rustle. Then an acorn dropped, and as I looked up searching for the bushy-tailed harvester who had loosed it, a half dozen ripened leaves followed the nut in zig-zag descent. A feeble gust of wind stirred the treetops into mild activity and the discharge of their summer garb, and suddenly the whole forest was alive with leaves lightly falling like giant varicolored snowflakes.

Then the sounds of country life invaded my moment of silence and solitude. Cattle lowed, a dog barked, a rooster gave forth at the top of his voice, a turkey gobbled, and morning class waited, still a mile's walk away.

***

Looking back, my years in Oklahoma were happy ones. In 1945, having business in Oklahoma City and Guthrie, I detoured via Pleasant Valley, crossed the Cimarron via the narrow, rickety bridge built in pre-automobile days and still resting on the oak pilings from our farm which I had helped cut, and turned eastward. Four-and-a-half miles from Coyle, in central Oklahoma, I found myself in our old village of Clarkson, empty now except of remembrance.

The mere sight of a few old landmarks occasioned a flood of long-forgotten names and memories. I drove slowly past the corner where the candy store had stood. The old stone building was entirely gone, but plain in my mind was the ancient counter with its curved glass top through which we used to eye gum drops, chocolate drops, and peppermint sticks. I could see myself puzzling over how to spend my three pennies to my best advantage, and recalled how we hung about until Ike, the proprietor, came to wait on us himself, rather than trade with his less-generous wife. The hog lot had extended under the store, adding a flavor to the place and at times music of a kind.

I drove north, and one by one, the old neighbor families came to mind as I passed their farmsteads. There was the church where I had spent so many hours of childhood and youth. There marched by many, many Sunday mornings when all the boys had stood around admiring the buggies, the spring wagons, and surreys or evaluating fresh-broken young colts and studying harness and saddlery until singing called us into the meetinghouse. The old benches were still there, the home-built pulpit, bare pine walls, splintery floor, and pot-bellied stove. The venerable church stirred a hundred memories to life.

Finally, there was my old home, but nothing looked like home. All was changed, and for the worse. Even the timber seemed to have been cut over, leaving only scrubby trees and brush. A few scrawny trees marked the orchards.

The empty site of Clarkson, shorn of its buildings, brought me some pleasant memories. But at the farm itself, the effects of time and careless use were all too plain.

3

Always small for my age, I recall yearning to reach seventy

pounds by my thirteenth birthday in January 1900. Perhaps the natural handicap of smallness forced me into a course of independent thinking. I was always the smallest on the ball team, the smallest in a wrestling match, the smallest in foot races. I came to realize that I must surpass my fellows in order to be regarded as equal. I learned early on to view myself in a dual light—as the person my fellows saw and as the person I knew subjectively.

I made up for being the smallest in the class by excelling in spelling, ciphering, or geography, the only subjects that really mattered in our country school. For several years, I was the school champion in ciphering, and on one occasion won a contest with a neighboring school—the only inter-school contest we ever had. Ciphering and spelling contests, mental competitions in which size played no part, kept up my morale.

It was with great difficulty that I kept my self-esteem from being buried under a constant barrage of reminders, by ridicule or by physical defeats, of my size. Consequently, my ego sought mental fields to conquer, but even here, I struggled against a handicap. I was extremely bashful.

Each trial, each error, seemed to toughen the core of my determination, or perseverance, or plain stubbornness, that probably became the governing feature of my character.

I had to wage inner battles and endure defeats in the small, competitive community of my boyhood, but there were always lives to be watched and facts to be discovered in the natural world that touched every hour and act of our rural existence. And sometimes, here was a source of personal victory for me.

***

It was summer, and plow time.[1] The cotton and corn showed promise of a bountiful harvest. The wheat had been cut, and I, who was in my middle teens, had the job of plowing the field under in preparation for the next crop. The usual growth of weeds in the stubble had been attacked by millions of ugly yellow-and-black striped caterpillars, each with a "deadly" horn surmounting its rear end. They were ravenous creatures, and almost before we knew it, they were devouring the "persly" and pig weeds, leaf and branch. Within a matter of days, they had grown to a length of two inches, and it seemed a foregone conclusion that as soon as the weeds were finished, the caterpillars would head for the cotton, corn, alfalfa, and other crops.

They were the sole subject of conversation outside the church on Sunday morning, and in the Sunday school classes, they were referred to as a plague sent to punish our community for some unidentified sin. From the pulpit, this anticipated calamity was compared to the biblical plagues of Egypt.

By the following Saturday, the repulsive creatures had grown to frightful size, and their food plants had just about disappeared from the stubble field. Disaster seemed imminent, for it looked like they would now march into our crops. Fear settled over the community. Fervent prayers were offered for deliverance from this awful menace.

On Monday, I returned to the plow. There was a certain satisfaction in plowing the ugly things under, and each round made with the plow added another fourteen inches to the barren plowed ground that the survivors would have to cross to get to our other fields, where the precious crops were still intact.

A great surprise awaited me. When I reached the field, the caterpillars were nearly all gone! Where? I hurried to inspect the

---

[1] Much of this recollection is based on my story, "The Mystery of the Little Brown Jugs," published in *Natural History*, September, 1956.

neighboring fields. No caterpillars! Had the prayers actually been answered? Had the Almighty withdrawn the pestilence? The community heaved a great sigh of relief. Whatever had brought on the pestilence had evidently been righted, and our crops were saved. To me, however, the matter seemed not so simple as that.

As I sat on the moving plow that forenoon, I pondered. Also, I observed. There were many holes in the ground, I noticed—holes about a half-inch in diameter, which I had not seen before. Also, a number of the caterpillars were turned up by the plow. They were inactive, however, and many of them looked misshapen and "sick." This I did not understand.

A few days later, I began plowing up odd-looking brown objects, pointed at one end and with a sort of pitcher-handle at the other. Generally these objects were inactive, but if touched, the pointed end would wiggle.

Things like this were not entirely new to me. I had often seen similar but smaller ones when the garden was plowed in the spring. They were commonly called "little brown jugs," and the story was that they turned into butterflies. But this sounded rather "fishy," and I had never found anyone who could prove it. In fact, everybody I knew seemed to regard the story as fiction.

Certainly nothing of the sort was taught in our little yellow schoolhouse, which had neither a library nor, in fact, even a bookshelf. The same was true of our home. Books on natural history were completely unknown to us. Caterpillars were not discussed in the Bible, or in the Montgomery Ward Catalog, or in the church paper.

Perhaps here was my chance to find out the truth of the butterfly story. I'm afraid I gave the horses more rest than I was supposed to during the next few days, for I frequently stopped to examine these "little brown jugs." I could not help wondering if they were related to the vanished caterpillars.

OPEN FIELDS

As I knelt in the plowed ground examining a "little brown jug" one day, I noticed a peculiar little lump of something lying near the small end of it. I picked it up and examined it. When I tried to pick it apart, it surprised me by stretching out accordion-like; and soon I found I held the skin of one of the caterpillars—head, horn, and all!

My excitement was indescribable. I began examining other "brown jugs," and, sure enough, there was a wad of skin at the end of every one. I had made a discovery!

But it would not do to talk about this either at home or to the neighbors. To explain the disappearance of the caterpillars in terms of a natural process would be to doubt the Almighty's intervention, for which thanks had been expressed sincerely and repeatedly. I myself was a bit frightened by what I had found. But the urge to find out things was too strong to stifle; consequently, I continued to investigate on the sly. When returning to the field that afternoon, I concealed a tin can on my person and returned that evening with it two-thirds filled with moist earth and a "little brown jug" buried in the dirt. I managed to hide it in the barn and then after dark took it to where there was an opening in the foundation on the back side of the house. There I placed the can out of sight.

It was about two weeks later when, on a Sunday afternoon, we had company as usual and had enjoyed one of my mother's bounteous Sunday dinners. After we had gorged ourselves with watermelon and were sitting out under the big mulberry tree on the north side of the house, conversation among the adults turned to the caterpillar plague and what had become of the ugly pests. Various explanations were offered, but divine intervention was strongly favored. Finally someone said that he had read that such caterpillars changed into butterflies. He said that they were supposed to go into the ground and undergo a change, after which they were said to get wings.

The "brown jugs" I had examined showed signs of rudimentary

wing pads under a shiny brown skin, but it was hard to believe that winged insects could come out of the almost inactive objects I had found. My observations *had* proved that the caterpillars had undergone a change, however, so I thought it was time for me to speak. I told them what I had found. All seemed surprised and interested.

Finally I got up the courage to go around the corner of the house and see if my specimen was still there. I took out the can and was surprised to find the "brown jug" on top of the dirt instead of buried as I had left it. Placing it on the palm of my hand, I walked around to where the group waited.

As I entered, hand extended, I felt the object suddenly move. I looked at it and saw that it seemed to be coming apart. Its skin had split at the larger end and along more than half its length. As the startled group looked on, a husky, large gray moth struggled out of that brown skin. The half-inch wing pads rapidly expanded as when air is forced into a crumpled paper bag, only these grew into flat, velvety wings. I was surrounded by a group of white faces and bulging eyes as the newborn, quivering, restless moth crept over my hand shaking its six-inch spread of new wings. It seemed to feel as much at home in its new world as if it had always had wings and had lived in that environment for years.

There are no words to describe the expressions of bewildered amazement and fear that I saw on the faces of the little group around me as they stared at this mysterious and dramatic apparition. For me, the experience was almost as if my mind had shucked off an imprisoning shell. I was enjoying the greatest thrill of my life.

Speechless, we watched the newborn creature exercise its every appendage. Then, with evident confidence, it fluttered away to a nearby mulberry branch. And from the little group of sober-faced onlookers came an audible whisper, "A miracle!"

***

In late summer or early fall, when I was about seventeen, I had been plowing in the northwest field and was returning to the house for the noon meal in response to the big dinner bell. I was aboard Old Min, the family favorite, with her teammate, Topsy, following behind. Mind and body relaxed, I rode homeward, but kept my eyes busy as usual. Suddenly, my gaze fixed on what appeared to be a star in the daylight sky. I blinked and looked again. It was still there, positioned about nine-thirty or ten o'clock, plain as could be.

Never had I heard of anyone seeing a star in bright daylight, except from the depths of wells, and I wanted no part of a family ribbing for a claim that could not be substantiated. On reaching home, I fed the team and checked again from the stable door, where I noticed my star shone more clearly from the building's shade. Before entering the kitchen, I checked from the northeast corner of the house, also shaded, and established markers among the tips of tree branches by which to show my discovery.

Mother's appetizing meal waited while I led the bunch outside. Exhibit of a daylight star demanded some time and patience from hungry folk, but as soon as one had verified it, the others too had to satisfy themselves.

The next day, I found my star again, at the same hour and in about the same position relative to the sun. All week I followed it, and on Sunday, the search for the star furnished a welcome diversion for the gathering crowd at church and earned me some distinction. But when the music began, we ceased our stargazing. Anything detracting from the service would have been unacceptable.

School had been in progress more than two months when we Nininger brothers entered for the term. I found opportunity to bring my "daylight star" to the attention of our teacher. She seemed less astounded than the rest of us but could offer no information. Only years later, when I entered the State Normal School at Alva, did I learn the object's identity. Dr. G. W. Stevens, who was to mean much to me in my student days and influence my life after, listened to my story with a smile. I learned from him that my daylight "star" was the planet Venus, visible much of the time by day.

Astronomically, my "discovery" was of no significance, but psychologically, it was important. This experience helped to make the history and development of science, as I studied it in the textbooks years later, come alive and become understandable for me.

4

School conditions were no better for us in Oklahoma than in Missouri. Students attended when there was no urgent farm work. The Nininger boys usually entered in November after most of the cotton was picked and stayed home on warm days to finish snapping the late bolls. In March, it was time to begin preparing the ground for the next crop to be planted in April.

When I was nineteen, we had a teacher named Tom Haight who had finished the ninth grade and had attended one or two summer courses at the State Normal College. About mid-year, he suggested I take the "county examination." He explained that if I passed, I would graduate from common school and would be eligible to enter high school. I knew full well that I would never be allowed to go any further than common school. I had heard the preacher say it many times: "Those big schools change people's opinions."

To this day, I don't know how Tom Haight talked me into taking that examination. But I did take it, in March, I believe, and promptly forgot about it. Spring work was on us, and six days a week were field days. Cotton field, corn field, bean field. One day in May, I was picking cowpeas in the orchard near the house when Mother brought a letter from the county superintendent informing us that I had earned a certificate of graduation and had passed with a "high average of 88 percent." So far as I know, I was the first ever to graduate from that school, though others soon followed.

Nininger farmstead near Clarkson, Oklahoma, several miles from Coyle, Oklahoma, c. 1906. From left: Harvey's father (James), mother (Mary Ann), and sister (Naomi). Harvey is on the right.

My two older brothers had married by this time. Father had leased a quarter-section of state school land that cornered with our farm, and when Jake married, the land was rented to him to live on. Then Jake became a preacher by the simple process of being elected by the church to fill a preaching mission. He had no choice but to accept unless he wished to ignore the will of the church, an act almost synonymous with going against God. When Jake was called, John and his bride moved to the leased land.

My school days were over, and my twentieth birthday would occur the next January. By Father's wish, I was preparing to take over the family farm. I began to think of putting in my first crop.

Then, that summer of my "graduation," the minister of our church persuaded me to go out as a solicitor for funds for an Orphan's Home Society, agreeing to pay me $1 a day and travel expenses with the understanding that I would lodge with church people along the way. Beginning after the end of the heaviest of the cotton-picking, I was to visit a number of churches in outlying areas of the state, some as far as 100 miles from home.

Since coming to Oklahoma, I had visited only one county besides our own. The first trip of my fundraising mission was to Muskogee. I was gone about two weeks and received my dollar-a-day, my train fare, and a brand-new conviction that I needed more education. I had found myself embarrassed for lack of it, and I made up my mind in those two weeks that I was going away to school.

How? I would attend the annual Bible Institute at McPherson College in McPherson, Kansas, 200 miles from home, held for one week each January. Here I would receive day-long instruction, rich educational offerings that would fill my need. Founded twenty years prior, the college was then little more than an academy and Bible school, but to a boy who had never read anything but the Three R's, the Bible, and the Montgomery Ward Catalog, every campus figure seemed an intellectual giant.

On my return home, I announced that I was going to "take two years off and get an education." Father now saw all his plans upset. Mother cried bitter tears and seemed to look upon me as a lost soul. All our lives, we had been warned that a college education was a sure road to perdition.

Until my attendance at the seven-day Bible School, I had to my knowledge met but one college graduate, a preacher who had visited our little church one Sunday, and he had not left a very favorable impression upon our community.

Though restricted as my horizons were, and limited as my

experience had been, I had seen enough beyond our rural barriers to make up my mind that I must have some "higher" education. I wasn't really needed on the farm, for at that time, we passed as "well-to-do," and Father could hire help. But my plan to desert the farm and my eagerness to reach college unsettled the family. The blow to family equanimity and conscience was somewhat softened by my decision to attend the Northwestern Normal School at Alva. There I could live with my brother Jake and his wife and child. Jake was preaching in a little church eight miles from the Normal School, where he was attending some classes—apparently his studies were made acceptable by his position as preacher.

The Niningers, c. 1907. Back row (from left): Harvey, Jake (Harvey's brother), Jake's wife Ella and daughter Alberta, Frances (John's wife), and Harvey's brothers John and Roy. Front row (from left): James (Harvey's father), Naomi (Harvey's sister) and Mary Ann (Harvey's mother).

Perhaps if I had never known Tom Haight, some other event or person might have steered me to higher education, but I doubt it. My other teachers shared the impression that dominated our little cotton-growing community: any learning past common school was not only useless, but dangerous.

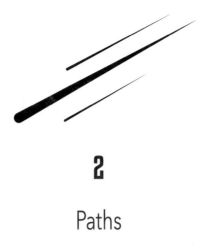

2

Paths

*What I am to be,
I am now becoming.*

—Benjamin Franklin

No rural school I attended in Kansas, Missouri, or Oklahoma had even a semblance of a library. A dictionary was the only book besides those in our desks. Discovery of the small college library was for me like discovering a new continent. It was but one room of perhaps 4,000 volumes. The librarian gave me permission to look among the shelves. I glanced at title after title. Most had no meaning for me, and it seemed evident that most of them could never mean anything. Finally, coming across Emerson's Essays, I took down a volume and began to read. Fascinated, I sat down and read it through, forgetful of time.

The Northwestern Normal School at Alva, eventually to become Northwestern Oklahoma State University, was then part high school and part college. It offered a "B" class, an "A" class, and the freshman and sophomore college years. I was entered in the "B" class and was soon allowed to advance to the "A" class.

When I entered the Normal School in the fall of 1907, I was twenty years old and had been out of school since my "graduation" more than a year prior. I had saved $250 and expected that my savings would take care of my education expenses for two years. Before the end of the first semester, I could see that I would be out of money before the first year ended, and went to the dean to ask about employment. All the campus jobs were taken.

Jake and I had been driving to and from classes in a buggy behind two Indian ponies. The winter was cold, and to reach our eight o'clock class on time, we had to be up before daylight and on our way by seven. On cold days we rode with a lantern under our heavy lap robe to keep our feet from freezing.

Toward spring, we moved into town, eliminating the long drives and creating time for me to take on a job. I concentrated on the problem of finding work, so enthusiastic about the great possibilities of learning that I was determined not to quit short of my goal of two years. It occurred to me that some of the lawyers and doctors in the town might appreciate having their offices swept out each morning and fires started in the potbellied stoves before they came to work, especially on cold mornings. Down the street I went, calling on professional men and offering this service for one dollar a week. I lined up seven customers; a boy could batch in a cheap room for $7 a week. Rising at 4:30 in the morning, I swept out each office, built the fires, and dusted the furniture, then headed for class.

***

My horizons were broadening. I began to think beyond my "two years" toward a full college course. Only a hundred miles from home, I found at Alva facilities for learning and enjoyment of which I never could have dreamed. The library had been a revelation. Now I found another world for exploration.

The biology department maintained a nice little museum, and I began spending there whatever time was spared beyond classes, my chores, and the library. Soon I had memorized the names of all the stuffed birds and mammals. I couldn't resist enrolling for zoology, although I had not even known the word until my interest in the museum had led me to it. I entered the class forewarned in my own mind to resist any radical teachings or hints of "evolution," but once in the class, I found myself at home.

Dr. G. W. Stevens was the most popular man on the faculty. Known for his tough courses, he was the kindest of men and would go to any length to help a student who needed his assistance, but he was merciless in his demands on those who sought to evade or bluff assignments.

His discussions were so interesting that students were delighted when he strayed from the assigned subject to draw upon his wide background of experience, but he was always conscious of what he was doing and why. As the class began to believe he had forgotten the assignment, he would suddenly announce that they would have the opportunity to write on the undiscussed matter in a special quiz.

Dr. Stevens would stand before his high school zoology class and dissect mythical powers and behaviors imparted to animal species with the same care and thoroughness he demanded of us in our dissection of cats and grasshoppers in the laboratory. He would get to the truth in some such manner as the following:

"Tell the class about that joint snake, Mr. Link," he would instruct.

And Link would tell with some gusto how he had beaten the joint snake with a stick and had broken it into a dozen pieces, and then, an hour later, how its head end had turned about and attached all of the other pieces in proper order, whereupon the creature had slithered away into the grass.

"You hear that, class?"

At this stage, Link would be mentally puffed up in somewhat the manner of a strutting turkey gobbling gayly on a warm April morning.

"Where did this happen, Mr. Link?"

"In the grass between the road and the cotton field."

"Did you mark the place where you broke it up, so that you would know where to come back to look for the pieces?"

"No, I didn't need to; I knew right where it was."

"Did you have any trouble finding the snake when you came back?"

"No, not much. In fact, none at all, because he had crawled to the edge of the cotton field, so I saw him before I got to the grass where I'd broken him up."

"Then you didn't actually see the head go around and couple up all of its pieces?"

"Not exactly, but it was wiggling along sort of in the direction of the rest of him when I left to take care of my team. But I knew that's what he was starting to do, 'cause Dad's told me lots of times how they do it."

"What do you think about it, class? Any questions?"

By the time the class had asked all of its questions, all that kept Link from defensive belligerence was the professor's calm mien.

"Next time you break up one of these joint snakes, I tell you what you do. Stay right there and watch till they're all hooked together again. And listen—now you take careful note of whether they come together in the same order, whether the front end of every piece is in front and whether they are all in the same arrangement as before."

Then, smiling, "Would be sort of funny if the tail somehow got set up in the middle, wouldn't it?"

Variations of this joint snake tale would keep coming up. Other members of the class would come in with reports they had found in various "reliable" books. Finally, when no more snakes were encountered, the teacher would lay the fallacious myth away with positive and indisputable evidence.

It was his thorough-going analysis of these legendary beliefs, together with the thorough but good-natured kidding with which he attacked them, that led me to a series of personal experiments that ultimately solidified an attitude of skepticism toward anything unproven by fact.

For instance, I allowed myself to be bitten twice by a blue racer snake. To prove that a centipede's sharp-clawed legs could not pierce the skin, I gripped the head of one of the ugly creatures between thumb and finger while it writhed and pulled to free itself. This was not a pleasant experiment, but as a demonstration, it was satisfying. Many times I have been served well by the background of knowledge gained from such trials, and by the habit of providing or demanding demonstrative proof of assertions.

***

When Dr. Stevens called me from the laboratory into his office one day, I was frightened, though to my knowledge, I had done nothing to incur his displeasure. He questioned me a few minutes and then shocked me almost out of my shoes by saying he would like me to serve as his museum assistant the next semester when the regular man would be away. I demurred that I wasn't sure I could handle the job. But through my mind was running a warning that to take it, much as I might want it, would mean outright alliance with things scientific.

Dr. Stevens assured me he had watched my work, particularly in the laboratory, and was confident of my ability. He added that he knew I was earning my own way, and this work for him would pay $15 a month for Saturdays and an hour or so a day during the week. Addition of this sum to my earnings from morning janitorial work would mean a real financial boon. After battling my conscience a few days, it was agreed I would report for work early the next semester.

On a Saturday morning, I arrived at the laboratory to start my job. It was a little before eight o'clock, the time Dr. Stevens had set. The senior assistant and he, the professor announced, were going into the field, but I was to work until they returned some time in the afternoon. Dr. Stevens pulled a box of excelsior from beneath a shelf and escorted me to the workbench. He fastened the handle of a pair of tin snips in a vise, then took a handful of excelsior and began to chop it up, letting the bits fall into a tray on the floor. My assignment was to chop the excelsior into lengths of about one-quarter inch, with permission to close shop and go home if the other two had not returned when I had snipped all that was in the box.

After I had worked at the wooden hay for an hour, I could see scarcely any effect on the size of the pile. My arm was tiring so from the repeated motion that my muscles were ready to rebel. The glamor of my new job had dimmed considerably by noon, but I was determined to stick it out and never let the professor know my feelings.

My flagging muscles became unreliable and finally I slipped and cut my right forefinger to the bone. Despite the pain, the time necessary to stop the bleeding and bandage the wound seemed a welcome interval. I almost enjoyed the rest, save for worrying what Dr. Stevens would think of a fellow so awkward he couldn't get through the first day with all of his fingers.

After eating a sandwich, I snipped again until about 3:30, when the two returned from their field trip. Dr. Stevens inquired about my finger, saw that it was sterilized and bandaged properly, complimented me on my box of clipped excelsior and told me it was sufficient.

During the months that followed, I had the questionable pleasure of seeing that box of excelsior clippings resting under the lower shelf just where I had placed it that first day. It never was used, and I finally concluded that it never was the professor's intention to use it except to test the staying qualities of a new worker.

Years later, I heard Dr. Stevens tell how he had tested a man whom he had selected as a hunting companion for an Alaskan trip. He had taken the candidate on an antelope hunt into New Mexico Territory, a bleak and sparsely settled region in those days, and so managed things that they were forced to spend one cold, snowy night on the plains without camping equipment. After his young friend had stood or marched about the campfire all night without sleep and without complaint, he was judged a satisfactory partner for the trip to the north.

All my subsequent assignments were agreeable, if not always pleasant. I was given opportunity to learn how to prepare bird and mammal skins, mount birds, and to try my hand at mounting a wild pig specimen.

Dr. Stevens's dictum was, "Observe nature."

One was not to depend on books, which sometimes might be incorrect, but to look to nature for facts and record them accurately.

"There are no authorities in botany," he would say. "Only plants are the authority. There are no authorities in zoology; animals are the authority."

Dr. Stevens was one of the finest scientists I have ever known. In all my travels among schools and colleges in subsequent years, whenever I have met past students of his, they have been outstanding citizens of their communities. He was the teacher they all remembered. There was something about his personality that rooted deeply into anyone who came under his tutelage.

He was a man of great ability and humility. Probably the one serious mistake of his professional life was his failure to publish anything extensive. His single great manuscript, on which he worked for years, he finally sold to another scientist in order to pay bills incurred while doing the research for it. I like to think of him in terms I once heard used to describe another great teacher: "He never published books, but he published a number of very successful young men."

***

An activity at the Normal School that helped diminish my bashfulness was participation in the debating society. After Jake and I were advanced from "B" status, we were invited to join a new debate group that had been started to counter the established debating society's policy of freezing out younger students. Within a year, our new group was a serious competitor, and by the end of its second year was considered by many—certainly by its members—as superior to the older organization. In this group of peers, my shyness faded and I soon was able to hold my own.

The differences of our rural and fundamentalist background receded for Jake and me as our stay at Alva lengthened. As our vision widened, we found it more and more difficult to defend the peculiar manner of dress required by our church. Home training had equipped us to go our own way regardless of the customs of others, but subjectively, as we attempted to rationalize our conduct, we found ourselves beginning to direct some of our new "independent thinking" to the teaching of our church.

In the midst of this mental confusion, I remained sensitive to the feelings and opinions of my parents. By the end of my second year at the Normal School, my mind was in turmoil. My present interests and enthusiasms were in direct opposition to all I had been taught from childhood. When I had left our home community to seek education, I had gone braced stubbornly against its "evils," particularly those of the demon science, but now, everything I was doing and that I wanted to do seemed in conflict with my conscience.

On the train headed for Alva that first year, I had met one of our local preachers, a cousin of my father, who felt it his duty to extend a warning. Education, he reminded me, was dangerous. It changed people, he said, changed their ideas and their manner of dress and action. I had assured him that nothing of the kind would happen to me.

And yet I felt my moorings loosening. My thinking was changing; I *did* dress differently. My mind questioned where this deepening interest in science might lead me. Would my religion suffer? Ultimately, I devised a compromise between the past I had lived with my parents and the future I hoped for myself. I decided that for my third year of study, I would transfer to McPherson College in Kansas, operated by the Church of the Brethren, a school that seemed to offer a safe shield against skepticism and especially that most dangerous of all heresies, evolution.

The decision was a painful one. My mentor and employer, Dr. Stevens, was deeply concerned, mostly, I think, for my sake. Characteristically, he said little. Later, I understood that he had developed great hopes for me and sensed the struggle I was going through. Near the end of the school year, Dr. Stevens told me he was to be traveling soon to our county seat, a 200-mile round trip from Alva, and would like to drive the additional eighteen miles to my home and meet my parents. An invitation was extended, and in due time, he drove from town to our farm by horse and buggy. He stayed the night and did his best to win Mother and Father over to the idea of my returning to Northwestern Normal School as his assistant. He was persuasive, but the decision for McPherson held.

My brother Roy, almost three years my junior, was pleading for his freedom to join me, and would be ready in another year. Our parents felt they could accede to his wishes only if we attended a church institution. Whether the final choice was made out of filial respect and duty or my own judgment in answer to inner torment, I am not sure. In any event, Dunkard training and parental loyalty prevailed.

2

When the fall term opened at McPherson College, as a transfer, I was still taking mostly freshman work. While my two years and a summer at the Normal School had little more than put me through the equivalent of high school, it had set me on my way. And I was surprised to find there was independent thinking at McPherson, too. Although faculty members conformed more or less to church traditions, they also saw beyond denominational bounds of thought. They exhibited an attitude at least as liberal as that of Northwestern Normal School, and it was all the more stimulating for having sprung from the same soil as that in which my own roots were anchored.

Now began in earnest that mental metamorphosis which accompanies true education. With the encouragement of professors whose backgrounds had been similar to that of my parents, I developed an independence of thought that disregarded tradition on all sides. During my student life at McPherson College, every facet of my religious and political upbringing was re-examined.

Unlike a few of my fellows, I did not throw off social and moral restraints as well. A critical look at the facts of life convinced me that a life in harmony with a liberal interpretation of the Christian philosophy was the most practical course to follow.

Several mottos and definitions came to have very strong meaning for me. I developed a habit that has followed me since of leaning on, and being influenced by, certain inspirational or wise phrases. I do not now recall the source of several that were so meaningful during my college days, but I can still feel their impact on my thinking. "Wisdom is knowing what to do next," for example. And this definition of faith: "The momentum of a righteous life, the distance into the future a soul can run who has been running right in the past." When my brother Roy joined me at McPherson and we first moved into the dormitory room we were to share, we found a motto hanging on the wall: "What I am to be, I am now becoming." "Dad" Eliott, a YMCA leader of those days who traveled among Midwest colleges, one day admonished us thus in assembly: "So live that you may look back on your life with satisfaction when you are forty."

My intellectual course had been set on "independent thinking," but mottos and clichés have their usefulness in crystallizing thought and compressing philosophy into capsule size.

***

In addition to the bits and pieces of man's intelligence I was absorbing and feeding into my own developing philosophy of life, I picked up smatterings of varied experience in the course of earning my way through an education.

Father had acquired a confectionary shop in Guthrie, Oklahoma, in a real estate deal, thinking he could resell it for a good profit, but having been a farmer all his life, his judgment on such matters was not infallible. When he found himself stuck with a bad proposition, we boys leaped at what we envisioned as a way to help Dad out of a predicament and at the same time set aside a stake for the next year's schooling.

We were young and optimistic. Oklahoma, in 1910, was young and unsettled politically. While I finished out my spring term at McPherson, Jake and Roy had no sooner started the confectionery business than the "Sooner" state moved its capitol from our bustling town of Guthrie to Oklahoma City. Most of our town's prosperity went along with it. So did most of the confectionery's profits, though, luckier than some, our shop did a little better than break even.

But only a little better. By fall, our school "stake" amounted to some $16 to $20, and in the sale of the shop, we barely managed to earn back what we'd invested. We sat at one of our ice cream tables, "celebrating" the end of our confectionery venture with a farewell soda. The college year was due to open in two weeks. Our few dollars wouldn't go far.

Having to work our way through school was expected, but usually there was a good running start from the summer's earnings. From there, we could janitor, press clothes, wash windows, clean old stoves, milk cows, clean chicken houses, split kindling—there were many methods to care for the year's bills. But to start from zero, with no jobs lined up in advance... Things did not look good.

We finally agreed that our newly acquired art of candy-making should be made to pay our way. We had purchased the previous confectioner's book of recipes, and Roy had taken to the business with a good deal of skill. Before we left that ice cream table, we had made plans for our year at McPherson—my second and Roy's first. We would either rent or buy on credit candy-making equipment, buy supplies on credit, and make and sell candy wholesale to the grocers of the college town.

Fortunately, I had enough money left from the previous year's work for railroad fare to the college. Roy was to stay on in Guthrie the remaining two weeks, earning enough to come along in time for the opening of school. I would stop in Wichita, call at the confectionary supply house we had been using, and try to open a credit account for our new venture.

Mr. Kevan alternately smiled and frowned as he heard my story.

I admitted we were taking a risk but insisted that he was not. I told him we would pay our bills even if we had to quit college and work our fingernails off to do it. Soberly, he admitted that he believed we would. We proceeded to the stock room to select supplies and equipment. I told him to ship the supplies in two weeks but hold the equipment. I had a hunch that we might be able to rent from another source.

School opened. We enrolled on credit, bought our books on credit, and rented the cheapest room we could find to batch in. We were able to arrange for the use of some abandoned candy-making equipment stored in the restaurant cellar of a man who had dropped the candy-making end of a restaurant/sweet shop in McPherson the year before. He said we could make our candy in the basement and pay the rent by delivering sweets to his sales counter at cost.

In that basement, we made two things: candy, and war on rats. We were watched by gleaming beady eyes from all dark corners while we worked. All supplies had to be stored in closed metal or glass containers during every minute of our absence. Everything was kept covered. Each time we used our marble slab, we had to scrub it down.

We made various fudges and creams, pulled taffies, and leaned heavily on peanut confections—clusters, plank, and brittle—from our purchased book of recipes. We also made a "cream peanut" fudge that was our own development and sold well. We had a good, hot furnace with a thirty-gallon copper kettle atop a brick well and gas jet, and an additional gas plate. We stirred our products with a four-inch-wide paddle as long as a canoe oar and blistered our hands slinging taffy over a great hook on the wall, pulling the sweet ropes until they gleamed.

From clothing stores we collected shirt boxes in which to deliver the candy. We had large labels printed—"Nininger Brothers' Homemade Candy"—again on credit, to cover the original box labels.

Grocers were slow to begin buying from two college boys a new product which, for all they knew, might not sell—and if it did, how long would the boys keep producing it? Finally, I persuaded each store to take at least one box on consignment with the understanding that it be placed where customers could see it. Before long, we were delivering 500 to 600 pounds a week to ten or twelve shops. We found it necessary to work not only all day on Saturdays but also on Friday afternoons.

In a few weeks, we were able to move from the dreary basement to a small building above ground, free of the rat menace. Our stack of bills began thinning down nicely, but with our full college schedule, extra-curricular activities, and the candy-making, we were putting in about eighteen hours every day. Rising time was 4:30 a.m., and bedtime was at whatever moment we no longer could stay awake over our books.

As Christmas approached, the merchants' shipments of holiday candies began arriving from orders made the previous spring. Meanwhile, the grocers found their factory candy stockpiling while our homestyle product sold. They told us apologetically that we would have to hold up our deliveries.

The candy salesmen from out of town had been hit hard. These grocers were old customers of theirs, and to keep us out, the salesmen would sometimes risk a shipment of "specials" without an order, knowing it would not likely be rejected. Such tactics held us at bay pretty much for another month after Christmas. The two-month recess in our business put us in a state close to desperation.

One Saturday morning, I changed my approach as I started on my rounds of the stores. Most candy was sold in bulk. The standard practice was to display the candies in six or eight long pans in a large glass counter case. I persuaded each grocer to assign about thirty percent of his candy pans to us. If any of our product failed to sell promptly, we were to pick it up and give full credit. Then, if the customers preferred the factory candy—well, we would have to take our licking.

In a short time, we again were delivering 500 to 600 pounds of candy a week and were working almost night and day to carry our business and college work. Finally, in the spring, an executive of one factory came to call on us. He wanted to buy us out. Our success at bargaining for what we considered fair competition had made us rather cocky, and we rejected his offer with an attitude of total independence.

We continued the candy venture through the remainder of our college years. As graduation approached, a leading grocer came to us with a proposition to build a real factory and back us financially. By that time, however, we had committed ourselves to careers in education.

3

If necessity mothered the invention that enabled us to support ourselves through college, the same was true when it came to developing modes of study.

It must have been during my second year at McPherson that I enrolled in Dr. Claude Shirk's physics class. I don't think any girl ever got through that course without having a cry in class, because Dr. Shirk, a brilliant man, was very strict and rigid in his attitude toward students, even rather ruthless in his way of driving them.

Despite an interest in physics, I didn't seem to be getting along very well. One day, Dr. Shirk criticized my recitation so mercilessly I made up my mind to quit the class, but before I got back to my room, I had another idea. His censure had made me feel there must be something wrong with my way of studying. Since I had an hour between classes, I went to my room, took the new physics assignment and read it over paragraph by paragraph, thinking each through carefully. Then I took a pencil and wrote out in a single sentence the meaning of each paragraph. I was going to show Dr. Shirk, or I was going to quit class. I went over my sentences again. Then I closed the book—I was so angry that I threw it—and told myself I would never open it again if the professor were not satisfied with the lesson.

I didn't even carry the book along to the next session of class, nor did I take my notes. When I was called on, I made no effort to recall the words of the text, but told what I knew in my own way. Dr. Shirk smiled and complimented me. From then on, I used that method in every class, though I did not always write out my thoughts. This experience gave me something that has lasted all my life. I had learned to master in twenty minutes lessons that used to take an hour or more to understand. I had learned to concentrate, to shut out everything and absolutely grab the meat out of any paragraph I read.

***

During the summer break after my second year at McPherson, in my annual effort to raise money for the fall term, I joined a photographic group that traveled about on motorcycles taking harvest pictures, a project that seemed to promise good returns but which ended in early and complete failure. In midsummer, I was returning to my brother John's farm in Nickerson, Kansas, and stopped at Alva to visit Dr. Stevens. He seemed delighted to see me, asked what I was doing and what my plans were for the rest of the summer. I explained I was going to join a threshing machine crew and hoped to earn about two dollars a day.

He then asked me to stay in Alva and work with him in the taxidermy shop, at $1.50 a day plus room and board. I went into a spare room and unpacked my bag. The next day, I started work in the shop with Dr. Stevens and Roy Chestnut, who was a close friend from my two years at Alva. As we worked, Dr. Stevens asked me many questions about my school work at McPherson. He mentioned that he was going back to Harvard that fall to finish his doctoral dissertation. The Ph.D. degree was not so common on faculties back then. I believe G. W. Stevens became the second on the staff at Alva to hold the doctoral title. Prior to his earning the degree that year, we addressed him as "Professor," commonly used in those days for any teacher.

Harvey Nininger and Harley-Davidson motorcycle, 1912. To earn money for school, Harvey took part in a failed photographic project in which photographers traveled about on motorcycles taking "harvest pictures."

He explained that the school president had agreed to grant him a leave of absence with pay for the first three quarters of the coming school year, provided that he hire a substitute to take his place. Then he surprised me by suggesting that I teach his classes while he was away.

I was flabbergasted and couldn't believe he was serious. As the days went by, however, he brought the matter up repeatedly, assuring me that he meant it. My objections and refusals could not dissuade him, and one day, he came to work saying that he had seen the president and that everything was arranged. I was hired. I still did not accept, but after a few more days, he convinced me, and I set about making arrangements for room and board and going over the program with him.

So I, a college junior, served that year as substitute head of the department of biology at Northwestern Normal School at Alva. Roy agreed to help keep from the students the secret that their teacher was but a third-year man himself.

That school year of 1912–13, I handled classes in botany, zoology, and physiology. We had some interesting times, and I managed to stay ahead of my students in the textbooks. What I lacked in knowledge and experience as a teacher, I sought to make up in hard work, both on my part and in the way I loaded it onto my classes. I was carrying some junior work by correspondence with McPherson College, and I often had to stay up late at night to prepare for the next day's teaching. This experience convinced me that the best teaching is by the teacher who has not himself gone too far beyond his subject matter; if the teacher has become too advanced, too complex, too specialized in his knowledge, he may no longer be able to find the best way to communicate with his students.

4

After so much exposure to Dr. Stevens's philosophy that nature and observed fact are the only authorities, and after a rather heady year serving as teacher to my peers, I became rather zestful at wading in and dispelling old beliefs and traditions that colored the lives of others not so advantaged by modern education as I believed myself to be. However, I leaned more heavily than Dr. Stevens would have approved on the testimony of my textbooks to attack and lay low old wives' tales and superstitions. Then, finally, as is so often the way with idols, I discovered that my most cherished text had feet of clay.

During my senior college year, back at McPherson after my teaching stint at Alva, I was given charge of an elementary laboratory class in zoology. Classroom assignments included dissection of grasshoppers and the making of anatomical drawings. In an attempt to assist a weak student, I started to sketch a grasshopper's mandibles. Insisting that the student refer directly to the specimen rather than to the picture in the textbook, I was alternately peering into the microscope and employing my pencil, when suddenly I realized that what I was seeing and sketching bore little resemblance to the reproduction in the book. I rechecked my work and verified that there was a serious discrepancy. This was startling, for I knew the professor, the head of the department, had absolute faith in the textbook he had chosen and held its author in the highest regard.

The laboratory period was soon over, and as was the custom for this course, the students left their dissecting pans on the tables. I settled down to examine all the partially dissected specimens. First I determined that our specimens were of the same species as that depicted in the book. Then I set about comparing all of the specimens. They were as alike as peas in a pod, but certainly quite unlike the picture in the text. Next I examined other books. Not only did I find the same error, I found that most of the representations were copies from older books and most had been copied or recopied from an old and respected text that had been published some sixty years earlier.

I examined my conscience. Surely the obvious error must be brought to the attention of the department chief, but I knew the respect he held for the author of our book, and I knew also of his temper. I decided to take the discrepancy to him, but determined first that I would come with an open-and-shut case.

It was early October, and there were still many grasshoppers about the campus, so I proceeded to collect them. I gathered some thirty different species of Orthoptera, including a good supply of *Dissosteira carolina*, the species purported to be displayed in our text. Transforming my room into a kind of specialized laboratory, I worked day after day, often slighting some of my classes, making drawings and more drawings, representing in minutest detail the slight differences between various species. I became so familiar with these "grasshopper teeth" that I could make a fair drawing blindfolded.

Finally, I approached Dr. H. J. Harnley, the zoology professor, carrying my proof in a cigar box and my courage in my mouth. I told him there was an error in the textbook. The professor indignantly told me about the author's qualifications, opening to the flyleaf of the book and reading off titles and degrees. How dared I presume to criticize? I was reminded of my status as an undergraduate. I was privileged to work in the laboratory to help pay my way, not to rewrite the zoology course.

I proffered my box of specimens, and my natural stubbornness asserted itself. I asked him not to accept me or my drawings, but to look at nature. The book showed mouthparts better suited to a flesh-eater than to a vegetarian, and the grasshopper is a vegetarian. I held out a mounted set of mouthparts and a small hand lens.

Grumbling, he brought the magnifying glass to his eye impatiently, and looked from the specimen to the book and back again. He examined and re-examined, turning his chair about for more light from the window. He asked if I was sure the specimen was the same species as the one illustrated in the book. I assured him it was and told him of the studies I had made of other species, and that they, too, showed a similar discrepancy with the textbook drawing.

Finally, with a kind of lament, he admitted that the book was wrong. Then he brightened, noting that this would be an honor for

44 IN SEARCH OF FALLING STARS

both me and the college. He instructed me to write up my findings. Published in a standard scientific journal, they would "change every zoology textbook in the world."

With his blessing, and with the resources of the laboratory, I devoted all my spare time to making the best series of drawings I could. I inspected the eating apparatus of every specimen of Orthoptera the town and countryside yielded—cockroach, katydid, locust. My roommate began to refer to me as "Mr. Grasshopper," and the name spread through the dormitory and across the campus.

Finally, I developed a series of drawings I was confident could be defended against any criticism. I sent off my paper for publication, and the year after my graduation, it was in print, in a "standard scientific journal." I waited for the predicted impact—the praise, the criticism, the flood of textbooks corrected so that students now might receive a true description of this important little creature of the insect world, Orthoptera.

The great majority of animal species have mouthparts built on the same plan as *Dissosteira* rather than the plan assigned to man and other vertebrates. Probably nine-tenths of the animal world has achieved survival by biting and chewing with jaws that move laterally rather than back and forth. *Dissosteira's* mouthparts were fitted for grinding of vegetation rather than for tearing of animal tissues.

For forty-five years, I watched zoology texts and saw no indication of awareness by anybody of the error I had so meticulously set out to correct. There were only the same old representations of a vegetable-eating grasshopper with the mouthparts of a carnivore. Finally, a 1960 textbook appeared that carried an amended representation of the mouthparts of Orthoptera. Whether my published paper at last had been heeded, or whether after a half century another writer had made the same finding, I had no way of knowing. Despite this somewhat shaking experience wondering whether science really does move forward on an expanding base of new knowledge and corrected error, I continued, whenever occasion presented, to debunk many of the cherished notions by which and with which people live.

3

Stopping Places

> *Optimism is the faith*
> *that leads to achievement.*
> *Nothing can be done*
> *without hope and confidence.*
>
> —Helen Keller

It was in Latin class that I was first attracted to Addie Delp. She was an outstanding student, she had intelligence in addition to physical charms, and she had been brought up in an environment similar to my own. Addie was sensible, efficient, knew how to make a home, loved children, and insisted that she loved me also. We were engaged for more than three years, but during two of those years, one or the other of us was teaching and away from McPherson.

Toward the end of May 1914, I received my diploma, and on June 5, we were married. A job waited at LaVerne College in Lordsburg, California, within walking distance of Pomona and Claremont

Colleges. I was to receive the princely salary of $80 a month, with room and board furnished in the college dormitory.

Immediately after our marriage, I took a job in the hayfields to earn fare for our coming trip. We spent the summer at Nickerson, Kansas, where my brother John farmed and kept honey bees. When I was not haying, I helped with the bees. Addie busied herself with cooking and keeping house.

The long train ride to California was a colorful experience for a pair of newly-wed Kansas youngsters. In the village of Lordsburg, in the midst of citrus and walnut groves, we were as happy as could be, with beautiful mountains beckoning on one side and Pacific beaches on the other. It seemed a veritable heaven for a young biologist and his wife. Days were not long enough for all of the climbing and beach excursions we so much enjoyed.

Harvey Nininger, Senior Photo, McPherson College, 1914.

Addie had not yet completed her college work and enrolled in certain courses for credit and in art studies for enjoyment. And I plunged into my work as teacher. At LaVerne, I collided, via the subject of evolution, with the Church. The college, previously known as "Palmera," was a sister school of McPherson. It had been only an academy and junior college, but in 1914–15, it qualified to present the four-year curriculum. My teaching duties included zoology, botany, chemistry, physics—thus I was not to occupy a chair, but rather a rough-hewn bench. After the second year, my attentions were to be devoted only to the biological sciences.

LaVerne was run by a very fundamentalist board of trustees. Other colleges operated by the Church of the Brethren had promoted a more liberal viewpoint by that time, but the mild California climate had attracted many of the elder brethren of the church to LaVerne, and the board of this school, unbeknownst to me, had continued to cling to many of the most strict and unbending views on education.

Among the courses scheduled for me in the second semester was one labeled "Bionomics." Whoever had written the catalog under which I assumed my duties had apparently copied the curriculum for the newly created department of biology from a similar track offered at the better-established McPherson College. McPherson offered a course in Bionomics, patterned after one that had originated at Stanford University, which was mainly a study of evolution.

As the second term approached, I handed my order for textbooks to Mr. Joseph Brubaker, business manager for LaVerne who was also in charge of the bookstore. Brother Brubaker was a successful and progressive orange-grower, but he was also one of the bulwarks of the local forces of fundamentalism. He was a man of about sixty and a fine, upstanding citizen of the community. He had been a good friend to me, but when he looked over my book list, he was plainly shocked.

His little goatee bristled. "We can't order this book."

"Why? What do you mean?"

He pointed to the title. "Well, it says here, 'Evolution.' We can't have evolution taught here. It is not according to the Bible."

I explained that my contracted course load included the study in bionomics—that bionomics involved evolution and, indeed, *was* evolution.

Mr. Brubaker proceeded to give me quite a lecture. I saw there was a complete misunderstanding. The administration clearly had no idea as to the content of some of the courses catalogued. I was sure, however, that Mr. Brubaker did not represent the views of the entire administration. It was evident that if the present difficulty were not ironed out, the conservative faction of the school and the local sponsoring church would really stir up a fight. When an attempt was made to persuade me to cancel the course and change my teaching methods, I insisted on a hearing before the board of trustees.

The college president was broadminded, but he was not in good health and had allowed others to pretty much run the school. He sat quietly at my "heresy trial," while the board chairman outlined the case and each trustee expounded his views. For my turn, I announced that if the course were canceled, a search should begin at once for someone to take my place, and that I would not be on duty the following Monday morning.

This got under Brother Brubaker's skin, for we had always been congenial until he'd seen that awful word, "evolution." To forestall my leaving, he proposed that I be authorized to deliver one lecture demonstrating there was no truth to the evolution theory and then never allow it to be mentioned again. Although some of the board viewed this suggestion as too restrictive, it was evident that the majority, aside from the college president, agreed with Brubaker.

I addressed the business manager. "Your two daughters have been in my zoology class for about three months. Have you noticed any ill effects on their religion or morals?"

"Oh, no," Brubaker replied. "Your influence has been most wholesome." He then went on at some length to extol the beneficial effects of my presence on the faculty.

I proceeded to relate that I had been discussing evolution, more or less, every day since the course had begun. I stressed that my chief interest was to teach students to think, that I did not try to sway them to believe the theory, but that I outlined the reasons for its acceptance by most zoologists as a law of life and also presented the arguments against it. And I further stated that so long as I remained at LaVerne, I would continue to teach evolution in that way.

Now the president spoke up. He reminded the trustees that teachers were hired for their knowledge of subjects to be taught, that their chief function was to help students learn how to evaluate various questions and think for themselves. After some further and more sound discussion by the board, it was decided that I should continue to teach in my own way.

2

Despite my very meager salary, Addie and I lived richly during the four years we spent at LaVerne. Some of our friends owned automobiles, and California was blessed with quite a lot of hard-surfaced roads for that time period, but we loved hiking into the solitude of the nearby mountains.

In addition to my teaching, I was doing some advanced study and research. I earned my master's degree in entomology for research performed under the direction of Professor William A. Hilton of nearby Pomona College. I was intrigued by the abundant opportunities for field work and lined up a number of projects involving the behavior and life histories of certain birds and insects that took me both to the mountains and the shore.

Addie shared and enjoyed most of these field activities. The San Gabriel Mountains claimed a considerable amount of our energy and attention. We made many hikes into the canyons and to the mountain tops, sometimes alone, but often with one or more companions. One summer week, we camped on a solid rock precipice that in the rainy season would be transformed into a waterfall some sixty feet high.

We chose this site to study a family of golden eagles whose nest was perched on the broken top of a large spruce trunk about ten feet lower than we were and about sixty feet away. Near the same aerie, we had the good fortune to observe two of the great, rare California condors.

Some of my studies concerned certain parasites that affect the development of the large, black carpenter bee, *Xylocopa orpifex*, and the ground bee, *Anthophora*. I made almost weekly eight-mile round trip journeys to observe carpenter bees that tunneled into the redwood frame of an old, deserted cabin and to work out the life history of a colony of ground bees whose behavior I believed had been incorrectly described in the entomological literature.

Altogether I made some eighteen trips to study the carpenter bee colony during autumn, winter, and spring, and the following summer, I continued my research with individuals I had brought down to the laboratory by removing and transporting the timbers they occupied. I observed them at the laboratory by opening the wood and arranging the habitats under glass, protected by dark covers when not under study.

The beaches were open and accessible, and most attractive to a young couple who never before had seen a body of water larger than the Arkansas River. The shoreline below Long Beach stretched open to the south for miles and miles, uninhabited except by the myriad of organisms that preceded man's existence.

During one summer term, I taught a zoology laboratory class for Pomona College. The course was given at Laguna Beach, and Addie and I set up our tent at a location which now is at the approximate center of a dense urban area. Throughout our four years at LaVerne, we frequently took classes to Laguna Beach and other shore points to study plant and animal life.

We spent two summers at the Berkeley campus of the University of California, beautiful and stately, and memorable for the fragrance of eucalyptus trees, the Campanile, and the graceful Greek theater. The natural environment of the campus was its glory. We loved the rustic bridges and grew familiar with the songs of the thrushes we heard on hikes up Strawberry Canyon and the call of the Gambel's sparrow which seemed always to be singing near the bell tower.

Harvey and Addie Nininger at camp in California. They moved from Kansas to California shortly after they were married, and lived there from 1914-1918.

In 1915, when we learned that Professor Loye Holmes Miller was conducting field trips to Muir Woods and other scenic destinations for bird study, Addie and I immediately enrolled in his zoology class. Dr. Miller was an expert in the art of recognizing birds in the field not only by color marking, but by shape, pose, flight, behavior, habitat, feeding habits, and song. His class, though consisting mainly of teachers, included persons of all ages, from teens to grandparents, and they were an enthusiastic and interesting group.

I have held a great fascination with birds as far back as I can remember. My mother had an abiding affection for the little hummingbirds that occasionally visited us, though no one else in the neighborhood ever showed interest in any of the great feathered population other than game birds and hawks, the latter because of their supposed depredations on poultry. At school, birds were never referred to outside of nursery rhymes and poetry.

From Dr. Stevens at Alva and from Dr. Miller, however, I learned that it is perfectly respectable for an adult to be interested enough in birds to devote time to their study; that there are books and periodicals devoted solely to supplying reliable information about birds; that birds are of great economic importance and the majority are beneficial to man; and that I could learn not only how to identify birds, but how to teach others to recognize, enjoy, and appreciate them.

I used the Berkeley experience to justify our request to the administration that we institute a course on ornithology at LaVerne. During the next fifteen years, wherever I taught, I always presented bird courses, both for college credit and for the pleasure of those who might care to sign up for the field trips alone without regard for lectures or laboratory. When former students comment that the bird course proved to be the "most practical" of their college studies, I recall the difficulties I had advocating for such an "impractical" course.

In 1927, I wrote *A Field Guide to the Birds of Central Kansas*, intended to enable the amateur to identify species in the field and utilizing identification aids similar to those taught by Dr. Miller. Over the years, Addie and I have "counted" birds for the Audubon Society and have made searches for rare species. We feed, water, watch, and cater to the Gambel's quail, towhees, juncos, wrens, bushtits, jays, and other birds that frequent the hospitality center just outside our sliding glass kitchen door. I have noted forty-two species in all.

***

September 1, 1914, marked a demise to make wise men hang their heads in remorse. The mourned was neither king nor president, performer nor sports figure. Rather, it was a single bird, the last surviving member of a race. The last known passenger pigeon of the United States died in a Cincinnati zoo.

As Addie and I tramped about with Professor Miller on his bird appreciation hikes, the fate of the passenger pigeon was still fresh in our minds. Recalling Dr. Stevens's admonishments regarding the balance of nature, I seethed inwardly that something had been so out of balance in this instance, and I embraced the cause of conservation with new determination. America's murder of the passenger pigeon is a chapter of shame in its national history.

A century earlier, millions of passenger pigeons had swarmed the forests of the Mississippi Valley. It was their habit to come together at evening to a common roosting place, darkening the sun as they flew against its fading rays, forming long streaming clouds. The whistle of wings produced a deafening roar. In the morning, they would flow away in all directions to feed upon acorns, wild fruit, and seeds.

But generally, they left many of the flock behind. Men had learned that their plump bodies made good pot pies. Fried pigeon was a common dish. No one thought it possible to kill off the wild pigeons. No bird was more abundant. For some families, they were a chief source of meat. They were an ideal game bird. Not only were they palatable, but they were easily obtained.

Their habit of roosting in close flocks made it easy to kill the pigeons in great numbers. This habit was their downfall. From far and wide, men would come with wagons and guns. Some carried nets looped to the ends of long poles. Others bore sticks and clubs. And as the birds swarmed in for their night's rest, they were slaughtered ruthlessly, gathered into stacks, and loaded into wagons for the markets. They were hauled away not by the dozens but by wagon and cart loads. Tons were sold for a cent or two each in the great city markets. Quantities were taken home by the hunters and salted down in barrels for the winter's meat. And thousands of wounded were left to die and rot under the roosting trees.

It took only a few decades of this one-sided warfare until the flocks were so decimated that the few remaining pigeons became a matter of public concern. Nature lovers got serious about the issue. Laws were passed. But it was too late. The flocks had dwindled to bands, pairs, solitary birds. And, finally, Man—modern, civilized Man—dispatched another victim to follow the fate of the dodo and the great auk. The prairies had already been stripped of their herds of bison; the beaver had been driven into remote retreat; now, in the twentieth century, another terrible blow had been struck against nature. This deed was final and effective, with no pardons and no second chances.

We look back today and say, "What a pity that our fathers did not see it, that not enough birds were spared to replenish the race." But rather than condemn the past, we might better look to today. The better to see today, look back again a century.

It is spring, a warm day in late March, a day when the soft sun caresses the cool, moist soil to arouse early vegetation from a long slumber, one of those first lazy days when winter clothes feel too warm and the thin shade of a leafless tree is welcome. Suddenly clear, resonant whooping sounds are heard in the distance. There are more and more whoops, and then there appear, high up, not geese, but cranes in a long line. The line breaks, the birds circle about on quiet wings, round and round without formation, yet not in disorder, while the clear resonant calls come louder, more frequent. The birds are snowy white, with black-tipped wings, huge, much larger than geese. Their flight is easy, graceful. They seem to bask in the atmosphere as an expert swimmer might skim the surface of a pool. Suddenly, a shot rings out nearby, a victim plummets, and the flock forms again and streams rapidly away.

Here is another beautiful species, one that once bred all the way from Nebraska and Iowa to west central Canada. The whooping crane was not especially rare during the 1890s. The last I remember seeing them during migration was in 1907 or 1908, on a sandbar in the Cimarron River of Oklahoma. There were several, in company with a large flock of sandhill cranes, all lined up in soldier fashion in the spring wind. The whooping cranes and several of the sandhill species indulged in spectacular dances, leaping into the air, flapping and kicking ludicrously.

The whooping cranes are not all gone, but they were very nearly exterminated by the 1930s and doubtless would have been completely wiped out had nature lovers all over the United States and Canada, led by the National Audubon Society, not moved heroically to save them. Now protected, those that remain are just a little better than holding their own, but the future of the species is still far from assured.

These species had been on the earth millions of years before man emerged from caves. Even a few thousand years ago, they constituted a far more significant part of the earth's population than mankind. They belong to the environment into which humankind was born. Surely no one generation has the right to eliminate from this planet any such member of the animal kingdom.

3

In the spring of 1918, I resigned from LaVerne to pursue further graduate studies. The University of Pennsylvania indicated interest in an extensive research program on tobacco smoking that I had outlined, and I was assured that some form of scholarship or fellowship would be established to support my work. On that promise, we prepared for a move east in the fall.

But the fortunes and misfortunes of war by this time had forced many research laboratories to focus on programs strictly related to the war effort. Word came from Pennsylvania that my proposed program would have to wait until the war was over. We therefore proceeded to the University of California at Berkeley, where I enrolled and was given a position as laboratory assistant, which covered our living expenses.

I decided to seek my doctorate at Berkeley and went to the acting head of the department of zoology to discuss a subject for my dissertation. I outlined the same program I had presented to the University of Pennsylvania on the subject of tobacco. The professor listened, then drummed his fingers thoughtfully on the desk.

"I do not think anyone here would be interested in that particular problem."

He then described what he thought would be a better one. He wished for me to work on cultures of *Paramecia*, recording details of their behavior under various conditions. I deferred giving an answer and left his office feeling utterly disgusted. I knew that he was writing a book, and it was plain that he needed someone to carry out a prolonged program of observation and record-keeping that would provide data for that book. Such work could be accomplished by any bright senior or junior student and would be almost completely useless in my preparation for teaching.

During the summer, tension built up among the students. Every day, men were answering the call of the draft, while others who had been exempted for their university studies were leaving to volunteer. It was evident that I would not be drafted for service. I was thirty-one, married, and my wife was expecting our first child. Addie and I together, earlier, had applied through the church for placement in Herbert Hoover's program of providing food relief to starving

Europe, an activity that would be useful in a warring world without running counter to our religious upbringing. But this had not materialized. I wrote the war department offering my services in a noncombatant capacity. I was promptly assigned to an insect control program in the state of South Dakota but was instructed to await further orders.

While waiting for these orders to come, we exchanged places with friends who resided in Reedley, California. They took our apartment in Berkeley and attended the university. We moved into their home in Reedley and signed up for work in the fruit harvest. We spent a few weeks picking grapes and cutting peaches for drying.

The waiting began to seem unnecessarily long. Our university pay had, of course, stopped. The harvest was over. Our meager savings were shrinking rapidly. I began pounding the sidewalks looking for work. Any worthwhile job was available only if I could assure the employer that I would not be rushing off to war, so all I could find was day-to-day employment. For a few days, I assisted a plumber, I cleaned old stoves for a secondhand store, and performed various other odd jobs. My earnings were only about two dollars a day, but that bought groceries.

The long-delayed orders from the United States Bureau of Entomology came at last, and October 1918 found us in South Dakota.

Arrival in new country always gives a special kind of thrill. Brisk, cool autumn air greeted us. The leaves had mostly fallen, and I loved everything I saw, even though by all indications, winter was approaching.

We were headquartered at the agricultural college in the town of Brookings. We found living quarters of a sort, and I began at once to line up my work, which for the most part was to consist of traveling among the county agricultural offices and farm organizations throughout the state.

South Dakota was raw and undeveloped. The state had one great resource, arable lands. I was being sent to set up a statewide insect control program to improve food production by eradicating the pests that were depleting the much-needed crops. It was a useful application of my background in entomology, although a far cry from hunting out and cataloging the more fascinating members of the insect world.

There were several of us serving as a team associated with the agricultural extension service: specialists on poultry, hogs, cattle and sheep, an expert on economics, a domestic-science advisor, and myself. For the most part, we did not travel to the same place at the same time. Each specialist worked out his own program from a shared office in the State College of Agricultural and Mechanical Arts at Brookings. I visited every county in the state. County agents eagerly outlined programs of meetings with farmers in areas of their counties that had experienced insect depredations. Grasshoppers were the worst pest, but cutworms, wireworms, potato and cucumber beetles, chinch bugs and various other destructive insects constituted a real challenge. I loved the work.

Travel from town to town was by train. There was a depressing aspect to travel in the winter of 1918–19. The war-time plague of influenza was raging throughout our country, especially in the military training camps. At each station where my train stopped, I would see unloaded from the baggage car long pine boxes in which the bodies of boys who had fallen victim to the flu were being returned to their parents. Civil hospitals were full of patients also, and in many homes there were seriously ill persons.

In order to attend farm meetings, usually held in rural schoolhouses, we would depend on truck, wagon, sled, or Model T Ford. There were no hard-surfaced roads.

For several weeks, Dale White, the Stanley County agent, and I attended meetings of farmers in the out-back portions of the thinly settled county. We were working in a region of barren plain, broken by ranges of low hills. In 1919, except for small, widely separated tracts, this was open grazing country. There were no fences, and houses were several miles apart.

Persistent early spring snows slowed us and finally marooned us. We were twenty-three miles from the nearest railroad point when we had to abandon our snowbound little Ford. The railhead was 200 miles from home. Trains were halted, telephone wires were down, the mail was not going through. I had been expected home thirty-six hours earlier. There was no means of sending a message to Addie, who was alone at Brookings, expecting our first child within weeks.

Dale and I had been driving along a ridge-crest trail when the left front wheel came off. The car sidled to a sliding stop.

"What the hell's the matter with the car?" he exclaimed.

I pointed to the wheel, rolling briskly off down the hill.

We hiked two miles to the nearest ranch. The following morning, Dale headed back to the car on horseback to retrieve a suitcase. I was to strike off two miles in the other direction, toward another ranch, to catch the delayed mail sleigh which surely would be on its way again and hold it until Dale could arrive. Then we both would ride the mail stage into town.

As I started out, wrapped in a fur-lined coat, I felt as though the tramp over squeaking snow on such a crystal-clear morning promised a pleasant diversion after several days of wading drifts, shoveling snow from in front of the Ford, and constantly striving to find the way through blinding blizzard conditions. Twice Dale and I had been lost, and once an elderly member of our party had narrowly escaped freezing to death.

All of this made me appreciate the fairness of the morning. As I reached the top of the first hill, I turned to look about, for the past few days had taught me not to trust too far a winter day on the South

Dakota plains, no matter how fair. On the northwestern horizon hung a narrow, bluish haze which seemed so far away and small that I felt somewhat cowardly even for marking it. After I had gone about a mile, I saw that the cloud had risen rapidly, and though it appeared thinner, I was convinced it was flying snow driven by wind.

Over the past week, I had learned that residents of this country took no chances in a blizzard. It was common practice to keep a rope stretched between house and barn during stormy weather to avoid getting lost going from one to the other.

I calculated rapidly. It must be as far back to cover as ahead, but the direction of the storm favored going on, for the wind would be at my back. As I forged ahead, however, I carefully studied the landscape for features that would help me to get along without my eyes should a severe storm overtake me.

About a quarter of a mile below me, on the southward slope of a ravine was an old, broken-down shack. The south and east walls had fallen away, but the remaining walls and roof appeared to offer a sort of shelter. I planned how to find my way to the shack, if necessary, in blinding snow.

As I passed the head of the ravine, the wind was rising, but still no snow was flying. I started climbing what I hoped would be the last hill before the ranch house. I had not gone more than a hundred yards when the breeze stiffened. A little snow flew. Only a few paces, surely, would carry me to the last downward slope to the ranch.

When I was nearly to the hilltop, the storm seemed to explode around me. The air was so instantly filled with snow that absolutely nothing was visible beyond five or six feet. I recalled with bitter understanding what the old rancher had said: "No man can keep his way in a blizzard. All this country looks the same in a blizzard."

I turned back, searching for my own footprints. When I thought that I must have reached the ravine, I walked back and forth a couple of times to check my bearings, dreading to leave the only landmark I had—my own fading tracks—then plunged down the slope to my right. The side blast of wind almost threw me off my feet. I landed in a plum thicket with snow up to my armpits, and was encouraged, for I had noticed such a thicket at the base of the steep embankment I had selected to mark the point of departure from my trail. Several minutes later, a moment's lull in the wind, or more probably its leap over the ravine, cleared the air sufficiently to give me a quick glimpse of the shack.

Once shelter seemed within reach, my mind began to struggle with another problem. How could I avoid freezing to death? I seldom carried matches, and I scolded myself for this neglect. Then there flitted into my consciousness a faint hope. Several days before, on seating myself in the Ford, I had heard a faint "snap," and on investigation had located a small object, presumably a match or a toothpick, which had worked through a pocket and found its way down to the corner of my coat. On the basis of reasonable conjecture, taking my habits into account, there were about nine chances that this object was a toothpick to one chance that it was a match, to say nothing of the possibility it might be neither, but I cherished that ten-percent chance.

When I reached the windbreak, the contrast with the fury outside its shell was startling. At first, it seemed I could live there comfortably for days. I covered my face with my mittens until I could loosen the icicles that had formed over my eyes. As I swung my arms to warm my stiffened fingers, the cold returned; even this shelter could not promise survival.

The shack, perhaps twelve by sixteen feet in its principal dimensions, had been enlarged by a small extension of some five-by-seven feet on the east end of the north side, and while all the other walls now were dilapidated, the northwest corner of the addition remained intact except for numerous cracks and holes. I heaped snow against these walls to reinforce my wind screen.

Then I took out my knife and cut a hole in the corner of my coat, and with great care coaxed out a match stick, almost the complete stem of a headless match. I searched again for the head off that stem, finally persuading a small, hard lump to the hole, where I triumphantly drew forth the match head. It seemed undamaged. I tucked it into my vest pocket and replaced my gloves until the exercise of banking more snow against the cracks of my shelter had warmed my hands well enough to be trusted to handle that solitary match head. I intended to do my very best to kindle a fire with that treasured bit of sulfur and phosphorus.

I gathered small boards, and with my pocket knife reduced them to a reassuring pile of finely cut kindling. In a snowdrift at the side of the shed, I found a crushed paper barrel which I carried in to house my fire. I placed the drum on its side, laid the kindling in carefully, and added several pages of the railroad guide I was carrying in my hand bag. A lone iron shelf bracket hung on the wall, and I salvaged this to use as a match scratcher.

Under the protection of my overcoat I then scratched my match, more carefully than I had ever before or have since. There was a crackle, a bit of smoke, smell of brimstone, the burst of a phosphorous flame and its dwindling and extinction, and at last the kindling blazed up.

I set about preparing to stay, however long, until the storm had passed. First, the needs of my fire and the supply of fuel had to be measured. I timed the blaze for one hour, measured the amount of wood that had been burned, and calculated there was enough lumber in the shack to keep me from freezing for six days without having to burn the corner which formed my shelter.

Having noticed a ragged strip of green window shade dangling from its otherwise bare roller, I detached it carefully, tacked one end to a floor board, ran it up through the stove pipe hole in the roof, nailed another board to the end of the first to extend the length, and thus ran a "flag" as high as possible to attract any searchers or passersby. Next, I contemplated how best to economize my energies. I would be able to maintain myself for a week, I was certain, and surely the storm would break in that time.

At first, I was troubled with anxiety about my wife and the worry I knew she must be undergoing. Then I feared that Dale White might strike out to search for me and suffer a worse plight than my own. Very soon I concluded that the best service I could render those two people would be to see to it that I did not worry away my capacity for endurance, and from that moment, I began to think of my situation as an interesting adventure.

I sat by the fire on an old broken chair rescued from another snowdrift and drank snow water which I melted in my folding drinking cup. At intervals I added a board to my fire, or to my shelter to stop up a crack, or piled more snow against the wall from the outside. My biggest worry was that I might fall asleep, allow the fire to die, and fail to waken. So I wrote an explanatory note, in case the worse came to pass, on both sides of a single sheet of stationary from the Portland Hotel, Yanktown, South Dakota.

The storm, however, waned by late afternoon. I thought of our hosts back at the ranch a mile and a half away. If it cleared enough before dark, I could climb to the top of the hill behind my little castle so that they might see me. Strong wind still blew, but all loose snow had been moved to low ground, and the air was clear.

But before I could venture out, my frost-bitten face needed protection, and I remembered how the wind had penetrated my fur-lined coat like needles. Pondering the problem, I saw a solution. The paper barrel that sheltered my fire was not so large that my overcoat could not be buttoned over it. I cut a length to reach from my armpits to hips.

Then I detached the bottom and cut eye holes in it, with another slit for my mouth. Broken down the middle and with a spare shoestring attached to hold it in place, it gave wonderful protection for my face.

As the sun approached the horizon, I made my way in my crude armor to the crest of the hill where I could see across to Edd Klopping's ranch. Almost at once, I saw him leave the back door, and a moment later, a horse and rider dashed from the barn and headed my way at a dead run. The horse skidded to a stop. Edd dismounted, grabbed me, threw me across the horse's rump, leaped back to his saddle, and spurred the panting animal to a gallop.

"Stop! Stop!" I yelled. "You're killing me. Edd, you don't know what sort of contraption I'm wearing." My hip-to-shoulder wind break seemed about to amputate my arms and hindquarters.

Edd slowed his horse to a walk. "I thought you looked strange, but I had so feared you were dead that I didn't stop to ask questions." When we reached the house and I took off my mask and overcoat, Edd saw my wind shield and nearly fell over laughing. He hung the barrel and mask from a nail on the west wall of the house as mementos.

Dale White had returned from our wrecked Ford just as the storm was reaching its height, stabled his horse, and waited until the worst had passed, then started out on horseback to find me. Failing in this, he went on into town with the horse-drawn mail sleigh.

The next morning, I took a horse and started for town. At noon, I stopped at a ranch to ask about Dale. He came out from the kitchen and hugged me like a sweetheart. Several men were with him.

"We were heading out right now on our way to find you in a snow drift."

***

A few weeks after my return from Stanley County, our first child, Bobby, was born in the Brookings hospital, big and bouncing at ten pounds.

In May 1919, the US Bureau of Entomology terminated its emergency field program. I might have renewed job negotiations with the Universities of California or Pennsylvania, but the termination found us with the added expense of a new addition to the family, and time was of the essence. For some time, there had been in my hands a request from Southwestern College in Winfield, Kansas, to head the department of biology. I decided to accept it. Then I wrote to the State Agricultural College in Manhattan, Kansas, saying that I had an empty summer, and was promptly offered employment for the season as an extension agent in insect pest control. I escorted Addie and the baby to her family home near Murdock, Kansas, and spent an extremely busy summer teaching Kansas farmers how to control cutworms and grasshoppers.

4

With the start of the school year of 1919–20, I settled into to my expected teaching career as professor of biology at Southwestern College. The following spring, I was called to my alma mater, which was then in the midst of a development drive. McPherson College had a new president. There were great plans to strengthen the faculty and build a new science hall. I hated to leave Southwestern, where I was just getting started and was well satisfied, but I handed in my resignation effective at the end of the summer session. The prospect of returning to the old school at its time of growth was intriguing.

The president at McPherson was Dr. Daniel Webster Kurtz, who guided the college from 1915 to 1925. He was a man of distinctive appearance—short, stockily built, with strong facial features, an impressive head of curly black hair, and a voice of rich timbre and wide oratorical range. He was a popular lecturer and an avid reader who could extract the message of a book on philosophy or religion and present it in a most forceful manner.

Those of us who had left the college but had kept in touch with its progress were much impressed with the program of activity and growth under Dr. Kurtz and his new, young staff. He had a way of inspiring his young teachers.

Addie and I settled down to home, family, and my teaching. We were to stay ten years. We had a small frame house about a mile from the campus. Addie's sister and mother lived next door. Our daughters, Doris and Margaret, were born in 1921 and 1925.

Life was carefree, if not luxurious. We romped with the youngsters in our small house and yard fitted with swing and sandpile, apricot tree, pet dog and cat, kiddie-car, wagon, dolls, and tea tables and all the other impedimenta of childhood. In the spring, we hunted daisies and flew kites on the hills across the "draw" that cut just past our house. In the winter, the youngsters were bundled up and onto the sled, laughing and giggling until one would tumble off and perhaps return crying. Then we would head inside, all of us, to warm up by the old base burner.

Ours was the home life of a typical young family, with the usual share of hurts and tragedies to bear and the gradually settling plans for the future.

***

A benchmark of Dr. Kurtz's presidency at McPherson was his willingness to support and to seek funds for new ventures by the college or members of its faculty. One such venture was the McPherson Scientific Expedition of 1921, an insect-collecting project organized by Mr. Warren Knaus, editor of the town paper and recognized expert in the study of beetles, in which I participated on behalf of the college.

Our purpose was not only to contribute to Mr. Knaus's already famous collection of beetles, but to assemble as large and representative a group of the insects as possible for the college, both for display and as stock for trade.

Warren Knaus demonstrated that it was possible to take a scientific hobby, master it, and become a recognized authority, even if the interest was only avocational. Despite the fact I had spent nearly a year earning a living as an entomologist, I owe most of my abiding interest in entomology to Mr. Knaus.

Born in Indiana in 1858, Warren Knaus received most of his schooling and his college education in Kansas and settled in McPherson. Here, he edited the *Democrat-Opinion* in a Republican county, served as president of the local Chamber of Commerce, was a leader in most of the important clubs and committees of the town, supported a family of eight girls, and all the while collected beetles.

He became a world-acknowledged expert on Coleoptera, the group of insects whose known species, numbering some 250,000, account for several times the total number of species of mammals, reptiles, and birds. His private collection contained more than 100,000 specimens, representing about 9,000 species of North American beetles, cataloged in a book known affectionately as "Knaus's Bible," which he kept centered on his desk at all times.

From Warren Knaus, I learned how complex and different each insect was from any other, whether it were one that gleamed like a polished gem stone or one with a creepy legginess, unpleasant protective scent, or ugly set of pincers. Mr. Knaus discovered some 100 species, of which about twenty bear his own name appended to their species designation.

"My children are all girls and cannot perpetuate my name," he would say. "It will be perpetuated by my bugs."

Mr. Knaus was a public-spirited, and also religious, man. He contributed regularly to support of his church, but I never knew him to attend services. He was made a fellow of the American Association for the Advancement of Science in 1921, was similarly honored by the American Society of Entomologists, and served as president of the Kansas Academy of Science.

He was a tall man. He walked with a long stride but stooped slightly, bending forward because one eye was completely blind while the retina of the other was so badly scarred he could see details only by holding an object within six inches of his face. In field work, he required field glasses (binoculars) even for eight- or ten-foot distances. For the 1921 expedition, Foster Hoover, my lab assistant, and I served as his extra "eyes."

As had Dr. Stevens, Mr. Knaus gave me lessons in stick-to-itiveness.

The trip occupied two months and took us as far as California, with the major collecting done in Utah and Nevada. Mr. Knaus had timed our itinerary carefully so that we would arrive at certain localities at the precise time and season when rarities had been encountered by collectors thirty or more years earlier. Some of these species had not been seen since.

Our first stop was at Medicine Bow, Wyoming. In 1872, on the shores of nearby Lake Aurora, Professor Williston had discovered the species of tiger beetle that bears his name, *Cicindela Willistoni*, which had not been reported since from any other location. Knaus was determined that we add a few specimens to his limited series of this handsome beetle.

Though it was early July, we disembarked from the train at Medicine Bow on a cold and drizzly day that gave us little hope for success in our search. After a seven-mile walk through sage brush and mud, we arrived on the lake shore cold and wet.

This particular beetle, which runs very actively over the moist shores on sunny days, takes cover when the weather is cool. We sought specimens under logs, boards, and drift of all kinds. We dug into beetle burrows. Foster Hoover and I worked very industriously for hours, but without success. Due to his defective eyesight, Mr. Knaus contented himself with caring for our catch—and this consisted of almost everything except the beetle we were hunting. Several times he advised us to give up, since the day was so unfavorable. But we were persistent, interested as we were in being on the beetle's native ground, and impressed as we were by the dollar price each specimen would command.

When at last we were ready to return to camp, Mr. Knaus walked over to where Foster and I had been turning up logs, stones, sticks, and clods. He stooped and turned over a very small piece of board. He knelt on his knees, his long frame doubled up like a jackknife, his eyebrows nearly brushing the ground and his hands fumbling for his insect bottle and catching forceps. Beside him, six of the precious tiger beetles stood all in a row, evidently waiting for a chance to get into the bottle of a real entomologist. Foster and I spent another hour hunting, but our efforts were fruitless.

Near Death Valley Junction, California, in the midst of great, bleak stretches of sand, gravel, and barren mountains, the nearest being the Funeral Range on the west, lies a small oasis watered by several large springs which flow out of the earth, cool and refreshing, in volume sufficient to form a small river which traverses a few miles of the parching soils, then sinks away or evaporates as do hundreds of other half-hearted streams of the desert. The watered area supports a growth of salt grass and a few other saline plants, for it is heavily impregnated with salts. This lonely place, known as Ash Meadows, is just the sort of isolated spot where one is especially likely to find new or rare species. *Cicindela Nevadica* had been described in 1872 from a single specimen found here and was not reported again until

1919, when another ten individuals were found at the same site.

Our side trip to Ash Meadows cost nearly a hundred dollars and seven days of hard work by the three of us. This was a substantial investment. We found and captured a number of the beetles, and they brought an extremely high price, enough to allay a good part of the expedition expense. But what if we had found none? Knaus taught me that no true naturalist worries over spilt milk, nor becomes despondent because the object of his search eludes him. Disappointments must be expected and accepted as matters of course; rare and valuable specimens often cost many a "water haul." In later years, when so many of my meteorite searches dissolved into disappointment, I appreciated having learned this lesson.

Knaus, then sixty-three years old, thrived on frustration and hard work. He seemed to look and feel stronger at the end than at the beginning of our eight-week expedition. Our average work day stretched to fourteen hours, and on one day in the Wasatch mountains, we walked more than twenty miles over rough trail.

Near Parowan, Utah, we were carrying our bedrolls and other equipment up a canyon when we were overtaken on the trail by two men riding on the running gears of an old lumber wagon. They had sacks of grain and boxes of groceries tied on the wagonhounds, coupling pole, and axletrees, but they offered to add our load and found a place for Knaus to ride while Foster and I walked. Soon we were engaged in a discussion as to whether or not we should follow the trail to its top. Our hosts urged us to take advantage of the opportunity to view what they assured us would soon become a favorite destination of every traveler on the Arrowhead Trail. An access road was just then being completed.

Though we had left camp prepared for only an overnight side trip and were little suited for a stay at an elevation between 10,000 and 11,000 feet, the scenery was growing ever more attractive and the description ever more glowing, and we continued to go farther and farther yet. We approached a great battlement of giant sandstone cliffs, beautifully carved and rising high above an exquisite stand of blue spruce. Surely this was the spot? The contempt with which this suggestion was received left us little doubt as to the virtues of the scenery yet ahead.

We camped by five o'clock, and our guides insisted we must see their promised land yet that night, so across a half-mile of rich, wild clover meadows, over clear brooks and through scattered blue spruce stands, we came to the north rim of Cedar Breaks, destined to be designated a national monument a dozen years later.

Harvey Nininger (left) and Warren Knaus (right) at Cedar Breaks, Utah. Nininger, Knaus, and Foster Hoover, Nininger's lab assistant, collected insects on the two-month long McPherson Scientific Expedition of 1921 sponsored by McPherson College.

We kept extending our unexpected stay, until ultimately we had spent twelve days collecting near Cedar Breaks, working much of the time at the 11,000-foot altitude of Brian Head Peak. Insect groups that we had found at desert level often were represented at the mountaintop by somewhat modified forms, apparently evolved by selective adaptation into species quite distinct yet recognizable as belonging to the same families or genera as those taken below.

I mentioned to Mr. Knaus that in several days of collecting, we had not taken a single specimen of the genus *Eleodes*. He agreed that it seemed probable that at least one representative of such a large family, although a typically desert genus, should have adapted to the higher elevation as had so many of the other beetles. Jokingly, I called to Foster Hoover to busy himself finding a new species of *Eleodes*. Foster began turning up every flat stone he could find, but his call came in just a few minutes. "Here it is!"

He had an *Eleodes* beetle such as I had never seen before. Mr. Knaus took the beetle between finger and thumb, brought it into focus for his good eye and, cocking his head far to one side, scrutinized it very critically. A grin of satisfied surprise spread over his grizzled countenance. "Well, I'll be jiggered. A million dollars to a penny, it's a new species."

We all turned up stones and located by our combined efforts another dozen specimens. After necessary procedures for verification, the new discovery was duly named *Eleodes parowana*, the 123rd member of its genus, christened after the Mormon town of Parowan at the mouth of the canyon below Cedar Breaks.

Our camping area was a naturalist's paradise. From the forbidding crevices in great black piles of lava, irate pikas and marmots fiercely asserted their rights and profanely maligned every intruder. White-crowned sparrows sang from mountain willow thickets along tiny streams, and purple finches from the treetops. The pine grosbeak whistled from deep in the forest, and the scolding bark of the Douglas squirrel echoed from among the tree trunks. Thousands of sheep fed in the wild clover.

***

Our Cedar Breaks experience reinforced my hope to somehow develop a summer school where natural history studies could be pursued *in situ*. On the way home, we made a brief survey of prospective locations near Colorado Springs.

Back in McPherson, memories of the refreshing atmosphere of Palmer Lake, Colorado, kept plaguing me to test the idea. It seemed there could be a no more inviting spot in which to offer true "refreshing" courses to public school teachers from the plains of Kansas. At this time, most summer session students were teachers.

I was aided in taking my mountain school proposal before officials of the college by B. E. Ebel, professor of modern languages, who enjoyed the highest respect of both faculty and students and was the embodiment of the term "gentleman." Many of our fellows held obvious views that the proposed summer program was a rather wild undertaking that would most likely soon need an undertaker, but permission was granted for a trial session, and the Rocky Mountain Summer School opened in Palmer Lake in the summer of 1922.

In addition to natural history, courses included education, languages, and history. A faculty of four was organized and a bulletin extolling the virtues of earning required credits in the cool and invigorating climate of the Rockies, far from the steaming Kansas plains, was circulated. The village fathers of Palmer Lake encouraged the project, helped to publicize it, and provided rustic quarters in what had been a YWCA camp. The two sessions of the first summer had a total enrollment of forty-five.

Although I held the position of director of the school, Professor Ebel was always my right-hand man. I depended on him so much, sometimes I came to think of him as both my hands. Together, we worked out the problems which arose during the nine years the Rocky Mountain Summer School operated.

One blow was the failure of the bank in a nearby town in which all our funds were on deposit. Another difficulty was severe criticism from other educators, perhaps out of jealousy of those of us who could teach in the delightfully cool climate of the Rockies while they sweltered in the heat of the plains. There were the usual student disciplinary problems, housing difficulties and other common ailments of institutional operation.

Professor Ebel was tall, straight and blond, with strong facial features which most of the time were organized into a winning smile. However, he was deeply serious, and on occasion could couch incisive criticism in language of poetic beauty. In spite of the close association required to address the constant problems of management, on top of caring for our own classes, I cannot remember that there ever passed between us a cross word.

In the early years of the camp, our dormitories were the simple tent-houses of the old YWCA facility. The miniature canvas-walled rooms, furnished crudely with a bed, a box, a candle, and two nails for wardrobe, made secrets among the occupants impossible. As one remarked, "the softer the whisper, the louder the sound."

A major feature of our school in the mountains was the emphasis placed upon outdoor excursions. Weekends were given over to picnics, hikes in nearby canyons or all-day drives to survey botanical and geological characteristics of the mountains or foothills. Once or twice each summer, a trip was planned up Pike's Peak, and both students and faculty were encouraged to participate. Some would go by auto, some on foot, some by the cog railroad and some by burro. It never was possible for all of the faculty to go on the same trip, and Professor Ebel usually found his work compelled him to omit the Pike's Peak excursion. One year, he decided to go, and though he was not young, he insisted on doing it the hard way—on foot. Some of us had serious misgivings about his undertaking to hike all the way to the summit of a peak above 14,000 feet in elevation, but he went anyway, leading a considerable party of students. He was determined to see that all of them made it to the top, and of course, that left him no choice but to go all the way himself.

The morning after their return, Professor Ebel was plainly feeling the effects of the trip. At lunch, I asked him what his opinion now was of Pike's Peak. With all the dignity that his six-feet-and-three-inches and his resolute face could muster, he told me, maintaining his usual formality of address, "Professor Nininger, make no mistake. Pike's Peak is a rugged, steep, genuinely tough, and big *he*-mountain!" Every year after that, he led the foot trek up the peak.

Each year, the Summer School enrollment and faculty increased. As in everything, Addie played a prominent and useful role. She served simultaneously as dean of women, manager of dormitories, and receptacle of all and sundry troubles and confidences, in addition to caring for our own growing children.

Bob had been three years old when the school was established and was twelve when it closed. Doris was eight months old when it opened, and Margaret was three months old at start of its fourth season.

A radical and romantic change in the setting of the Rocky Mountain Summer School took place in 1926, with the acquisition of the old and lovely Estemere estate. This beautiful building, set on broad, green lawns against the rugged side of Sundance Mountain, was a relic of the early grand days when Palmer Lake was the summer resort dream of General William J. Palmer, a Colorado pioneer and railroad giant, and his friend Dr. W. Finley Thompson, a wealthy physician and dentist from the Midwest.

Palmer and Thompson, attracted by the surroundings of what in the 1880s was the little-known rail station of Weisport, transformed the area. They dredged the little lake known as Loch Lomond, established boating and swimming and built a hotel, the Rockland, with a deer park, a bandstand, and a sprinkling of resort cottages. Reservoirs were constructed in the rocky canyon, and a carbide gas light system was developed.

Dr. Thompson initiated the name change of the town, which had also been known as Divide or Lake Station, to Palmer Lake in honor of his friend. When the general turned his attention toward the founding of Colorado Springs, the doctor stayed on, selected fifteen acres of mountainside, and in 1887 and 1888 built the turreted and gabled mansion he named the Estemere. The walled and terraced grounds were fenced with wrought iron, posted with lamps, guarded by bronze lions and graced by fountains and a summer house. The fabulous mansion eventually passed into the hands of Palmer Lake's citizens, who made the estate available for the Summer School.

Very few changes were made to the spacious home. Only such utilitarian pieces as additional beds and chairs supplemented the luxuries of the past. The billiard room was transformed into a library, but the felted gaming table remained. The red room, the violet room, the carnation room, the blue room—each with its characteristic wall covering and its own fireplace of individual design, together with a few gems of old furnishings—became the summer headquarters of Kansas girls. Soft gaslight glowed in every corner of the big home, caught and reflected in a glorious collection of crystalled and chimneyed chandeliers. The servants' cottage was converted into a men's dormitory, and the huge old carriage house became "Pioneer Hall," fitted with classrooms and auditorium. The great, old mansion was a never-to-be-forgotten background for student summers.

For five summers, 1926–30, the Estemere was host to youth and enthusiasm, echoing to story hours, home dramatics, dormitory parties, tennis and croquet on the lawns, and recitations. At the ringing of the dinner bell, students crowded into four big tables in the dining room and ate family-style. Townsfolk participated in many of the lecture and entertainment activities. The two sessions of 1930 had a combined enrollment of 150—more than three times the number of 1922.

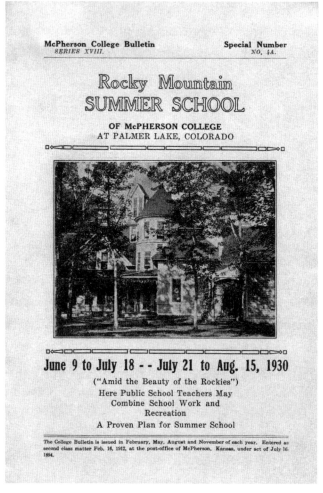

Rocky Mountain Summer School brochure. Harvey Nininger was the founder and dean of the Rocky Mountain Summer School in Palmer Lake, Colorado, which operated from 1922 through 1930.

Despite its success, the Rocky Mountain Summer School drew resistance. The Kansas State Board of Education, made up of heads of the various state schools, questioned the extension of a summer school to a neighboring state. Two times, our standards were put to question and then approved. Finally, the State Board simply adopted a rule requiring that all accredited courses be given on the home campus, or that any off-campus extension program be financed solely by the home school. The Board of McPherson College decided this could not be done, and the Rocky Mountain Summer School became history, only an experiment in education.

4
The Challenge

*If a man does not keep pace with his companions,
perhaps it is because he hears a different drummer.
Let him step to the music which he hears,
however measured or far away.*

—Henry David Thoreau

As America settled into the nonchalant years of the 1920s, I settled into the academic and community routine of small college faculty life. My specialty was biology, but I was also teaching a field course in geology. I enjoyed my work. Any problem that demanded critical observation intrigued me. I thought of myself as a *naturalist* in the older and truest sense of the word—one who critically observes and describes facts of nature.

About the middle of August 1923, in *The Scientific Monthly*, I found an article by Professor A. M. Miller of the University of Kentucky. The subject was meteorites. I cannot remember ever reading anything that so completely captivated me.

All during my childhood, meteors were regarded in about the same light as ghosts and dragons, mentioned rarely and never discussed seriously. During my Oklahoma and Kansas student days, I associated with field and laboratory scientists, yet I am sure I never heard the subject of meteorites mentioned more than once. On that occasion, it was simply dismissed as "irrelevant." The word "meteorite" was scarcely part of my vocabulary when I graduated from college and was almost equally unfamiliar after years of graduate study and teaching.

Now, reading Dr. Miller's article, I found there was a body of knowledge of which I had read nothing previously except a few scattered paragraphs in the public press. It was a source of actual embarrassment to me for some days, for I had been in the habit of thinking that I possessed a pretty fair education.

Only a few months before, while attending a meeting of ornithologists in Chicago, I had walked with a noted scientist from a Philadelphia museum through the mineralogical hall of the famous Field Museum. We came upon the great meteorite exhibit and paused before a case of beautiful specimens.

My friend broke our silence. "I wonder if they really know that these things come to the earth from space," he remarked seriously. He had spoken my own thoughts. As I was to learn through succeeding years, he had spoken the thoughts of many scientists, including not a few geologists.

I reviewed Dr. Miller's article for our Faculty Science Club. Subsequently, I spent many hours pondering the phenomenon of meteorite falls and their uniqueness. It seemed to me that by any fair appraisal, the arrival of a meteorite from space must be one of the most basic and fundamental of natural events. Here was a source of information concerning the universe beyond our atmosphere of a kind that astronomers were making no special effort to utilize. Here was a more or less constant contribution to our planet from space which geologists seemed to be making no effort to evaluate except as accidental recoveries were thrown into their laps.

What I would not give for the opportunity to witness the fall of a meteorite to earth! According to Dr. Miller's estimate, my chances were very slim, perhaps one in a million. Nevertheless, I could not resist considering the steps one might take to determine the course and the approximate place of landing, and how one might go about obtaining community assistance for a search. Dr. Miller had supplied no blueprint for such an undertaking.

Unknown to me of course, in August, while I was first reading Dr. Miller's article, there was out in space a meteorite traveling in an orbit which crossed that of our planet at a point some millions of miles ahead of us; and the meteorite's speed and timing was such that we would reach that crossing at the same time. Most important for me was the fact that the side of our planet on which I lived would be turned toward the stranger at precisely the right moment to receive it. Of our hundred-million-square-mile hemisphere, only that half-million-square-mile area within which McPherson College was located would be permitted to experience the light display which would mark the end of the meteorite's flight.[2]

---

[2] Nininger used the terms "meteor" and "meteorite" as defined in the glossary beginning on page 553 of this book throughout his career. Meteor: "the light phenomena caused by a meteorite's passage through the atmosphere." Meteorite: "a mass of solid matter, too small to be considered an asteroid; either traveling through space as an unattached unit, passing through the atmosphere, or having landed on the earth and still retaining its identity. He did not use the term "meteoroid," the commonly accepted term now used to describe an object smaller than an asteroid traveling through space. For consistency and simplicity, "meteor" and "meteorite" are used herein as Nininger intended. *Ed.*

On the evening of November 9, 1923, faculty and students gathered in the chapel for a lecture, play, or other program common to college communities. At the close of the program, I walked toward home with my friend, Professor E. L. Craik. We paused in front of his house to chat. Astronomy didn't enter our conversation. Had anyone asked, I could have told them that we were sailing along in our orbit at about eighteen-and-a-half miles per second, that our town and college along with all the rest of the earth's surface were in normal rotation about its axis at a little less than a thousand miles an hour. But I surely could not have told that a meteorite only a few thousand miles away was following a course that within minutes would mean the extinction of this minor member of the solar family before our very eyes and at the same time would start a metamorphosis in the life of a young biologist.

Suddenly a blazing stream of fire pierced the sky, lighting the landscape as though nature had pressed a giant electric switch. The blade of light vanished with equal suddenness, leaving a darkness seeming thicker than before.

Momentarily, Professor Craik was speechless. Then he saw that I was bent over, making a mark on the sidewalk. He asked me what I was doing. I remember telling him I was going to find that meteorite and that I was plotting its path from where I'd seen it. He laughed, but I was serious. I will never forget the conversation that followed.

"Do you really think that something has actually come to Earth?" he asked.

I told him I was certain of it.

"Well, where do you think it landed?"

"Probably within 150 miles," I estimated.

He laughed again. "Now I know you must be kidding."

"No. I'm serious, and I'm going to hunt for it."

"Has such a thing ever been done?"

It was the question that has followed me all of my life, thrown at me whenever I have proposed something new.

Professor Craik and I didn't talk long that night. I wanted to dispatch a message to the principal newspapers of our state. I asked the editors to publish a request to all who had seen the great meteor that had passed toward the southwest at 8:57 p.m. Central Standard Time to please furnish to me the following information:

1) The location from which the observation was made.

2) The exact direction of the fireball in relation to the observer when it disappeared, naming, if possible, some town in Kansas or a neighboring state which was in line with the path of vision.

3) How high above the horizon was the fireball when it disappeared or was extinguished—this to be stated in degrees, remembering that from horizon to zenith is 90 degrees.

4) Was any sound heard in connection with or shortly after passing of the fireball?

Reports piled up, and so did my difficulties. As my file began to bulge, I called in Professor Charles Morris of the physics department, who had surveying instruments and knew how to use them. We met in my laboratory on a Saturday to digest the reports that had accumulated.

A hundred and fifty miles from McPherson, near the end of the November 9 meteor's flight, people had been frantic. They had leaped from their beds and rushed from their houses in the belief that their homes were afire. Men on the highways stopped their cars, blinded by the flash of light. Livestock crouched in fear. Children cried and women prayed. A few moments later, the sounds of blasts and a thunderous roar seemed to confirm fears that the end of the world had come.

One can readily imagine that a farmer, merchant, or other person dazzled for the first time in his life by the spectacle of a great fireball might have some difficulty reporting the experience with accuracy. People wish to be helpful on occasions of this kind. So instead of merely supplying the simple information requested, the writers offered more. Why should a man satisfy himself with saying simply that the object disappeared in the direction of a town sixty or eighty miles away, when he could tell me the site, within a hundred feet or so, where the thing actually had struck? Or so he thought.

This was precisely what most of the reports set out to do. When Morris had waded through a dozen or so such letters from correspondents representing every corner of Kansas and localities in Nebraska and Oklahoma, each claiming that the meteorite's landing place was within sight of his house, he threw up his hands. There was no use wasting time on such a hodgepodge of conflicting reports, he insisted. He was used to working with accurate data that could be measured and weighed. Certainly no good could come of struggling with this packet of misinformation.

Subsequently I learned that other scientists during the past century had, like Morris, turned away from such a confusing mass of reports and given up somewhat half-hearted attempts to trace down such a fall. I reasoned that one cannot expect accurate data on such an exciting and even frightening event, coming as it does, unexpectedly and lasting only seconds. However, one can, by careful screening, select from a large number of witnesses those who will be most able to supply additional useful information through personal interview.

With an objective of such importance, one had better use whatever reports he has as best he can. I also reasoned thus: meteorite falls are important events—not mere incidents—and are rarer than comets. Meteorites are tangible astronomical objects, not mere mysterious objects viewed from afar. In my opinion, they were far more important than comets so far as either could be studied in that day.

So I used the correspondence as a guide to a mere rough approximation, judging from each letter whether indeed there was useful information that could be gained by talking with the writer on the spot where he had stood at the time of this exciting experience. By selecting the best half-dozen accounts and plotting on a map the reported observations, it was possible to build convincing evidence that the fall had taken place within a fairly constricted area, perhaps forty miles across. The indicated location was not that expected by any of the observers but was where the lines of all the sightings taken together converged.

A commander of artillery has difficulty in scoring a good percentage of hits on a target ten miles away, even though he knows the exact location of the target, knows atmospheric conditions, direction and velocity of air currents, shape and weight of his projectile, and velocity with which it leaves the muzzle of his gun. Trained technicians carry out his firing orders and trained observers correct his judgment.

In Kansas, in November 1923, no one had expected the arrival of the meteorite. No trained observers were scanning the skies to record its passage. I had to depend on the memories of startled laymen, who recorded their observations or were interviewed several days after the event. Most observers were 100 miles away from the passing light. The light never reached a point closer than eight to ten miles above the earth. Somehow, the course must be ascertained and projected to a striking point on the ground. I did not know the speed of the projectile, nor its size, shape, or specific gravity. I could only approximate velocity, height, and course. I did not know if the path changed somewhat after burning out. I could assume that the meteorite would fall a little more directly as it slowed. But I knew nothing of air currents or atmospheric conditions. Its entire lighted path was limited to those parts of the atmosphere which were above the region of exploration of meteorology of that day. The meteorite was likely fifteen to thirty miles from its target when last seen.

I had marked the line of disappearance of the meteor, as I had witnessed it, as just behind a branch "next to the top of the pine tree in Mr. Price's yard down the street." I added to my own observation the written reports that seemed useful. Then, in interviewing witnesses, I found that by having them point out stars, nearby objects, or perhaps the branches of trees through which they had looked at the fireball, I was able to determine a consensus as to the altitude of the meteor's disappearance. By studying the course, the seeming velocity, reports of an explosion and description of accompanying sounds, I reached the conclusion that one fragment had fallen in the vicinity of Coldwater, Kansas, and that another, though with less likelihood, might have fallen to the north near Greensburg, Kansas.

The best that I felt could be hoped for was the designation of a township (six miles square), perhaps half a township, in which fragments might have landed. But how to search half a township? One fact favored me. Meteorites were known in many cases to burst when passing through the air, scattering in fragments over an area of up to fifty square miles. This multiplied the chances of finding a part of the fall. Even taking the optimistic view and assuming there were a dozen or fifty fragments, how could one go about searching such an area?

Everyone has suffered frustrating experiences in searching around his own premises for some lost or mislaid article. Consider expanding the area of search to even one square mile. If I were to take a piece of a brick and bury it a few inches in the ground, announcing it was buried on a certain square mile, how many searchers would come forth even if, instead of a piece of a brick, it were something of value? Most persons would see little use in searching. In spite of the slender chance of making a find by direct search, I couldn't resist but try, and I took three or four volunteer students on a weekend search. We found nothing.

I conceived a plan to take the public into my confidence and form a sort of partnership with residents of the area in which I believed the meteorite might have fallen. During late 1923 and into the autumn of 1924, I made several excursions into the vicinity of Coldwater and Greensburg, visiting schools, asking cooperation of the press in printing appeals. I explained the nature and behavior of meteorites and the reasons they are valuable for study, then offered to pay a good price for any specimen found. I counted on this incentive to alert the whole community and generate interest in the meteorite. So far as I know, this method had not been used before.

The plan I followed in 1923 is the one I have used ever since. And until such time as provision shall be made to secure instrumental data by means of automatic devices I am convinced that this is the only method for successful "chasing" of fireballs.[3] Ultimately, there were two meteorites recovered in the vicinity of Coldwater. I now credit neither of them to the fall I witnessed. Rather, I credit their finding to my alerting of the public.

On one of my first visits to the general area, covering parts of two counties, where I was concentrating on gathering data in an endeavor to pin down the fall to a smaller target, I decided to use a Sunday morning to try for an opportunity to speak to a church group. I got to the church just as the crowd was coming from Sunday School classes and gathering for the preaching service. An usher carried a note for me to the minister asking permission to make a three-minute

---

[3] Since the original publication of *Find a Falling Star*, several camera networks have been established in various locations around the world to observe meteors. The widespread adoption of "doorbell cameras" and "dash cameras" has also assisted in tracking fireballs. Additionally, Doppler radar, used for weather forecasting, has aided in the recovery of meteorites. *Ed.*

announcement. He kindly extended me the courtesy. In exactly three minutes, I requested those who had witnessed the great fireball on November 9 to see me at the end of the service. Then I joined in the order of worship.

At the close, Deacon A. M. Brown came to me to say that he had not witnessed the fireball, but that based on my description of meteorites, he thought he might have one in his back yard. He took me to his house and showed me a forty-one-pound meteorite, an oxidized nickel-iron, that he had plowed up four years earlier. He said he had always wondered what such a heavy rock was doing in a field all by itself.

It was evident from the oxidation that this meteorite had lain on the earth a very long time, but the thrill it gave me can never be described. I purchased Deacon Brown's meteorite then and there, paying one dollar a pound, the price I had determined was customary.

Eleven months after my 1923 fireball, a man plowing his ground bumped into a stone which seemed very unusual to him. He brought the rock to town to show it to the local newspaper editor, who wrote to me immediately that someone had found "a piece of the meteor that I had been hunting."

On my next opportunity, while on my way to collect fossils in Clark County with Handel T. Martin of the University of Kansas, I stopped in Coldwater to examine the fragment. It proved to be an eleven-pound stony meteorite. It was found fifteen or so miles south of the course of the meteor as I had mapped its flight, and I had reservations that it might belong to a different and earlier fall, since it showed some evidence of weathering, but I tentatively assigned it to the fall of November 9, 1923. The newspapers, however, for the most part did not repeat my qualification, being more satisfied to say that the wild star chaser had made a catch.

In these days, one would not need to be so puzzled as was I. The

90   IN SEARCH OF FALLING STARS

approximate time of the meteorite's stay on earth could be determined by isotopic studies. Even without such modern aids, I soon learned to recognize the effects and degree of weathering in a rough way. It was not long before I determined that this was not a fresh stone, but one which had lain in the soil for many years before it was struck by the plow.

***

H. T. Martin, my companion on that visit to Clark County when the smaller Coldwater stone was obtained, was a respected fossil preparator and a congenial associate on fossil-hunting expeditions. We were checking out a new fossil discovery that had been reported to me during my meteorite-hunting forays. During the trip, Martin related a most interesting account of a meteorite find he had made some twenty years before.

He was at the American Museum of Natural History in New York City, visiting with the curator of minerals, when Dr. H. A. Ward walked in, having just returned from a trip abroad. Dr. Ward, as usual, was dressed in his solemn Prince Albert coat, in the tail of which were pockets. Ward drew forth from one of these pockets a fist-sized stony meteorite which he offered to sell to the museum curator.

Martin, too, examined the specimen, and said, "I know where there is a big one out in Western Kansas."

His story was received with some misgivings, but Martin was a careful observer and, now that he had learned something of the value of meteorites, he decided on a course of action. He wrote to a young man who had assisted him in field work a few years before, telling him just how many miles to go south from the town of Gove, then east to the corner of Farmer Brown's northwest forty, through the wire gate into the pasture and down the fence to a small branch of

THE CHALLENGE 91

such and such a creek, then northeast along the north bank of the stream about 200 yards. There he was to find, lying half buried in the buffalo grass sod, "a big, dark brown rock."

Martin's final instructions were just as explicit. "Load it in your wagon and take it into town and ship it to my home address."

Thus, the sixty-five-pound Jerome, Kansas, meteorite was recovered and preserved. It is now in the Yale collection.

2

Do with your life
something that has never been done,
but which you feel needs doing.

That November night in 1923, when the sky suddenly was cleft by a shaft of light half as wide as the moon and far brighter, I did not suspect that the event would change my whole life, but almost immediately I felt certain effects. Apart from the excitement of trying to plot the course and recover the meteorite, I found myself completely engrossed, during all the time I could spare from my college and family duties, in an effort to learn as much as I could, as fast as I could, about this new subject. I consulted all the references available, ordered literature, and spoke with all my professional acquaintances. The search for information was less rewarding than had been my search for the meteorite itself.

I wrote to the Field Museum in Chicago for information and was rewarded with a copy of Dr. Oliver C. Farrington's *Catalogue of the Meteorites of North America*, published by the National Academy of Sciences in 1915. Here was a great fund of information on all of the falls and finds that had been collected previous to 1909. I studied it avidly and found that a generation before my time, in the 1880s and

92   IN SEARCH OF FALLING STARS

early 1890s, the state of Kansas had yielded more than her normal share of the meteorite finds made within the borders of the United States. Among these finds had been the largest stony meteorite recorded in the world up to that time and also, the greatest pallasite ever collected. Surely the University of Kansas at Lawrence would be a fount of knowledge.

I gathered together enough cash for train fare and a night's lodging at a third-rate hotel and set out fully confident that I would return with useful information. During my frantic search for the November 9 meteorite, I had come across two farmers who had in their possession small meteorites from a recognized fall for which they had failed to find a market twenty years before. These I had managed to purchase with borrowed funds. They gave me the reassurance of first-hand knowledge, and I carried them with me in my briefcase.

The train's wheels could not turn fast enough. Every station stop bored me on that trip to the university. Professional pride demanded that I fill the wide gap in my knowledge of the world around me.

I was dismayed when members of the geology department at the university were unable to answer any of my questions. They not only professed ignorance of the subject; at first, they showed total lack of interest. They showed me one meteorite specimen on loan from the astronomy department where it had lain unlabeled for years. They believed it to be the Tonganoxie, Kansas, meteorite, but were not quite certain. I looked at the forty-pound specimen and saw instantly that it was a pallasite. Even I had learned the conspicuous difference between pallasites and siderites and knew that Tonganoxie was a siderite. This could not possibly be a part of any recorded Kansas meteorite other than the Kiowa County (Brenham) fall that had been discovered in 1885 and described five years later.

I headed for the astronomy department. When the professor emerged from his office to learn what he could do for a young teacher from one of the state's small colleges, I stated my case with great confidence that here at least would be a source of facts.

Somewhat apologetically, the astronomer told me he had but one meteorite specimen, a small slice he personally had purchased from a supply house some years before. He produced an etched slice of one of the most common nickel-iron meteorites and gave me a rough approximation of its chemical composition.

"But I would like to learn more about stony meteorites," I told him.

"Stony meteorites? I didn't know there were any."

I assured him that stony meteorites reportedly were more abundant than "irons" among witnessed falls. He seemed skeptical. I reached into my briefcase and drew forth the small specimen of the Modoc fall which I had purchased from Mr. J. K. Freed, on whose Kansas farm it had fallen twenty years before. The professor took it in his hand, looked it over, hefted it, turned it about.

"You think this is a meteorite?"

I nodded.

He shook his head. "What makes you think so?"

I told him the story of its fall.

"How do you know this is one of the stones that fell?"

I recounted how Mr. Freed had collected some sixty or more stones from the shower, that he had sold most of them to the great museums, that both Smithsonian and American Museum scientists had examined this very one. By this time, he was becoming genuinely interested, and he questioned me further. Then he told me of a student who a few weeks before had brought in a stone for inspection.

"This looks for all the world like the one she brought in, only hers was much larger. She said her family had seen it fall on their farm. But I told her it could not be a meteorite, since meteorites are composed of nickel and iron." Now he was on his feet. He grabbed his hat. "Let's go over to the laboratory."

And off we went, downstairs, across the street, and into another building. This was a Saturday and the laboratory was empty. The janitor was called. Had he seen a black rock, the size of a cantaloupe, on one of the tables? The custodian reminded the astronomer of standing instructions that anything left on the tables was to go into the trash barrel.

"Where is the trash barrel?"

It had been emptied. The clean-up truck came three times a week to haul everything to the town dump.

Our hope of finding the specimen was shattered. Dejectedly, my host led me back to his office.

On my way home, I considered the matter. My first reaction was great disappointment that these men had proved to be ignorant of this unique and, by its very nature, fundamental aspect of our universe. But on second thought, I realized that inasmuch as these men had been trained in the best universities of our land, they must be fair examples of a situation that pervaded our whole educational system. The subject of meteorites simply wasn't being taught.

I was both appalled and challenged. Newly introduced to the subject of meteorites, I was finding that in such a state of affairs, and in comparison with others then on the science scene in Kansas, I was not far from becoming an "expert," due only to my avid reading of whatever I could lay my hands on and the fact that I was making some personal field investigations.

As nearly as I could find out, the principal reason for the dearth of information was the scarcity of meteorite finds. Although Dr. Miller, in the first article I had read on the subject, had written interestingly and informatively about meteorites, in no place had he indicated that the subject was of real importance. Since initial recognition of meteorites by science in 1803, only fifty-three falls had been recorded for the entire United States—an average of one in a little more than two years. If I could peddle my enthusiasm to enough people in enough places, I believed I could modify these statistics considerably.

It had taken much of my time for almost a year to locate and recover the two Coldwater stones. Whether I could speed up the recovery process in some way, I could not tell, but I hoped that as I gave lectures in rural high schools and grade schools, the farm children who attended those schools would spread the word to their parents.

Dr. George P. Merrill had written that most meteorite finds had been turned up by the plow. I had been a farm boy, too, and to me it seemed the most important factor of all in where and why meteorites came to light was being overlooked. In the land that our family had cleared of timber in Oklahoma, rocks were scarcely known, and there was only one thing to be done with rocks when they were encountered by the plow—get them out of the field. Meteorites? We had never heard of such a thing. Neither had almost anyone else. Here, perhaps, was an answer.

The literature was full of stories of meteorite finds made purely by accident. Surely a planned strategy of educational propaganda would increase the number of such finds. If the residents of our state of Kansas could be kept alerted as to the nature and importance of meteorites, it seemed to me that a far greater number would be brought to light, especially if an attractive bonus were offered for each find.

But I was a biologist, I reminded myself. Meteorites should be the concern of astronomers and geologists. But if nobody had been trained to observe, collect, and study them, then a biologist, I decided, might just as well start working on the subject as anyone else, especially since astronomers and geologists seemed for the most part to be otherwise occupied.

How could the population of a state of 81,000 square miles be alerted properly? How could such a program be financed? Surveys and searches were expensive. When I took $41 from the budget of our growing family to purchase Deacon Brown's meteorite, I started spending sleepless hours wondering how I would finance this new hobby that had become my consuming interest. I ran into financial walls. All small colleges were struggling to maintain their meager programs. Universities were better financed, but their funds had been committed. This idea of mine was so unorthodox that no one could, or would, justify its receiving attention while many other programs, some of them widely recognized and considered of far greater importance, went unsupported.

Any program I could carry out would have to be self-supporting. I turned to my experience as a natural-history lecturer. I estimated that I could schedule one or two appearances a week at points not too far distant from McPherson. President Kurtz allowed me to carry a light teaching load and devote as much time as I considered advisable to my research program. Lectures seldom paid much more than travel expenses. My hope, though, was to arouse interest that, in turn, would result in the discovery of new meteorites. The catalogs of scientific supply houses showed that quite high prices were charged for meteorite specimens.

In addition to providing a source of support, my search for specimens was intended to test what I considered an unwarranted assumption as to the scarcity of meteorites *in the soil*, and the rarity of *falls*. The experts had said that the scarcity of meteorites was a fact, and that was that.

Had my two finds at Coldwater been a mere case of beginner's luck? I could not believe so. There just was no logical reason for believing that meteorites fell more frequently on one part of the earth than another. Yet intensive search had yielded two falls of unknown date in a county where I had cause to believe that still a third meteorite lay undiscovered.

Here was the other side of the coin or, perhaps, here was another application of the old chicken-and-egg riddle. Did lack of knowledge really come from a paucity of specimens? Or did the extreme rarity of meteorites reflect a paucity of knowledge?

The complete neglect of the subject of meteorites in all curricula had rendered the population helpless to recognize them except in extremely rare instances where some farmer or ranchman had inherited an unusually large measure of curiosity and then luckily had met a scientist who knew and told him something about meteorites.

Deacon Brown's meteorite and the smaller Coldwater stone, found in a period of less than eleven months, represented twice as many meteorites as had been recorded for the entire state of Kansas during the fifteen-year period just prior to 1923. The interest and publicity attendant on my survey and search must have been responsible for their recovery.

I came to believe that the discovery of the older stones was more significant than would be the finding of the actual meteorite of November 9. The two Coldwater finds demonstrated a fact which I had then already suspected and have since come to recognize as a rule—meteorites are widely distributed over the earth, and any considerable search of a territory of considerable size will probably result in the discovery of a new meteorite.

5

Ways and Half-Ways

*These Americans are not to be discouraged by any new difficulty,...
they never worry about a problem beforehand,
but when it develops that is the time to take care of it.
This is the pioneer spirit of America that I have heard so much about...
and it is the quality that I admire most in men.*

—Bernt Balchen, *Come North with Me*

By 1925, I had taught for most of eleven years, winter and summer. For a year or more, I had been thinking that teaching was tending to become routine. I felt unfit for routine, unable to go endlessly on in set patterns. I had always been happiest when tackling new problems, and though I enjoyed teaching, I felt that I had reached a point in my life where a change to research was in order. How to enter upon such a program and support my family at the same time posed real problems, however. Meteorite investigation, as I envisioned it, would offer a real challenge, but so far, such a program remained untried, and I had yet to mention it to anyone who thought it might succeed. Certainly I could not hope to find institutional support.

Biological research was a wide-open field, but there was little chance to realize more than a bare living unless one achieved prominence. I was thirty-eight years old; the time for establishing an advantageous association with a university staff was in one's early twenties. Starting at my age, I could hardly hope for a really rewarding position until about retirement time.

The lecture field offered possibilities. The natural history presentations I had worked up, bolstered by "magic lantern" slides, had proven popular. The talks on meteorites I had made in connection with the Coldwater search had been received with more than ordinary interest.

With each year, the leash that was holding me became stronger: more responsibility, preparation of a proposed zoology textbook, the Rocky Mountain Summer School. These recurrent obligations weighted the days and months with a burdensome regularity. For several years, I had been noticing in my own life a slow but appreciable decrease in the quality which enables a man to attack the difficult and to try the untried. Our civilization softens and tones down too much those attributes of primitive man which enabled him to cope with problems of life in the wilderness. With the disappearance of these characteristics of strength and dare, other virtues are liable to go as well.

On a cold December night in 1924, I sat by the old base burner, receiving the warm glow from its isinglass window, reading an absorbing account in *National Geographic Magazine* of the recent explorations of Carlsbad Caverns.

Addie was busy with the babies. When she had a free minute, I made an announcement.

"We're going to see those wonderful caverns next year!"

"The whole family?"

"Yes, and including the baby due in March. We'll take the whole year off and travel. The baby can celebrate its first birthday on the road."

Addie was eager but ever practical. What would we use for money? How could we afford to travel as a family?

"I'll lecture in schools as we go. We can live as cheaply on the road as here and if I can get only ten lectures a month at $25 each, it will equal my salary. Besides, we may find saleable fossils or meteorites. If worse comes to worse, I'll pick cotton. I can still do some of the jobs that got me through school."

Just talking out loud about it was bringing back the courage I had been so afraid of losing. We had talked many times about a house-on-wheels. Now we would build one.

The trip I had in mind would have to be a gamble, but I could engage in this kind of gamble with a clear conscience, since I saw it as a venture for self-improvement, and by my philosophy of life, any effort at self-improvement was worth whatever it cost.

Addie loved to travel, and the idea looked attractive, but she was the cautious one of us, and she was surprisingly good at thinking up objections. She voiced some of them. And then she admitted finally that she would rather camp than anything else. She had hardly more than reached her decision when she gave birth to our third child, Margaret, and began adjusting her ideas toward ways of traveling with an infant.

We had seen magazine pictures of a few vehicles fitted with living quarters and had examined one example. We added ideas of our own to the good features we had observed. Once our house had been constructed satisfactorily in our minds, the next problem was on what to mount it.

Since our itinerary was to carry us into thinly settled and even uninhabited country, the best known makes of cars were the only

The Harvey Nininger Family, 1925. From left: Bob, Addie with Margaret on her lap, Doris, and Harvey.

ones to be considered seriously. At the same time, the car must be as dependable as a car could possibly be. Ford parts could be purchased at all garages and, so the story went, at ten-cent stores and soda fountains. Our beast of burden would be "Henry."

For $900 and our old car, we acquired a new Ford truck chassis and, with another $900, contracted to have firmly mounted on it a six- by six- by sixteen-foot "house." Our German carpenter believed in sturdy construction. Whatever shortcomings the structure had, lack of strength was not one of them. Our "house" rested on two sixteen-foot, oak two-by-tens, and every joint was mortised solidly and held together by screws or bolts.

When I approached the president and board of trustees of the college to ask for leave, they plainly thought I was making a foolish move, but they decided they could "permit" even if they couldn't "recommend" my venture.

When we started out on our eight-month tour, we had $19.67 in cash. We owed $300 at the bank, and some $50 worth of finishing still needed to be put to our wheeled home. I had a half-dozen lectures scheduled ahead, at $25 each, the nearest about 200 miles and two weeks from home.

***

Kansas roads of 1925 were unpaved. We had not gone ten miles from McPherson before rain began. Just thirteen miles and two hours from home, we had to pull off of the impassable mud ways and park beside the Smoky Hill River.

Ten days later, we were still there.

While we waited for the roads to dry, I managed to line up and deliver a lecture, traveling by railroad. The local fishermen kept us supplied with carp at ten cents a pound, and we started out again with

Custom-made house-car (the "Nininger Runabout") used by the family on an eight-month journey through the southwestern United States in 1925 and 1926.

most of our $19 capital intact.

Addie began her diary of our life in the "Nininger Runabout" with a notation that our house proved to be neither fly-proof—though its windows were screened—nor waterproof. One night, she noted, the family slept between rows of pans.

After our escape from the bank of the Smoky Hill River, we spent five days at Gaylord, Kansas, where we found several specimens of ammonites in the fossil beds but not much in the way of lecture proceeds.

We hurried to the next stop, Paradise, late on a Friday evening, to be in time for a lecture, but no crowd came, so there was no lecture.

However, four other lectures in the Paradise vicinity paid better than we knew, for nearly five years later, a boy who had attended one of them furnished proof of the correctness of my recovery theory by finding the first meteorite that I could trace directly to my educational efforts.

Addie's diary recorded our financial history. "Gave two lectures, so we're ahead." But at another town, there was no auditorium; at another, the school had chartered a train to attend a football game in the city—there was no one left to hear a lecture. Still, on the whole, our lecture experience during the early days in Kansas was fairly good, and we turned south with light hearts.

In southern Kansas, we were lucky to obtain a twenty-pound meteorite, a good specimen of the old, well-known Brenham fall. At once, I sent it to the National Museum for cutting, on a basis which would give them half for the cutting cost and return the remainder to me in the form of saleable slices. Here was a resource for ready cash, should it be needed somewhere along the line. Our pack already included several nice meteorite specimens which were brought from home to help illustrate lectures. These, too, could be sold if we needed to be bailed out financially.

Making camp along the way on the house-car trip of 1925-1926.

Our southward course took us through the Oklahoma and Texas panhandles, and in these areas as in Kansas, we were able to arrange for lectures. Then we entered cotton country. Here, we found, education did not prosper so well. I was reminded of my own childhood. Schools were poorly financed. Any program of outside talent had to be supported by admission charges—usually a nickel, sometimes ten cents.

At the outset, my plan was to charge a fee of $25 for a single lecture, or $50 for a series of three, including a talk on birds, one on fossils, and another on meteorites. But as we went deeper into Texas, we found it rarely possible to schedule anything except on an admission basis. Now, instead of dealing with school boards or administrators for my fee, I had to sit down after the lectures and help count out the nickels and dimes, receive my cut, and go home with a pocketful or less of small change. Often my receipts were $10, $5, or even less.

Soon after starting out, Addie began to keep our funds in a little pewter syrup pitcher we had received as a wedding present but had never used. Now it became our little piggy bank, and as we followed the difficult, rough roads farther into Texas, the little pitcher lost weight rapidly.

Our intent was to hold to daylight travel, but we found ourselves entering Fort Worth in supper-hour traffic. Since it was dusk, and we had to search for a campground, we pushed on through the city. This was our first experience at guiding our cumbersome gray bungalow through city congestion. At a red light, I stepped from the house-car and ran to ask directions of a policeman, rather expecting a bawling out for leaving our unwieldy vehicle double-parked as far out of the traffic as I could safely steer it.

"Is that your outfit?"

I admitted that it was.

To my surprise, his answer was that he must "see that rig." He crossed to the house-car, greeted the family, and asked to step inside. Addie led him on a tour from cellar to attic.

Near the front was a clothes closet and a writing desk, furnished with compartments above and shelves below for reference books and a typewriter. Each side of the house-car opened out in wings, and cushioned seats unfolded into two full beds. While traveling, the bedding was strapped to the walls above the seats. The baby's bed was on wheels and could be moved from one end of the house-car to the other. A small rocking chair could ride either in back or next to the driver's seat.

The kitchen had a sink with running water—we could carry sixteen gallons—a gas stove, work table, drawers and cupboards, and an ice box. A trapdoor in the floor opened to storage for vegetables and canned goods. A second trapdoor was for laundry. Out the back door were lockers which opened from the outside.

His tour complete, our cheerful policeman waved us on with careful directions to our destination. Meanwhile, the evening traffic had thickened and horns were being sounded in the grand cacophony that one encountered in those days whenever more than a half dozen cars faced any sort of road block. The Fort Worth campground was bright with autumn colors. Addie listed the trees in her diary: black jack and post oak, sumac, black haw, live oak, green briar, pecan, walnut, umbrella. The standard fee at most large campgrounds was fifty cents a day. A first-class facility would have a shower and washroom, a reading room, and cabins. Motels and trailer courts were unknown.

Most of the roads south of the Texas panhandle had been paved, but near Austin, we travelled some terrible ones. First attempts at paving had proven unsatisfactory, and the roads were being torn up for miles in preparation for rebuilding. We negotiated several hundreds of miles of freshly graded road beds, loose-piled gravel, unworked rocky hills, culverts and bridges under construction, and ditches that had to be bridged before crossing, without seeing the slightest evidence of attempts to provide detours. Travelers were expected to find their own way across or around such stretches. Once, we used railroad ties to make a bridge. We had many experiences with mud-holes and grades too steep or curves too short to negotiate, but our truck proved sturdy, and where it didn't fit, we managed to adapt our plans accordingly.

The birds, the flowers, the trees, the names of the towns differed, but the delays, the hurries, the changes of plans were always with us—only the time and the place and some of the details varied. In Austin there were redbirds, robins, mockingbirds and Mexican doves, and Bob, who was trying to keep up his second-grade work under his mother's supervision, started keeping a notebook of pressed tree leaves.

Addie's entry for Austin, November 26, ends, "We shall leave after dinner." The next notation is dated December 4. Location: Austin. "We are still here. After talking to the superintendent of schools, we decided to stay this week and put on the lectures in the schools. Gave six lectures which averaged $9. Not as much as we hoped."

As our Ford engine pulled us along toward the coastal town of Corpus Christi, we grew more than a little worried. We had been on the road three months. Christmas was only two weeks away. My wallet was flat, and only a few nickels and dimes remained in the pewter pitcher, barely enough to buy milk for the baby, but it was the thought of the holiday that really stung.

On a Friday, we approached Corpus Christi just before noon, hurrying for a chance to see school officials prior to the weekend. At 11:30 a.m., we pulled up before the high school, and I hastened inside.

The superintendent was a meek-mannered, middle-aged woman who seemed to see value in my lectures but told me that all such matters could be decided only by "the board." I asked who headed "the board." The superintendent told me the chairman was a Mr. Clark, president of the local bank.

Was he interested in science? Yes, but he was very stern and most hard-spoken unless favorably impressed at the first meeting.

I started for the bank in order to reach it before noon. Planning my approach as I hurried along, I stopped at the house-car to take a choice meteorite specimen from my little black bag and put it in my pocket.

In an office behind the small lobby of the bank, I found Mr. Clark, a man with a heavily lined face under a thick shock of wavy dark hair, bent intently over his work. I planted the black, lemon-sized stone on the desk before him.

"It won't explode," I said.

"What is that? Who are you?"

I told him the stone was a meteorite.

"Sure enough? It's the first one I ever saw. I've read about them and always wanted to see one. Are you sure it's a meteorite? Who are you?"

I explained my mission, and he instructed me to come back at

1:30 that same afternoon to present my proposition. "And bring along some specimens."

I returned on time, presented my case briefly to the school board chairman, exhibited a dozen specimens, and began to wrap them up.

"Wait a minute. What is that specimen you are wrapping worth?"

"Seventy dollars."

"I'll take it."

"Well, I didn't exactly mean to offer it for sale. I thought you merely wished to know the value."

"You're not backing out, are you?"

So I agreed to sell the meteorite to the school board chairman, along with two others, for a total of $270. Five minutes later, I was told that arrangements would be made for seven lectures to be given the following week.

We were gratefully busy. We did some remodeling on the house-car, purchasing new curtains and two more chairs. We went to the woods and found a small bay tree for Christmas. Decorated and standing in the cab of the house-car, next to the steering wheel, it added a cheery, if incongruous, note.

The town of Corpus Christi holds memories of tender, fried oysters, mountains of fish, and a happy Christmas. Daily we purchased oysters, delivered at our door, for twenty-five cents a quart. On Christmas day, I fished. It was one of the few marvelously successful days of all my fishing experience, for ordinarily, I am an unlucky fisherman. Standing on the rugged boulders of the breakwater, I would cast my shrimp-baited hook and draw in eleven-inch sea trout—correctly called weakfish—from the Gulf waters as fast as I could transfer them to the creel. I hauled in forty-four. Fried to a delicate brown, they were delicious.

***

Christmas was foggy and cold, but the day after was wonderfully warm as we left Corpus Christi to drive to the little town of Kingsville in the middle of the million-and-a-half-acre King Ranch.

That evening, the weather changed and it began to rain, the thermometer dropping steadily. The next day and the next it rained, all the while growing colder until the rain changed to snow. Then for three more days it rained, the water freezing as it fell until our house-car was covered with icicles, some as much as a yard in length. Our camp at Kingsville was among palms, orange trees, figs, and eucalyptus, with roses blossoming in December— the subtropical vegetation encased in a coating of ice making a curious spectacle.

We moved on to Brownsville, where my birthday was marked by a small box of chocolates and a bird-watching experience that stays bright in memory. Our camp was on the edge of a small narrow marsh or *resaca*, the remains of several linked lakes. Our front window looked directly out on an open pool a few hundred feet wide. Our front wheels were only a few yards from the shore. The shallow lake bottom bore a heavy growth of water grasses burdened with algae and teeming with small minnows and other choice morsels over which the waterfowl chuckled and quacked in evident satisfaction. Occasionally pied-billed grebes, soras, ring-necked ducks and bluebills would join the crowd. It was the herons, however, warily keeping back in the brush thickets and cattails, that I wished to see up close.

I arose and dressed before daylight to make a pre-breakfast tour. I flushed Louisiana herons and little blue herons but could never approach closer than fifty or sixty yards. Just at daylight, flocks of small waders flew overhead on their way from their nocturnal feeding ground to their diurnal resting places, probably on nearby Padre Island. Then as it grew lighter, Mexican cormorants circled over the lake, ducks dropped down, and six glossy ibises flew swiftly in from the direction of the river and disappeared in the brush along the opposite shore. Canvasbacks and ruddy ducks swam out in the middle of the pond. A kingfisher chirred along the farther shore, and a little sora scuttled to cover as a Harris's hawk swooped down near me.

After breakfast, Addie and I maneuvered through a brush thicket and surprised a host of land birds: warblers, finches, titmice, wrens, sparrows, pipits, a ruby-crowned kinglet and many others in their winter haunts, some in winter garb. A phoebe called her name impatiently from a low branch, gesticulating with her tail for emphasis. But the herons remained far away.

Just as we were returning, I stopped at the edge of the grassy marsh to watch through my field glasses a little blue heron perched on a post some sixty yards away. He came flying and commenced to feed only about twenty paces from me. In water a half-boot deep, I stood as quietly as I could, breathing as silently as I could, studying every move of the feeding bird. It stalked slowly about, nabbing a minnow here and there, pausing often to satisfy itself that I was but an extra stump, and harmless. Suddenly there was a swish of wings, and a Louisiana heron alighted directly in front of me. I focused my glasses and stared until my eyes watered, fearing each second's opportunity would be the last. For three-quarters of an hour, I watched the Louisiana feed. A sora swam within a few feet of me, apparently not noticing my presence. A reddish egret approached the marsh briefly, and a green heron arrived, squawked and departed in ungainly fashion. Twenty glossy ibises, somber and scaly-necked, came to feed with the herons, rapidly jabbing their long, curved beaks into the water, so differently from the deliberate strikes of the herons.

I felt not a day older.

***

We spent two weeks in and about Brownsville and Weslaco. The latter town, Addie lamented, seemed to be "our fate." The public camp was dirty, and though the fee was only twenty-five cents a day, it wasn't worth that. We felt fortunate to have our "house."

The lecture series did not meet with great demand in the Texas towns, although at McAllen, one lecture netted $30 on the ten-cent admission plan.

During the month of February, we stayed at various of the great Texas ranches—the half-million-acre Kennedy Ranch, the mammoth King Ranch, the Larellis Ranch, the Schreiner Ranch. The area abounded in wildlife. To reach a camp site on the Kennedy Ranch, we left the main road and lumbered slowly along a dim, crooked trail through broken sandy grassland toward a ragged live oak forest in the distance. The shadows lengthened rapidly as we turned from the dim road to a dimmer one and struck off southward through still sandier soil.

Just before sundown, we picked a stopping place where a stand of mesquite and cat's claw bordered on a heavier growth of live oak and cactus. With sixteen gallons of water in our tank and plenty of provisions in the larder we could sojourn comfortably, for several days at least, in the southern wilderness. While Addie prepared supper, I examined our campsite. The live-oak forest spread north and east. To the south, mesquite and cat's claw thickened into a dense tangle a mile away. Westward was an irregular admixture of this brush and live oak with open grassland. To the southwest, a quarter of a mile away, lay a quiet shallow lake surrounded by a heavy growth of swamp grass. Perhaps a hundred ducks were feeding on the water, and from somewhere on the shore, there arose periodically the loud resonant call of a sandhill crane. Here, we were far from the devastating heel of the white man's civilization in a situation which, were it not for calendar dates, might just as well have been a thousand years ago.

Darkness settled heavily around us. All was quiet, save the fervent hooting of a great horned owl a few rods away, answered faintly from the distance by his mate. Mosquitoes drove us inside, but I stayed listening for a time through the open screened window.

Our days in the Texas wilds were filled with surpassing pleasures. We had the delightful privilege of living with the conveniences of modernity in a primitive wilderness. Such a combination was rare then and is even rarer now. A century ago, no continent furnished more of the primitive than did ours. But never has the primitive been hustled out of man's environment faster than during those hundred years.

***

We accepted an invitation to return to the King Ranch to view the world's last remaining colony of the great whooping crane, then totaling but seven birds. The ranch manager, Mr. Kleberg, offered cars and guides to lead us to the cranes.

The whooping, or bugle cranes, seemed about to go the way of the passenger pigeon, the Carolina parakeet and those other vanished races whose mute, motionless remains are preserved to stare out at us through the glass windows of museum cases, mocking our blunders and recalling the beauty we've destroyed. The active campaign waged since 1941 by the National Audubon Society to save the whooping crane has brought hope that the species may live on and even produce in numbers, but in 1926 the only known specimens were that handful on the King Ranch in south Texas.

The whooping crane is the largest of the American waders, standing more than four feet tall, with snow-white plumage and long, slender legs and a long neck. The tips of the great wings are black, but when the wings are folded, this border is almost entirely concealed beneath the whiteness of the general coat. A strangely lengthened trachea, about twenty-five inches of which is coiled up in cornet-fashion and imbedded within the sternum or breast bone, is responsible for the resonant clarion call which has earned the bird its name. The bugle-like voice can be heard for miles under the right conditions.

Early writers tell of large flocks of these cranes feeding during the autumn months in the fields and marshes of Illinois, Missouri and other parts of the Mississippi Valley and of their nesting grounds throughout the northern states and Canada. Though the cranes were wary and resourceful, their large size and white plumage rendered them an easy mark for gunmen, so that as the central states were settled, the numbers of the species dwindled rapidly until at the end of the nineteenth century, there were but a few known bands remaining.

Mr. Kleberg put us into the helpful care of the jovial ranch superintendent, a Mr. Cody, who entrusted our mission to one of the ranch guides. We were taken by automobile over a tortuous course through the broad, low coastal plain dotted by thickets of mesquite and clumps of live oak but for the most part clothed only in a thin stand of various grasses and bearing a luxuriant crop of the large southern variety of prickly pear cactus.

As we entered a low, muddy flat, a large flock of sandhill cranes took flight with a trumpeting disapproval of our intrusion. We alerted three Mexican eagles, several black vultures, and a great horned owl. Then flocks of Brewer's blackbirds swirled like clouds of wind-driven smoke above the horizon, tossing and swerving, sharing the skies with long, white lines of snow geese rising, circling, and settling to the grassy feeding grounds. Flock after flock of pintail ducks, together with a sprinkling of other species, rose from scattered shallow ponds and marshes.

As we mounted a low, long ridge, we came upon some 600 snow geese standing at attention in close formation at the edge of a small, marshy pool bordered by tall grasses. The landscape beyond was mottled by prickly pear cactus thickets. We had admired the great flock for some minutes before we had become aware that a "cactus thicket" some fifty paces ahead of us was no thicket at all, but a flock of large Canada geese, statue-like in their pose of alarm with necks up-stretched like weed stalks. As we looked about more carefully, many of these flocks could be discerned interspersed in perfect camouflage with the actual cactus clumps, and we knew we must be at the very edge of an immense aggregation of these geese extending in broken lines for miles across the plains.

All morning, and again in the afternoon, we searched for the whooping cranes. We saw flocks of long-billed curlew, more geese, more ducks. We seemed to see every bird but the *grúa blanca*, the local Spanish name for the whooping crane.

The second day, we were driven out to search again, this time by Porferio, a middle-aged ranch hand whose job it was to keep in repair the twenty-seven windmills of the ranch. He knew every corner of the vast acreage. Porferio drove us to the edge of an immense marshy flat known as Laguna Larga. As we approached, flock after flock of snow geese arose from in front of our car, and in broken formation drifted away to join other thousands feeding in the marsh.

Suddenly we saw the great white cranes towering above the geese. They saw us immediately and stood motionless for a moment while we hastily shared the binoculars so that all could view this scene—possibly the last act of a drama in which the final fall of the curtain would mean oblivion for the actors.

Then, leaping up, the cranes spread their enormous black-tipped wings, and with measured stroke, flew toward the middle of the muddy marsh. Having doubled the distance between them and us, they settled again, but they did not resume their feeding. Across the broad expanse of friendly mud, they stared nervously, then rose again to widen still more this muddy barrier. They waded briefly, snapping up a morsel here and there, but persistent in eyeing us intruders. A third time, their distrust grew to fear, and away they flapped until they were separated from us by more than a mile.

Still I watched through the field glasses as they settled, newly reassured, to quiet feeding. They stalked slowly about on their long legs, jabbing at the ground for mouthfuls and then lifting heads high to look around. All attempts on our part to approach any nearer were answered by further flight.

In the afternoon, we returned and found them in almost the same spot as we had first seen them in the morning, but they showed still that same fierce fear of man. One was tempted to wonder if these last remaining individuals of their kind somehow sensed the impending doom of their race. Seven lonely birds stood in the broad, bleak wilderness with two thousand miles lying between them and their Canadian breeding ground, almost every mile of it occupied by their human enemy.

Only a century earlier, these beautiful birds had been the regal class of the bird world. Before the rise of man, these snowy knights of the marshland had come to be an old and respected tribe. For a thousand millennia, they had fed in the lagunas of the south and had gathered on the shores of northern lakes to strut and cavort in their wedding dances. Their numbers and strength rendered their rookeries comparatively safe against furred and feathered enemies, but when a new enemy had come armed with guns, the warfare was brief and one-sided. In a few geological minutes, the sun of their day of life had set and the impending night of destruction had settled upon them.

We had come 1,000 miles, but in the brief moments of their fading dusk, man no longer was tolerated. Only from the safe distance of mid-marshland did we have the privilege of viewing what appeared to be the last few remaining days of a murdered race.

During man's few thousand years on this planet, he has blotted out forests, drained lakes and swamps, and flooded deserts. He has carved his way across plains, turned rivers from their courses, and where it has seemed good to him, he has moved entire mountains. Man could direct the same ingenuity that enables him to harness waterfalls and lightning and send emissaries into space, and the same love of beauty that compels him to construct parks and gardens and exchange fortunes for painted canvas or carved stone, toward saving the dwindling species and the dwindling wilderness, should he so choose.

***

The wealth we were gaining from our journey was mostly observational; we were not doing very well financially. Addie for once was really downcast. When she wrote in her diary that we were "clear out of funds," it was no exaggeration. We had a letter to mail, applying for a lecture date ahead, but not the three cents to buy a stamp.

We stopped near a creek bordered by giant, wild pecan trees. The pecan season was past, but I knew from boyhood that there were always a few late nuts that fell after the regular harvest. Walking about in the deep carpet of dry leaves, I found a pecan or two. Returning to the house-car, I asked Addie to keep the children busy hunting nuts while I walked the half mile to town. I was sure I could return with some lecture money.

Carrying my bag of sample specimens, I climbed the long, sandy hill to the edge of town, spotted the school building, and headed for it. Earlier, I had met the high school principal of a town a few stops ahead on our itinerary, and he had given me an advance order for a lecture when I reached his community. This gave me an introduction of a kind to the head of this neighboring school, and this gentleman acted promptly. At the end of the class period, he called the students to the auditorium. I delivered my lecture, it was well-received, the principal was pleased, and I returned to my family with $25. We posted our letter, gathered in our pecans, and were on our way toward Carlsbad Caverns—the goal that had inspired our trip.

2

Carlsbad Caverns now has trained staff, elevators, electric lights, lunchroom, and restroom facilities, and boasts a well-engineered trail, toured easily by millions of visitors. On our visit, there were five of us in our party, and we five constituted the entire roster of visitors for the day. Each of us carrying a powerful gasoline lantern, we were led over the partial, very crude 1926 trail by Jim White, the first known discoverer and explorer of the caverns. The Park Service had named White the first guardian of the recently created national monument.

Jim White was an unschooled western mountaineer of sturdy character, undaunted courage, and iron will. For twenty-three years, he operated the guano mine at the cave for a California fertilizer company. He spent much of his spare time ferreting through the gloomy labyrinthine tunnels of the great Carlsbad underworld. He later guided Dr. Willis T. Lee, who conducted and reported a more scientific exploration of the caverns, but it was Jim White who first faced the dangers and difficulties and overcame them with his own ingenuity.

Some areas that were open on our first trip to Carlsbad Caverns are now closed to the public; other new areas are now accessible. The walk is easier, the light is better, the total area to view is greater. But no trip since has equaled our first for bewildering spectacle and impressive encounter with Nature in her secret haunts.

***

Traveling on to El Paso, we met snow again, but the sun soon returned, and we crossed the White Mountains through a beautiful and easy pass. We descended into a hundred miles of desert, where the road wound in and out among sand hummocks as high as our car. This hot, dry, townless strip was unexpected. Water was scarce and we spent a trying day. But at El Paso we found Camp Grande to be a camp well worth its fifty-cent fee, with hot showers, an electric washing machine, modern dining hall, kitchen, and recreation hall.

We followed the Rio Grande toward Las Cruces, gathering screw beans from the mesquite along the river. We encountered a sand storm, when at times so much sand was in the air that we could see but a few feet ahead of us. We drove off onto a side road and staked the house-car lest the wind blow it on its side.

On to Las Cruces, Deming, and Lordsburg, New Mexico, then Benson, Arizona, and finally Tucson. We traveled the desert through miles of white, yellow, gold, and orange poppies, and cacti clothed in green.

Our Ford engine was getting balky, taking hours to start in the morning, and once again we were counting pennies. We reached Tucson at noon with just enough money for a loaf of bread to accompany such supplies as were left in the larder. Lectures at the university fattened the wallet sufficiently to care for our needs for a few days.

Bob and Margaret celebrated their joint birthday—Bob's seventh and the first for the baby—in Sabino Canyon in the Santa Catalina mountains, before a backdrop of giant cacti punctured with holes made by woodpeckers and tenanted by owls.

We toured the Indian ruins at Casa Grande, visited Phoenix briefly, and then drove on to the mountains. We camped for the night at Congress Junction. It took a full day to drive the forty miles across two mountain ranges to Prescott. Rain dogged us. We were determined to visit Jerome, the site of Arizona's greatest copper mine.

From Prescott, we drove through Granite Dells, across Lonesome

Celebrating Bob and Margaret's joint birthday on March 28, 1926, in Sabino Canyon, Arizona, on the house-car trip. This was Bob's seventh birthday and Margaret's first. Doris is on the left.

Valley, and wound our way up Mingus Mountain. As we reached the summit, there was heavy rain. The road was narrow, crooked, and muddy with no safety barriers, and our cumbersome 6,000 pounds posed a real threat on some of those sharp, steep curves. I shifted into low, low gear, and we crept down the mountain.

It was noon when we left the summit, but we thought it best to go on to Jerome for fear the road would soon become impassable. The rain became torrential, and we silently watched the road grow more and more slick. Our three youngsters were becoming hungry and restless. The steep, muddy miles lengthened terribly.

When finally we reached Jerome, we were no better off. Hanging on the slopes of the mountain, Jerome's streets are so nearly one above the other that one can step from the sidewalk to the third-story porch of a house built on the street below on one side or cross the fifteen-foot street and enter the basement of a house on the other. There was no place to park our big contraption. Water was pouring down the steep streets.

As we were about to leave the town, I saw a fenced-in school-yard. The big gate was open. I brought our house-on-wheels to a stop on a graveled patch of the school yard.

While Addie prepared lunch, I dashed through the rain to the school office. The principal, Mr. Schrepple, decided there on the spot that all of the schools in Jerome should hear my lectures.

I went out to the house-car to eat lunch with the family in an atmosphere of optimism. This engagement was just what we needed. Then I rejoined Mr. Schrepple. He told me of several things he wished to show me in the community and excursions he would like our family to make into the surrounding country. On his outline was one item that intrigued me greatly, and we agreed to make it the first project on the list.

The principal reported that at the store in Camp Verde, some twenty-five miles distant, there was on display a slab of limestone bearing two big bear tracks. These were said to have resulted when a bear stepped into an irrigation ditch leading from Montezuma Well, whose waters were rich in lime. The theory was that the heavy solution of lime had solidified the tracks.

The explanation didn't ring true, but the described display

demanded to be visited. We arranged to drive the twenty-five miles to Camp Verde on Sunday. There in the store window was a slab of compact limestone with two deeply impressed, distinct footprints, large imprints such as one might see after an African lion had been sent too soon into a cage where fresh cement had been laid. My first glance told me these were no bear tracks, nor were they made in the lime currently crystallizing out of the water from Montezuma Well. For once, a run-down rumor proved better than its story. These were fossilized tracks from millennia past.

We photographed and measured the specimen and inquired after the source. The tracks had been left behind by a man who had since moved away, but there might be one person who could give us the information: Mr. Chick, the postmaster at tiny Cornville.

The next morning, Bob and I set out on foot for Cornville, twelve miles away. The Chick family of parents and four daughters comprised half of Cornville's total population.

Mr. Chick scratched his head and pondered. Finally, he said the stone was brought in by one of the Rogers boys. Several Rogers boys lived down Oak Creek a few miles, although the most likely brother had moved to Oregon.

Bob and I walked on. At the first Rogers house, we found no one home, so we inquired at the next place. Here there was no information, either, but we were directed to still another brother. An offer to pay $10 if he could show me a similar track in place brought agreement to make a search.

We would have to wade Oak Creek, which was running high from melting snow and was hip-deep, swift, and roiling over a bed of boulders. We took hand poles to brace ourselves, and loading Bob onto my back, I asked our guide to walk on my downstream side to assist me if I should stumble. Across we went and into the hills.

After a walk of a mile or so, we entered a ravine that nourished a stream in wet weather. There were no tracks. My guide thought they had been either washed out or covered over with sand. With spirits and clothes thoroughly dampened, we headed back, again wading the creek. As we emerged, we were met by another of the brothers, who informed us we had been in the wrong ravine, but said that he could lead us to the right place before sundown.

We crossed the creek a third time and headed farther back into the hills, descending into a gully called Bee Canyon, where numerous colonies of honey bees inhabited the honey-combed limestone near the top of overhanging cliffs. A long ladder stood against the bee-infested wall awaiting the next honey season.

A deeper part of the canyon was floored by limestone of which the center portion had been eroded away, lateral shelves remaining at either side. I brushed the sand from a small depression. There was a perfect imprint. As we all scratched like hungry chickens in the sand, we laid bare at least twenty of the ancient footprints. Further investigation revealed more. Altogether, there were fifteen lion, fifteen antelope, three moose, and five or six bear and camel tracks, all made by animals that had lived at least a million years ago.

There is something very impressive about footprints; they seem so fragrant of life. For me, no fossil brings quite such a thrill as does a clear, clean-cut impression of a foot that has never been seen by man.

Shadows were lengthening. We carefully removed one slab bearing two good tracks and carried it up on the bank, lest unexpected high water rob us of our prize. Then, much less weary than when we had come, we crossed the creek once more.

My lecture series was spread over a couple of weeks. Between appointments, Bob and I, with the help of Charlie Risinger of Jerome, worked at removing tracks. Dr. Barnum Brown, famous paleontologist of the American Museum of Natural History, was notified, and an order to ship the tracks fattened our piggy bank quite effectively.

The badly fractured rock required patience for its removal, and the heavy slabs had to be lugged up a steep cliff to the nearest point reachable by car. Day after day, we removed slabs and fragments of the impressed stone, wrapping each separately, packing them carefully in strong crates for shipment.

Fossilized lion tracks recovered near Cornville, Arizona, during the 1925–1926 house-car trip and sent to the American Museum of Natural History. Doris is on the left.

**

Addie celebrated her 34th birthday at the Grand Canyon by taking the Bright Angel Trail via muleback down to the Colorado River, along with Bob.

When I learned that Dr. Gilmore of the Smithsonian Institution was excavating fossil tracks a thousand feet below the canyon rim, Bob and I hiked down the Hermit Trail for a look. We found Dr. Gilmore and his assistants removing beautiful slabs of flagstone, marked by long series of perfect amphibian footprints. The long trails of tracks were as clear and distinct as if the animals had walked or run there the night before.

Dr. Gilmore was puzzled that the only footprints uncovered were those of vertebrate animals, while it was known that insects and other invertebrates were present in the age in which the tracks were made. In two seasons, he said, they had uncovered thousands of vertebrate tracks, but never a single invertebrate.

Without really being serious about it, I said that Bob and I would find some invertebrate prints. We had gone not more than 200 yards along the canyon wall before I saw a loose block of sandstone about a foot long and about four inches wide with a beautiful series of tiny footprints running right down the middle of it.

"Look, Bob," I said. "These are invertebrate footprints, the kind Dr. Gilmore has been wanting."

We laid the block on a boulder until we finished our walk, and then returned for it and carried it to the great authority on fossils. "There you are."

"You brought that down with you," Dr. Gilmore said accusingly, until his disbelief passed. "Well, it's the most amazing thing that you should come down here and in an hour find what we've been seeking for months."

The following winter, I received a copy of Dr. Gilmore's paper, in which was a photograph of my invertebrate footprints bearing the name Triavestigia Niningeri, assuring my name a small dusty niche in history by its attachment to a fossil some 200 million years old.[4]

3

As we passed the halfway point from Flagstaff to Winslow, we began to watch for the first sign of one of the chief goals of our trip, the great hole in the earth dug long before memory by a tremendous meteorite. We identified the broken-rimmed crust rising above the level plain of the desert, watching the elevation take on, as we grew nearer, the appearance of a low cone, a cone that differed from ordinary volcanic peaks in its being light-colored and composed of fragmental sediments rather than the usual volcanic products.

We drove our mounted cabin to the base of the abrupt slope leading to the rim of the yawning "Meteor Crater," pitched camp for the night, and then before turning in, climbed the few hundred feet of steep rise to look into this strangest of all earthly pits, where massive beds of limestone and sandstone had been shattered by a blow from other worlds. That evening, the sun bowed away through a filmy gossamer veil that a northbound zephyr wafted earthward, leaving soft, streamlined folds to catch the departing rays, spraying them across the sky in delicate blending tints like some background of an artist's canvas.

No other spot on the earth's surface could furnish more inspiration for imaginative conjecture than the crater, this spot where a belated planetesimal arrived sufficiently late in the process of earth-building to leave for man's contemplation the stone-shattered

---

[4]In 1966, a new mineral, a meteoritic sulfide found in certain chondritic meteorites, was christened Niningerite

entrance to its mountainous burrow. We know not whether in that far-off time a night was momentarily changed to day or whether a day was doubly brilliant, but the record is there, written unmistakably in the faithful tables of stone to a depth of a thousand feet, the record of one moment of transcendent brightness when a gleaming shaft split the sky and pierced the stony crust of our spinning planet.

After a night on the wall of this memorial to a cataclysm, we studied the pit, the rim, and the surrounding plain by daylight. I did not know then that this crater would become as familiar to me as a man's own garden, that for a half-dozen years, a half-dozen miles away, I would live in its figurative shadow.

On our way again, we exclaimed over the Painted Desert and Petrified Forest, examined a large meteorite on display at Navajo Station, and then traveled on east through the Navajo country of New Mexico. I purchased a small meteorite in Albuquerque. We drove on toward Santa Fe.

As far back as Jerome, we had been warned that two difficult mountain passes lay ahead of us. One was La Bajada Pass north of Albuquerque; the other was Raton Pass straddling the New Mexico–Colorado state line.

La Bajada looked steep, narrow, and crooked. We had been cautioned that we might not be able to negotiate the sharp turns with our wide rig and that the road was so narrow that if we were to stop, we would block traffic, should there be any. At the foot of the grade, I shifted into our lowest gear, which slowed us to a top speed of eight miles an hour but gave us ample power. Except for stops to cool the engine, we reached the top without incident.

From Bernalillo, the road had been desolate, and from the top of the mesa, the way ahead appeared desolate also. After a few miles of this emptiness, there loomed in the distance on our right a lone Indian, running toward us and carrying a gun and waving. It was evident from his course that he would intercept our vehicle, then cruising at ten miles an hour. I stopped. All he wanted was a drink of water, and then he headed back the way he had come, probably to a flock of sheep.

Santa Fe was a fascinating blend of Indian and early Spanish cultures. At Raton, while some work was done on the house-car, I made inquiries about the mountains looming rather formidably before us. Men would look our equipment over and be slow to speak. Some shook their heads. Others predicted that we could make it if we were cautious, and if we didn't meet traffic in the wrong places.

Raton Pass proved a long, hard pull, with rests for our faithful little motor wherever we could leave room for any opposing traffic that might come along. Once over the pass, we felt we were nearly "home," even though 200 miles still lay ahead to Palmer Lake, all of it dirt or gravel road, just as all our roads had been ever since we left the Rio Grande Valley, except around Tucson and Phoenix.

It was May 19, 1926. We had left McPherson in September, 1925. Summer School was due to open at Palmer Lake in ten days. Addie and I felt an exhilarating lift, and we were already plotting ways to travel more.

As we left Walsenburg and Pueblo behind, we counted wildflowers to speed the last miles: yellow ragwort, white evening primrose, purple vetch, blue penstemon, white loco weed, puccoon, western wallflower, wild verbena, flax, bunch flower, paint brush, astragalus. At Palmer Lake, we greeted old friends, set up camp on the village flat, and prepared for another session of the Rocky Mountain Summer School.

4

When we returned to McPherson, we had been away nearly a full year. The children went back to catch up with their schoolmates, and Addie and I returned to our home and college chores, but we had tasted roving independence and wished for more of it.

I still had hopes of somehow making a livelihood out of my interest in meteorites. No new meteorites had come to light as a result of my lectures during our house-car tour of Kansas, Oklahoma, Texas, New Mexico, and Arizona, but I had been able to buy a few specimens. One—the small Ballinger iron purchased in a mineral store in Manitou Springs, Colorado, at the end of our trip—proved to be a new fall.

The only visible result of my lectures was the very keen interest shown by the audiences, but I could not doubt ultimate success. My immediate problem was to add to my store of marketable material while continuing in the employ of the college. Dr. Kurtz continued to make it possible for me to carry on with the meteorite program by keeping my teaching load light.

Addie worried somewhat about the great amount of time this "hobby" of mine was consuming—for until then, I considered it nothing more than a hobby—despite my growing determination that I would somehow, sometime, make a career of it. We both wondered just where this sideline would take us.

$$***$$

In the 1920s and even in the early 1930s, there was no recognized science of meteoritics. Dr. Farrington had suggested the term as a name for the study of meteorites, but it never seemed to catch on. Actually, those who did most of the writing about meteorites did not regard the study as a separate science. Any paper dealing with meteorites appeared as merely incidental to the disciplines of chemistry, physics, or astronomy.

The men who in the 1920s were working with meteorites were

mineralogists, astronomers, chemists, or geologists who only performed meteoritical studies as side projects. The leading students of meteorites had told me that a science of meteorites was a wild dream, that the most elementary phase of such a science would have to be a plan for greatly multiplying the quantity of meteorites available for study.

There were several points on which past investigators agreed that I questioned seriously. One was the belief that there was no quantity of meteoritic material to recover. Another was the notion that the great majority of meteorites were of the nickel-iron variety, an idea accepted as fact by the most prominent American researchers.

I read and re-read my standard reference, Dr. Farrington's *Catalogue of the Meteorites of North America*, the giant Volume XIII of the *Memoirs of the National Academy of Sciences*. It carried descriptions of 247 North American meteorites: 161 irons, 76 stones and 10 stony-irons. Only three of the 161 irons and one of the ten stony-irons had been seen to fall, but 56 of the 76 stones were witnessed falls.

The discrepancy between the number of finds of irons and stones and the number of witnessed falls of the two groups was obvious and had been discussed in the literature by different writers. Various explanations had been advanced, but there was no satisfactory solution.

I continued learning what I could, rounding up for my own small collection such neglected specimens of recorded meteorites as I could run down and talking, talking, talking about meteorites.

The fact that I was earnestly concerning myself with a subject to which great institutions of learning saw fit to pay slight heed did not go unnoticed. A good deal of fun was poked at the "young star-chaser" who had jumped the track of his professional career and gone off on an ephemeral tangent. Was it not a little ridiculous for a young scientist of a small college to assume the role of critic and insist that such a subject receive more attention?

I could not discount the fact that although meteorites may constitute only a minute fraction of the materials comprising our planet, they are in a class all by themselves as extra-terrestrial. They are cosmic; they do not belong to our Earth, yet they join it. And if perchance this coming of matter from space has been the means by which our planet was formed, then indeed meteorites are entitled to top priority in man's investigative program.

No matter who said what, I was determined to continue to devote myself to meteorites.

5

As we approached the close of our house-car trip, Addie and I realized that the year's travel had done something for us that was in the nature of a rejuvenation. I had regained the zest and dare that had characterized my life after breaking away from cotton farming and embarking on an educational career. And Addie, in spite of the tremendous workload created by travel with three small children, felt that she wanted more of this kind of life.

We talked about what the year had meant to us and searched for a method of repeating such an adventure, the one greatest prerequisite to any further travel being a way to make it pay for itself.

I could not live with monotony. Some people can endure, even enjoy, the same activities of life year in and year out for long periods of time. But for others, such a program of repeated experience means a slow death of the spirit. I could not surrender the adventure of youth because of a mark on the calendar. The world was too large, too challenging, too brimming full of the sparkling elixir for me to turn to the stagnant waters of monotony. I would rather have a life of economic uncertainty with plenty of daring than have security with routine.

Our break with routine to wander the Southwest in our wheel-mounted home only intensified my yen for being on the move, trying something new. My plan for long-term, large-scale recovery of meteorites still loomed in my imagination, as did a longtime, cherished hope of taking college students on a tour to some of the sources of the natural histories they studied in class.

Perhaps another year of leave from my regular teaching, a year devoted to the student tour I had in mind, would also provide an opportunity to learn more about meteorites and to test further my own theories for their recovery.

Addie and I decided that after another year on campus, we would be ready to undertake a natural history trek with a class of students who would be working for college credit, *if* the college authorities were willing to give me leave for such a project. The board of trustees was persuaded. I was granted a traveling professorship and volunteered to forego salary provided I was allowed to collect a $75 tuition fee from each of the students.

The trek was to cover the school year of 1927–1928. The students would work the entire school year for eighteen hours of credit. About mid-year of 1926–1927, we began receiving applications for membership in this unusual class. We scheduled meetings at regular intervals to brief prospective students and prepare them for what we knew would be a difficult year. Students were to furnish their own transportation, camp equipment, and sustenance. There would be lectures or conferences on call, and notebooks were to be kept outlining field work accomplished.

Credits could be earned in zoology, botany, and geology. The study of zoology would include field observations of any kind of animal life; botany and geology likewise were interpreted broadly.

Thirteen students comprised the class as it was ultimately approved. They were organized into three groups; our own family made up the fourth.

Each group purchased a new two-door Pontiac sedan equipped with a large trunk. An additional cabinet was built along the left side of each car.

One car carried five men, another four, the third carried two coeds and a married couple, and the fourth bore our family of five, our children then being eight, five, and two years old.

Each group cooked and ate as a unit. There were five tents, the two girls and the married couple each having their own. Rules were adopted for the purpose of keeping together in travel, and all were to set up camp within calling distance.

The Rocky Mountain Summer School term was to close the first week in August. The date set for launching our Natural History Trek was August 22. The group would gather at Palmer Lake as the point of departure.

During preparations for the trip, we had been showered with a hundred queries—what were we doing, why, and how on earth could we? My plan to take along the family was seen as the limit of absurdity. The opinions of people around me, including my friends, always concerned me, but seldom changed my thinking on plans of action.

For some years, I had been growing impatient with the tendency to make college life too easy on the students. The practice of trying to feed students their education in lecture form rather than having them find out things for themselves had just about crowded out individualism as I saw it. The educational program had come to consist too largely of the exercise of memory. Our program would challenge the students' ingenuity, powers of observation, thought, and endurance.

As for my family, I could never hope to succeed in such an undertaking without the methodical mind of my wife to keep the machinery moving smoothly. When I have studied pictures of the two long trips we took with the family, I have thought back to the miracle that was

Addie, the miracle that any woman would have the nerve to undertake such journeys and the courage to see them through. The older children would miss a year in school, but much larger sources of knowledge would be opened for them to experience firsthand.

It was a happy, almost hilarious group that gathered at the Estemere to arrange the car packs. Each student had a roll of bedding, a suitcase or two, and for each carload, there were a tent, a stove and cooking utensils, a table and chairs, and a small basic library of science texts, altogether a mountainous pile to stow in an automobile.

Before day's end, several packages of excess were mailed home—a ritual that was repeated periodically during the trip as, parcel by parcel, we stripped our gear down to essentials and then, by collecting as we went along, would build it up again.

Four days away from Palmer Lake, we camped among the spruce and pines of Yellowstone National Park. We spent a dozen

Caravan of cars on the "Natural History Trek." Harvey Nininger, accompanied by his family, led students from McPherson College on a 19,000-mile trek throughout the United States for the school year of 1927-1928.

days in this great combination arboretum, zoological garden, and geologist's paradise. We studied the geology of the area, including its volcanics, the petrified forest, hot springs, geysers, and boiling mud pots. We cataloged much of the plant life, the birds, and the mammals within view.

From Yellowstone, we went across into Idaho and south to Salt Lake. We had arranged with the National Audubon Society to make observations of bird life throughout the trip. Our group was permitted to use the clubhouse of the Salt Lake City Gun Club at Duckville as headquarters for a study of water birds on the salt marshes.

At Cedar City, Utah, we had a day of frustration. Our arrival at the town was delayed by stops to look for squirrels, to study Indian paintings on the cliffs, and once, when we ran out of gas. Our stay was to feature oil changes and lubrication for the cars, a visit to an agricultural college, the examination of a ranger's collection of flowers, and to follow up on a meteorite report. The college turned out to be only a high school, we couldn't locate the ranger and his flowers, and the meteorite expedition was a twenty-mile wild goose chase over terrible roads, ending with a fruitless hunt on foot.

If one day was bad, the next had its rewards. We drove to Cedar Breaks through colored cliffs dotted with bunches of bright shrubs. Then we emerged atop a heavily forested plateau watered by streams bordered with bright dogwood. There was a background of spruce contrasted by aspen and oak and the deep, beautiful canyon of Cedar Breaks. We were permitted to camp on the lawn of a small hotel on the highway. The next day, we made a memorable tour of another canyon featuring marvelous examples of erosion, Bryce Canyon.

We moved on to spend two days on the north rim of the Grand Canyon, hiking about and admiring the chasm, dim and mysterious in cloudy weather. Coming out through the Kaibab forest, we counted eighty-seven deer, and twice stopped to photograph the rare white-tailed Kaibab squirrels. These little creatures would leap from the tops of trees when our boys climbed after them, seeming almost to float down with legs and tail extended.

We took a shortcut through the Kaibab Indian Reservation to Zion Canyon and spent several days there in perhaps the loveliest camp site of our trip. Wild grapevines climbed the trees, with great arbors framing a private camp spot for each party. There was a bountiful harvest of wild grapes.

Some of the group went to church the day following our arrival at Zion while others held a discussion group. This was typical of our observance of Sunday. Sometimes we simply held a period of silent meditation followed by an exchange of thoughts or recitation of a favorite verse.

The canyon was shadowed so deeply by the steep canyon walls that we were able to observe Venus, the star of my long-ago farm adventure, by day. Some of us climbed to Angel's Landing, 2,000 feet above the canyon. Addie gave up on this jaunt, when the couple of miles I had promised seemed to stretch out and straight up, and instead went back to camp and made five quarts of jelly from wild grapes. The next day, Addie, Bob, and several others climbed to the top of Lady Mountain, 1,200 feet higher than Angel's Landing and requiring, Addie reported, climbing 1,400 steps cut into the cliffs, with 2,000 feet of cable hand rail. Not to be outdone, the rest of the party and I climbed Lady Mountain the third day.

As we left Zion, we noticed that the piñon nuts were ripe and the cones were opening, discharging their crop of nuts. I suggested we do some gathering. Tarps were spread under the trees and the branches flailed with poles, bringing down showers of nuts and cones. It was a sticky job, but it paid off in more than fifty pounds of the delicious, tiny nuts.

We developed a good bit of family spirit as our tour progressed.

Our two small girls, when the hikes got a little long for their short legs, were treated to many a ride astride the back of one or another of the boys.

Our camp cookery became relatively sophisticated, as it needed to be for meals not to become unbearably dull over a period of nine months. By necessity, we stuck to inexpensive foods, but we expended some time and effort in their preparation. We found that making a batch of fudge or peanut brittle or popcorn balls over the campfire or cookstove would take the edge off of a bad day or sour dispositions.

St. George, Utah, was a mail stop for us. Here, as in every town where we expected mail, our first call, usually at a dead run, was on the post office. Addie shopped for candles for a birthday cake for Doris, who would be six in two days. At the next stop, Las Vegas, Nevada, a rather quiet place in 1927, we located a cake for the celebration. The whole bunch of us shared it and then made a batch of candy together.

A campsite during the Natural History Trek of 1927–1928.

Amid all of our mountain climbing, sight-seeing, campfire pleasures, sojourns in forests and deserts, and occasional mingling with civilized communities large and small, our group was not allowed to forget the principal object of the trek. Questions and discussions concerning our surroundings kept everyone alerted to watch for and record the varieties and adaptations and relationships of plant and animal life that are the concern of ecology. The canyon walls constituted an open geological text and the seashore offered the richest combination and intermingling of biological adaptations to be found on our planet. The student's efficiency could be measured by his or her reactions to these environments as reflected in campfire discussions and in notebooks.

California was a much-anticipated destination, but our welcome to the Golden State was something of a let-down. At the quarantine station, we were required to open all the gear of our whole outfit. Then, when it was all laid out for inspection, the officer glanced at it only briefly, making no real examination. And we, who had set up and disbanded camp several times weekly, had to undertake what was, so far as we could see, an extra and useless chore of unpacking and repacking our gear in scorching desert heat.

We spent several weeks in and about LaVerne and Los Angeles. I delivered a number of lectures in the vicinity, and the students put on several programs for college, high school, and church audiences.

The first week in November, our married couple had to leave the group for medical reasons. Our two coeds received permission from their parents to continue with us and bought a Dodge coupe for their transportation. Our caravan now numbered sixteen persons, counting our children, in four automobiles.

At Laguna Beach, we set our tents on a cliff overlooking the ocean with sandy beach and rocky shore. We spent afternoons on the rocks collecting seashore specimens at low tide, and the forenoons studying our catch.

We spent five weeks in California, then headed east again. Near Yuma, Arizona, we stopped among the small sand dunes and mesquite to search for sidewinder rattlesnakes and meteorites while the children dug tunnels in the sand. Later, we paused among the large sand dunes to see remains of the old plank road used to cross the desert sands before the pavement was put in.

When we reached Jerome, Arizona, we were covering ground familiar to us from the house-car trip. We camped near the little post office of Cornville. We made a quadrangle of our four tents, driving the stakes into the sand. The evenings were cold, so we went to the mesquite thickets and gathered brush for a fire in the center of the quadrangle. There we would sit until bedtime talking over the day's experiences, writing notes or letters, or singing to the accompaniment of a ukulele. Occasionally someone became eloquent and broke forth in a reading or poem, classical or otherwise.

We toured the mine at Jerome, the smelter at Clarkdale, the salt mine near Camp Verde. To celebrate Thanksgiving, we visited the Montezuma Castle cliff dwelling and had a wiener roast, then returned for a candy-making session in the evening. We passed up Montezuma Well, for there was a fifty-cent fee. There were a number of side excursions we got along without during the trip for the same reason as we were traveling on a slim budget.

We were camped within a few miles of the site where in the spring of 1926, I had found the fossil footprints. Two years had passed, and I thought it quite worthwhile to revisit the site. Just as I had hoped, weathering had exposed several new sets of tracks. We wired Dr. Byron Cummings of the University of Arizona at Tucson, who agreed to purchase whatever tracks we might recover, and proceeded to remove and ship the tracks to the university.

It was heavy work, wading the creek, meticulously uncovering and chiseling around the stone blocks. We hired a local man to help us blast them out, then carried the blocks on our backs across the creek and up the steep canyon wall to a point reachable by vehicle to carry them to camp for crating and shipment.

After eleven days at Cornville, we moved on to the Boyce Thompson Arboretum at Superior, Arizona. There we camped on the grounds for eight days at the invitation of the staff. The arboretum furnished an unparalleled opportunity to study hundreds of species of cacti.

Good fortune waited at Tucson. The University of Arizona had a check for $380 ready in payment for the tracks shipped from Cornville. We all gathered round the fire in the reading room at our camp, and I paid each in cash for the work they had done and then divided up the surplus.

I made a two-day side trip to the noted copper mine at Cananea, Mexico, accompanied by one of our students, Alex Richards, who was fluent in Spanish. The mine superintendent, with whom I had corresponded, was most helpful. He introduced us to a Dr. Hoaglund who, in turn, brought us into contact with a local resident who owned a sixty-eight-pound meteorite. I was able to buy this specimen of the Arispe meteorite with practically my last dollar, and with the certainty that it would repay me handsomely.

When we reached El Paso, we stayed again at Camp Grande. We were plagued with more cold weather complete with snow, and we were plagued again financially. I tried hard to arrange to have checks cashed for the students but could not. Finally, and with much difficulty, I wired home and was sent $100 by our bank.

It was nearly Christmas, and all the students, some of them a little homesick, were receiving packages—cakes, cookies, nut bread, candy, and butter. Addie and I took money my mother had sent and used it for one gift for each of the children. We drew names among our group and exchanged small gifts. We placed our packages under the camp's community Christmas tree and celebrated Christmas Eve with the other campers.

We attended Christmas morning church services, cooked a roast chicken dinner in the community kitchen, and then packed to move on toward Carlsbad the next day. We found the great caverns even more beautiful than before, with some areas being lighted by electricity. Carlsbad was an excellent study in geology for the students. They were given special attention, and they made the most of the opportunity.

Group photo at Carlsbad Caverns, Natural History Trek, 1927.

Near Valentine, Texas, we followed fossil tracks again. Evans Means, a cowboy on the Quinn ranch fifty miles from Valentine, whom we had met during our house-car trip into that extreme southwestern corner of Texas, had written to me to describe some fossil tracks on the ranch. I took the students out to view the tracks, which occurred in a layer of hard, flinty rock covering perhaps forty or fifty acres. There were hundreds of small bird tracks like those of sandpipers, some that resembled crane tracks, whole series that looked like those made by a house cat, and others like dog or wolf prints. There were also tracks of a three-toed animal, which we mused might have been an ancient tapir. The oddest of all were a series of two-toed, bird-like tracks about the size of those that would belong to a large turkey, about an inch long, with a stride of seven inches, the main toe pointed straight ahead and the second thrust to the outside at an angle. The formation in which these footprints were preserved was estimated to be at least thirty million years old.

Evans Means was an affable host. We celebrated New Year's with venison tenderloin steak for breakfast and roast for dinner, and a horseback ride into the hills to gather resurrection plants to send home.

The weather alternated between cold and sun, the roads varied from good to bad, the days were punctuated with flat tires or running out of gas, lecture dates were obtained or refused. We visited canneries, mines, smelters, museums, lumber yards, universities. Bob did his homework, and Addie maintained her notebook of bird observations for the Audubon Society and completed the same credit courses in ornithology as the other students.

At the King Ranch this time, Porferio not only escorted us to a vantage-point from which to see the whooping cranes, but set up an armadillo hunt. At night, miles out in the mesquite thickets, dogs hunted out the little armored fellows, and at the first yelp, everybody ran leaping over bushes and dodging under branches until, with flashlights, they reached the dog with his armadillo captured by the tail. We caught two of the odd little creatures, unharmed, and observed them for several hours.

At the University of Texas in Austin, we visited Dr. Hermann J. Muller, later to become a world-renowned biologist, who outlined for the students some new studies he was then undertaking involving flies, the effects of x-rays, and the mechanics of heredity. I had read of Dr. Muller's achievements in experiments with x-rays on the fruit fly *Drosophila*, which already had produced many new species of that fly, and for which he was awarded the Nobel prize twenty years later. At the very period when his revolutionary achievement in genetics was first being recognized, he took time to give my traveling class a lecture and a guided tour of his laboratory.

Since the start of the trip, I had carried with me a forty-pound Canyon Diablo meteorite from the great Arizona crater. I received an order for a slice of that meteorite, and in El Paso, I went into a large machine shop to inquire about having a piece cut off that would measure four to five square inches.

I appeared at the office of the superintendent at 8 a.m., just as he was opening for the day. When I described my need and asked if the shop could handle it, he examined the specimen and told me yes.

"How long will it take?" I asked.

He shrugged and replied that it would be fifteen or twenty minutes.

I turned the specimen over to him and said I would return in a couple of hours.

Upon returning, I looked into the shop, where the man stationed at the power hack saw was wearing a rather sour expression. There was a considerable pile of worn-out blades, but the cut had reached a depth of only about three-quarters of an inch. Inquiry brought a somewhat glum remark to the effect that the job was not going well.

I offered to manage the cutting job. I kept the saw going about an hour, then told the operator a new blade was needed. He said he was afraid there were no more blades, but when I assured him I was prepared to pay for however many blades were required, he brought a dozen more, and the work was continued until closing time. The cut was not quite finished, but by using a hammer and chisel, prying the slice out and pounding it back a dozen or so times, we finally forced off the piece.

When I went to settle with the superintendent, he refused to let me pay for anything more than the cost of the blades, and he asked a good many questions about meteorites.

A few weeks later, I took the same specimen into a machine shop in another city to have a second slice removed. This time I tried to warn them in advance how difficult meteorites were to cut, but I was assured that the job would be finished in a half hour. I delayed my return, and after about two hours, I found the saw operator working away, again with a number of worn-out blades lying beside the saw. When I returned in the afternoon, I found he had improvised a seat on the beam of the large hack saw and was being carried back and forth somewhat like a cowboy on a bronco.

By closing time, the two-by-three-inch cut was near enough to completion that I paid for the job, took the specimen with me, and later finished the remaining bit with a hand hack saw.

We drove through rain from Roanoke, Louisiana, to New Orleans, riding through marvelous live oak forests hung with dripping Spanish moss. Everywhere in the swamp country we marveled at the cypress trees, flaring at the base with great roots anchored in the swamp or river banks.

Ferries became commonplace, from large Mississippi River steamers to little rafts just big enough to carry one car across a stream by means of pulleys.

Heavy rains marked most of our stay in New Orleans. We obtained permission to view the books containing Audubon's original paintings of birds. The six large volumes, about three by four feet and four inches thick, were guarded very jealously. We were told at that time they were valued at $35,000.

As we drove through Florida, we passed through town after town, mostly those that "weren't"—mere lamp-post cities, relics of the Florida "boom and bust." At Key Largo, we spent the low-tide hours gathering shells and studying the abundance of animal and plant life in the water. Then we bought a week's supplies and headed for Big Cypress Swamp in the Everglades to search for the rare and elusive ivory-billed woodpecker.

We camped on the shore of beautiful little Deep Lake, eighty-six feet in depth, surrounded by trees and thickets except for one small beach with palm trees where we pitched our tents. We never did see the ivory-bill as a group, although two of the boys on a hike into the deep forest believed they had. While many ornithologists thought that the great bird had already been exterminated, the existence of a few surviving birds at that time was later verified.

We hunted the ivory-bill again among the many small islands and lakes of the great Okefenokee Swamp. We stocked supplies for a week and headed toward the edge. Here, we located a typical South Georgia swamp native, Hamp Mizzell, who offered camp space for the night, amused us with his fiddle and songs, and identified for us various trees and plants of the swamp. The following morning, he put us on the right road to Jones' Island, near the edge of Billy's Lake.

Caravan of cars at Okefenokee Swamp, Natural History Trek, 1928.

We were slowed by detours, rough roads, mud, and water, and it was about supper time when we reached our camp spot, almost at the rim of the lake. A boardwalk raised up on poles over the swamp led through the trees to the boat landing on the lake, a quarter mile away.

We arranged for another swamp man, Jackson Lee, to guide us into the interior. Nine of us, including Bob, set out well before dawn on a two-day search for the ivory-bill.

John Lehman's diary describes our introduction to the swamp:

The morning was just beginning to break as our four boats nosed into the lake. We paddled for two miles through the narrow lake, then plied our way with long poles through shallow waters covered with lily pads and everywhere surrounded by cypress trees. After five miles of this, we shouldered our supplies and bedding for a tramp to drier ground. The earth often gave way beneath our feet and we sank to our knees or thighs. We cut a path through the dense thicket, often following alligator runs to avoid cutting a trail. After an hour of wading and stooping and resting, we had advanced about three-eighths of a mile.

Beside a pine tree, we found a spot of ground that was scarcely higher than the soggy muck nearby, but which our guide informed us was "as good as we will find." So we relieved our shoulders of their burden, built a fire of deadwood nearby, and dug a well with a hatchet. We spent the remainder of the day scouting for such birds as we could see and hear and, of course, the ivory-billed woodpecker.

The rain showed little respect for the crude shelters we made of branches and bark. When we awoke the next morning, daylight was breaking beyond a heavy veil of clouds. We gathered our half-soaked clothing in our hands and dressed by the roaring fire, then hung our bedding to dry from the branches of trees. We breakfasted on salmon cakes and hard-boiled eggs.

Then we went out again to explore the forest, picking our way over logs and through brush, cutting our way through tangles, wading ankle-deep water or sloshing through spongy leaf beds with an almost constant dripping of water from branches above. Several times we stopped and sat silently to listen for bird notes. The extreme quiet was impressive. Once we heard on three sides of us the loud coarse bellowing of large alligators, but we did not see any but the smaller ones.

Soon after noon, we slopped our way back to the boats with all our duffel and started on the return journey. Then the real rain came, harder and harder, the giant drops seeming almost like hail. We had to bail water from the boats. Finally the sun burst through, but still the rain continued in torrential spurts. The large drops looked like pearls as they danced upon the surface of the lake, so close together they seemed to be contending for space. With the sun on us, we suddenly were not at all chilled and were having fun.

Wind clouds appeared ahead in the west, black and ragged, chasing each other furiously. We paddled vigorously to beat the windstorm to our pier before the fretted surface of the water should develop real waves, carried our duffel ashore and hastened over the long boardwalk back to camp.

Okefenokee marked a certain personal victory. Just before reaching the swamp, we experienced the only near-revolt against our leadership of the entire trek. On our way in, we had spent the most difficult day of travel of the trip. As we rested between assaults on seemingly impassable mud holes and trestle bridges, several of the boys called me over to where they sat in a small circle and announced their joint decision that we should abandon our plan to go into the swamp. I reminded them of my repeated warnings at enrollment time that difficult situations would arise and stated firmly that we would go ahead as planned.

Though we came close to the break-up point, we mustered sufficient spirit of agreement to go on to Okefenokee. Our two days in the swamp, with water and ooze below us and rain streaming down over us, proved to be a completely miserable experience, but the kind of ordeal which we would look back upon later with smug enjoyment. Somehow, the acute physical discomforts drowned out the misunderstandings, and we came back out, as Addie described us, "wet as fish in the water, but a jolly bunch."

Later when we disbanded, while the students were saying their goodbyes, two of the boys stated that our visit to Okefenokee Swamp was the highlight of the trip. Both had participated in the attempt to dissuade me from going into the swamp.

We had a long, hard pull through thirty miles of mud back to Waycross. One car, with a broken axle, had to be pulled all but the first seven miles. As we battled the mud, we passed chain gangs of Georgia prisoners working on the roads. They pushed us from holes several times, and once, guards ordered the prisoners to pick up one of our stranded vehicles, together with its four occupants, and place it on firm ground.

From Waycross on, we were prepared to join the crowd of sightseeing tourists and maintain a more civilized schedule and appearance with less casual ways. As we started north, spirits lightened. Driving through Tennessee, we found the first signs of spring. Yellow jonquils were blooming everywhere in dooryards and on hillsides.

It was a welcome change to transfer our interests from natural to social history, touring the scenes of our nation's greatest turmoils and highest accomplishments and visiting the former homes of the country's great leaders of the past. Colleges, universities, museums, libraries, research establishments, historic points, and industrial plants were the new subjects of our studies. During the following eight weeks, we visited the most important educational centers of the east.

I decided to let each student plan his own program for our two-week stay in the nation's capital, except for certain assignments involving the US National Museum and a few suggested points of interest. Then I turned the group loose with the requirement that each keep notes on his or her own activities and hand them to me at the close of our stay.

From Washington, we traveled northeast—to Gettysburg, Philadelphia, New York City, and Boston. Our year's work was now finished, and from here on, it was optional for the students to stay together or to separate. Addie wrote a somber note in her diary:

We had our last meeting of the group together. The next morning, four of the boys left us. There was to be a wedding in Ohio, and they needed to finish the trip a little earlier than the rest of us. We had camped together so long and had learned to know each other so well that we felt like one large family, and I think of them as belonging to us as much as our own children. Many a time we swapped an egg for some crackers, borrowed a skillet, or loaned an eggbeater or potato masher. Sometimes it was a recipe for drop biscuits or omelet. So it was with a feeling of regret that we took the last pictures and said goodbye to the four boys there in the camp overlooking the city of Boston.

The rest of us went on northward to Niagara Falls and for a look into Canada. At Detroit, the two girls left us to continue on home. By the time we reached Coldwater, Michigan, the remaining five boys were traveling slowly to keep the rear end of their car from going to pieces. They were about out of money and food and were feeling blue.

Though the final days of our nine-month outing lacked the gaiety of the start, later memories emphasized the brighter times and minimized the hardships. Looking back from thirty-five years, Kenneth

Rock described it thus:

What a year! We thrived on cereals, pancakes, soups, and fruits. We slept well in our tents night after night, on deserts and in mountains, in snow and in rain. We drove Yellowstone bears from our larders by night, checked on the seven whooping cranes near Roosevelt Station in September, and found the same seven on the King Ranch in January. We chiseled, blasted, and crated four-by-six-foot slabs of four-by-six-inch cat tracks in Arizona's Verde Valley for Dr. Cummings and the University of Arizona Museum. We blistered hands hacksawing meteorites....

We went deep-sea fishing off the coast of California, and we poled our way through and over the lily pads, alligators, and cottonmouth water moccasins of the Okefenokee. We were awestruck by the giant General Sherman, the largest living tree in the world, in Sequoia National Park, and the colorful vastness of the Grand Canyon of the Colorado; we reveled at the booming breakers at La Jolla and were depressed by the devastation of the Mississippi floodwaters in the Louisiana delta country. We studied the living deserts of the Southwest and classified plant life in the major arboretums of the East. We wrote voluminous scientific notes in the field, and later spent a week checking them in the Smithsonian. We pried abalone from the rocky coast of Laguna, cut our feet on the coral of the Florida Keys, and were robbed of some of our belongings outside the Field Museum in Chicago.

During the year, we heard the pin drop in the Mormon Tabernacle, and we sat across the aisle from John D. Rockefeller, Jr., in Riverside Church, New York City. We spent several days being helpful in the Edwin Carewe filming of Ramona with Dolores Del Rio and Warner Baxter in Zion National Park. We saw the premier of the King of Kings in Grauman's Chinese Theater, Lon Chaney, Sr., in The

150 IN SEARCH OF FALLING STARS

Hunchback of Notre Dame in the Mosque Theater of Richmond, and we thrilled to the ballads of Sir Harry Lauder in the Chestnut Street Opera House in Philadelphia.

We angered the spitting cobras in Balboa Park and rode the three-hundred-pound Galapagos turtle in the New York Zoological Gardens. We climbed Pike's Peak in Colorado, and we went underground through unopened portions of Carlsbad Caverns with Jim White himself. We "burned up" on strongly seasoned barbecued venison on Cowboy Means' ranch at Valentine, Texas, and we almost died from vaccinations received on the International Bridge at Laredo.

We were a motley crew. We were immature, but we learned to give and take. During the nine months all of us grew in certain personal virtues such as patience, tolerance, loyalty, and cooperation.

Kenneth estimated the cost of the entire trip for him personally was $300, not counting a little money he had earned by writing newspaper articles, his share of the sale of tracks, and his money from the resale of the Pontiac.

During thirty-six weeks, we had covered 19,000 miles. We had visited the nation's capital, thirty colleges and universities, thirty-four museums, fifteen state capitol buildings, numerous libraries, a half-dozen mines, a score of industrial plants, and a dozen institutions of scientific research. The mettle of the group had been tested by three weeks in the desert, a week in the Everglades, a week in Okefenokee, and several weeks of freezing weather under tents. Field notes and museum notes were maintained with detail and regularity on birds, plants, mammals, reptiles, and fishes, with additional field observations on interesting geological phenomena.

It would be next to impossible today to capture the spirit of adventure and independence that marked our trek from coast to coast. We took whatever byway we chose, were able to do just about what we chose, with almost unlimited opportunity for solitary camping or exploring. America's wilderness today has shrunk for the most part to a few favored areas set aside as examples for future generations of what once was common.

6

Finding a Way

*The three great essentials
to achieve anything worthwhile are
first, hard work;
second, stick-to-itiveness;
third, common sense.*

—Thomas Edison

Although I wanted to give meteorites my full attention, for a while, I knew it would be necessary to hold my teaching job, lecture for fees, do as much field investigation as I could manage, and sell specimens as I obtained them. I had been able to pick up a few bargains in meteorites by purchase, but what I had counted on being *found* as a result of my educational program had just not materialized. Painstakingly, and sometimes painfully, I investigated lead after lead, only to have each turn out to be a dud. Some of these abortive reports sounded so infallible at the outset that it seemed investigation could not possibly fail to produce great finds.

During my early survey of the Coldwater fireball in 1923 and 1924, a former resident of Hodgeman County, about sixty miles north, had told me that when she was a small girl of eight or nine, she and other children used to play on a large meteorite that lay in the bottom of a hole some three or four feet deep—a meteorite that her father and neighbors had seen fall. Her description of the object was convincing. Various members of the family had made futile efforts to take samples from the mass, but failed, and since they knew of no use or value of the thing anyway, nothing further was done about it. The family moved during her teens, and she knew no more.

In later years, I would have placed such a story on a waiting list for future follow-up when I happened to be passing that way, but in my inexperience, I viewed the report as a *must*. Such a find might enable me to balance my budget, which was wearing thin from all my running around. So I made the trip to Jetmore, the county seat, via train and hired car. By eating cold lunches and sleeping in cheap rooms, and by delivering a ten-dollar lecture in a high school en route, I managed to pay nearly half the outlay, but the remainder meant more red ink added to the cost records of my meteorite hunts. I was able to find absolutely no trace of substantiation for the story.

Travel itself was difficult. There were practically no hard-surfaced roads, and most of the rural ways I was forced to use were not even graded. Cars were slow, tires were untrustworthy. The possibility of rain storms made use of these back roads very hazardous for one who must face classes Monday morning. I made most long trips by train, but meteorites, and especially the duds, seemed always to lie in the most out-of-the-way places.

In 1922, I had purchased an already old Model-T. It had no arrangement for carrying a spare tire; instead, I carried a good set of tire tools and patching materials, and certainly made good use of them. I would mend tires in ankle-deep mud while a raw wind numbed my fingers, or creep in low gear through mud for hours toward a remote farmhouse, only to find an ordinary rock instead of a "solid iron meteorite."

Were the old scholars right? Was my idea wild and unworkable? Many a weary hour I searched my plan for flaws when I would rather have slept, but I always arrived back at the same conclusion. I went back through a hundred years of records of meteorite discoveries. If those records told me anything between the lines, it was that the only reason meteorites were not found more frequently was that people just did not recognize them. I continued to trust that one or more pairs of the thousands of eyes I was training in the rudiments of meteorite recognition would come across a specimen in the course of routine duties on a farm or ranch, but I had to support and further my interest in meteorites somehow until specimens began coming to me. So I arranged to take extended leave of my classes for the fall semester of 1929 in order to go to Mexico, where I hoped I could acquire a number of specimens for stock in trade.

Because I had found the literature on the subject of meteorites so inadequate, I decided the best preparation I could undertake was to acquaint myself as thoroughly as possible with known American meteorites. Mostly by means of our two extended family trips, I had managed to examine the major collections at Washington, Chicago, New York, Harvard, Yale and Amherst—nearly all the collections of any note on the continent, with the important exception of the one in Mexico City, which held five of the greatest meteorites in the world.

Since Castillo's 1889 *Catalogue of Meteorites of Mexico*, there had been almost no contributions to the literature on meteorites by Mexican scientists, although a valuable body of meteorite writings had been previously built up in that country. Geographically, a greater incidence of meteorites had been recorded in Mexico than in most parts of the United States—a reflection of active interest.

At some time between 1884 and 1911, when he was the dictator of Mexico, Porfirio Diaz had issued an edict that all meteorites found would be considered government property. Since that time, quite a few meteorites had been found in Mexico and had been reported in the United States and elsewhere, but they had not been reported to Mexico City. I felt certain that more had been discovered and could be obtained by purchase, and I was confident I could secure a permit to remove such finds if I had attractive exchange specimens to offer.

In some ways, 1929 was not a very propitious time to go into that country—there had been a revolution the year before—but Mexico offered a clear challenge and a possible resource, and revolutions back then occurred in Mexico with almost the regularity of elections in the United States.

So I began to plan a trip. We were in debt as usual, but there was nothing crowding us, and we had credit at the bank. I would take out extra insurance, and my teaching salary would continue through the semester I would be away. I approached Alex Richards about going with me. He had been one of the stalwarts among the students on the Natural History Trek, had accompanied me on the side trip to Cananea, and was as eager to go further into Mexico as I was. Alex spoke Spanish fluently. He was twenty years old, resourceful, personable, and unafraid of hardship. Alex set about getting together $200 to match mine.

Alex had mechanical experience and aptitude. His favorite reading material seemed to be *Popular Mechanics*. I asked him to construct an automobile of the most rugged sort, made up of parts of other cars. I directed him to make a really husky car, a very ugly one, that nobody would want to try to steal. He succeeded better than I thought anybody could. No "hot rod" ever looked quite as tough as that car. It had seven speeds ahead and five backward, extra clearance, and a

156 IN SEARCH OF FALLING STARS

skid plate to protect its vital parts.

When we reached the border town of Laredo, Texas, in September 1929, we went to the Chamber of Commerce and asked for advice and information. We were told nobody drives to Mexico City, and nobody camps in Mexico. We were warned of a veritable scourge of banditry in Mexico. We would be robbed and then murdered to prevent the crime from being reported.

The Chamber of Commerce man went on to tell us we would not see a signpost or a road after leaving Monterrey. He warned, lectured, pounded his desk. Finally, he marked on our map—not a road map, just a common geographic map—the areas where banditry was reported to be the worst. He cautioned us about food, water, and disease. He inspected our Winchester automatic rifle and told us we would need it, that is, if the authorities would permit us to take it along. He tried in every way he could to be helpful, at the same time gesturing resignedly and clucking with concern. We would never, he declared, reach Mexico City. Or if we did, we would never return.

Alex and I mustered our courage. We had made so many preparations, we simply would not turn back. We carried camera equipment to record our experiences, a selection of meteorites for exchange, and a great cargo of food supplies and other paraphernalia. We laid out everything in the glaring sun on the International Bridge across the Rio Grande for the inspection of Mexican border officials, who looked over each item, down to the salt and pepper shakers, with courtesy, consideration, and delay.

The first 150 miles into Mexico was on a gravel road. The Laredo–Monterrey leg was the only strip of Mexico's national road-building program of 1929 yet opened to traffic. As we climbed from Monterrey and emerged through a mountain pass onto the great desert plateau that occupies most of northern Mexico, we passed the last of the road workers.

Just as we had been told at the Chamber of Commerce in Laredo, there were no signs and no roads. There were only oxcart trails, and no one we asked ever knew which trail we should take to bring us nearer to Mexico City. Everyone knew only that the city was very far. We would pull out our map, locate the village we were in, and select the next village in the general direction of our destination. Then we would ask how to reach that little town. If the way seemed too complicated, or if we were told there was only a foot trail or burro path, we would choose another village, or another, until we found one that could be reached by oxcart, and then try to negotiate the way with our jalopy. When we reached that village, we would repeat the procedure.

At camp en route to Mexico City, 1929. Alex Richards stands beside the vehicle custom-made for the trip. The pet hawk perched on his shoulder was purchased along the way.

The rough ruts of most of the oxcart roads failed to accommodate the standard wheel base of our car. A cart with two wheels five feet in diameter and drawn by oxen at the rate of one to two miles an hour was able to traverse a roadway that could practically wreck any vehicle attempting ten miles an hour. There was not much irrigation, but such ditches as existed, maybe two feet deep, simply crossed the road, with no attempt at bridging.

Though some of the roads were said to have been in use for as long as 400 years, the majority of them showed no evidence of any maintenance. When a traveler reached an entirely impassable section, he simply found a way around.

Towns with a population of 10,000 or so were blessed with "improved" streets whose paving consisted of small boulders, head-size to bucket-size. Such cobblestones lifted traffic out of the mud. Our first city of any size past Monterrey was Saltillo, with a population of about 60,000, surrounded by mountains, and spiked picturesquely by several cathedral towers. Tile-paneled lampposts lined the principal streets, and the richly adorned cemetery spoke eloquently of a glorious past.

Our route to Mexico City led south from Saltillo, via San Luis Potosí, but we wished to visit Torreón, some 200 miles across the desert to the west. Our first three days out of Saltillo were spent in a dreary search for the right trail. Then the thirsty desert surprised us with a truly beautiful oasis, an amethyst pool set in the dryland matrix, fringed by marsh grasses and guarded by a lone, stately cottonwood tree. This watering place was so welcome that we stayed a full twenty-four hours, filling our radiator, catching up on laundry, and collecting jars of insect specimens. From Torreón, we visited Jiminez, where large meteorites had been found a century earlier, then returned to Saltillo by another way and continued on toward the capital city.

We averaged forty miles a day, some days covering a hundred, some days as little as ten. We would stop by the roadside, usually near a village, and light our gasoline stove. Most evenings, we prepared flapjacks. By the time our fire was going well, a crowd of onlookers would be thick around us, not unfriendly, but curious, watching with some fascination the tossing and fielding of flapjacks in mid-air.

At noon or evening, if we stopped far out from any habitation with not a soul in sight, it was disconcerting to bring out our meal and then look up suddenly to see a couple of men standing among the desert shrubbery just a few steps away. Perhaps if we turned, we would see another visitor staring at us from behind a bush. We would usually manage to lift our rifle into view while ostensibly searching for some item among our supplies.

One day, we were interrupted at noon by a proud and tough-appearing caballero, heavily armed, who sat staring while we ate and then followed us for several miles when we started on. We were glad when he dropped behind.

One sundown found us weary in bandit country, and as we had seen no house for a long way, we picked a spot almost clear of brush and began setting up camp. Suddenly a very unfriendly looking character approached, then another showed up, and another. We managed to catch their interest with some trifle from our pack, then quickly replaced our gear and drove on. Long after dark, we reached a hacienda surrounded by a high wall with a guard at the gate.

We showed our credentials. The guard hesitated, then sent us on to a second gate that opened through another wall. Here we waited while this gateman took our papers to yet a third gate near the big house. After some time, we were allowed to pass on to the final gate that opened to the ranch headquarters. A man from the big house, who spoke some English, came out to us. He examined our papers carefully, looked over our pack, and invited us to drive into the passage that led to the central court. We parked according to his instructions and being terribly hungry, had begun eating a cold supper when a big, old-style Buick drove up, loaded with police, all armed and

160 IN SEARCH OF FALLING STARS

with machine guns mounted on the running boards of the heavy car. Two officers jumped from the car, demanding in Spanish to know who we were. We referred them to our host, who spoke to them at some length. Again we showed our papers. Alex answered questions. After inspecting our gear, the police went on their way. Our host told us they were searching for bandits who had killed a man that afternoon on a neighboring hacienda. We were not questioned further.

As we entered the Toluca valley, approaching Mexico City, our road led into a sort of quagmire and our car bogged down just as darkness fell. We selected a reasonably dry spot, set up our tent, had something to eat, and bedded down. About midnight, we were awakened by men talking and slapping at our tent. Alex's orders in Spanish to "go away" got no response. Alex took the rifle and sat at the closed tent flap with just enough aperture to see out. Evidently there were several men, apparently drunk. Alex rested the Winchester across his knee with the muzzle glinting in the moonlight and talked, gesturing with the gun from time to time. Meanwhile, I stirred about, making plenty of noise and carrying on muffled conversation with several nonexistent companions. Alex told our visitors repeatedly there was no room in the tent, we were already overcrowded. We kept this up for an hour or more until finally our visitors left. At daybreak, we loosened our car from the mud and went on our way.

***

The first place Alex and I visited in Mexico City was the *Instituto Geológica*. We knew no one there and carried no letters of introduction. When we asked about meteorites, we were directed to Dr. Frederick K. G. Mullerried. The short, stocky German geologist was a great field investigator who had traveled all over Mexico and later explored New Guinea. He had discovered a hitherto unknown volcano and a new species of fossil in Chiapas, a Mexican state bordering on Central America.

I explained to Dr. Mullerried that I had read accounts in the scientific literature of the large number of iron meteorites found in Mexico and asked if he could tell me how to learn something more about Mexican meteorites.

"I think you have come to Mexico at a fortunate time for you," he told me. "No one in Mexico is studying meteorites. There are a number of meteorite falls that never have been classified."

He arranged for me to see all the specimens in the National Museum. Immediately I saw that some were mislabeled, some unlabeled. During times of revolution, things had a way of getting mixed up, mishandled, even lost. I offered to write descriptions of some of the undescribed meteorites, correct the errors in labeling, and help put the collection in order. The museum officials were glad to have this done and arranged also for exchanges of their excess materials for the trading specimens I had brought with me in hope of just such a possibility.

For several weeks, Alex and I worked in the museum, cutting samples from unidentified meteorites, polishing and etching them to establish identification, labeling, and making notes on various specimens, including the history of each find. Here was a use for the skill in which I had been training myself, the ability to identify the correct origin and classification of nearly any meteorite specimen by surface features and, in the case of iron meteorites, by the etched Widmanstätten pattern.

In the course of my readings about meteorites, my attention had been captured by the story of the so-called Toluca fall of Mexico. When Spanish settlers first visited the little village of Xiquipilco near Toluca and Mexico City in 1766, they found the natives making implements out of native iron. At this time the existence of meteorites

162 IN SEARCH OF FALLING STARS

as matter from space had not been affirmed by the scientific world. The Xiquipilco community had no manufactured metal; the natives told visitors they "always" had made their tools from iron picked up in the fields.

A quarter of a century later, when the arrival of meteorites on earth had been confirmed, collectors remembered that distant Mexican village. By 1824, visiting scientists had obtained specimens. The natives were then still using iron fragments from their fields for making implements, but the creation of a market for the meteorites soon led them to obtain man-made steel for their tools, and they saved the natural iron for sale to visitors. In the latter part of the nineteenth century, great quantities of the material were shipped from the area. Dr. Jose Aguilera, a noted Mexican geologist, told me in 1929 that no less than twenty-two tons had been shipped out prior to 1906.

One of the burning aims of my developing interest in meteorites was to visit this village of Xiquipilco. Foote and Ward, two mineral dealers who had done most of the collecting in that area, had died shortly after the turn of the century. There was nothing in the recent literature about the Xiquipilco meteorites, and Mexico had been in revolutionary turmoil. These three facts indicated there might be much to gain from a visit to Xiquipilco. Perhaps this village, apparently unvisited by any collector for twenty-five years, would yield specimens at a cost that would enable me both to augment my personal collection and add to my exchange stock.

I spoke to Dr. Mullerried.

"While I am here, I would like to go over to Xiquipilco where they found so many irons."

"Oh, that's true," he said. "They did find many meteorites, but I haven't heard of anything coming from there for years. Besides, that's dangerous country."

He was not sure it would be wise for us to try to go, but he promised to make inquiries.

From time to time, I would ask him about it, but he never offered any encouragement. When I mentioned that I should like very much to locate one of the implements that were reported to have been made from meteorites at Xiquipilco, this interested him somewhat more, but still he fostered little hope for the trip. I told him I was set on going even if I had to go alone.

At last, one day he announced he had made arrangements for the trip if we still wanted to try it. At this time, Alex was seriously ill with amoebic dysentery, so Dr. Mullerried himself prepared to go with me.

The little village of Xiquipilco is only about thirty airline miles from Mexico City, but it lies across extremely rugged mountains. We were to go by train to the city of Toluca, take another train to Ixtlahuaca and hire some sort of conveyance from that point.

Before we left Mexico City, I cashed a fifty-dollar American Express Travelers Cheque so that I would have some negotiable money to buy specimens. In 1929, all Mexican money was silver. A peso was worth twenty-eight cents, and by the time I had loaded up with pesos from my check, together with the pesos I had already, my five pockets were bulging. The coins weighed me down, and every time I took a step, they jingled.

"Can't you keep them from rattling?" Dr. Mullerried would groan. "We'll be robbed here going down the street with such a jingle of money." He managed his silver supplies better. His own pesos were rolled up in his field clothes and stowed in his knapsack.

We stayed the night in Ixtlahuaca. Dr. Mullerried scouted around, inquiring about the possibilities of locating meteorites in Xiquipilco. He came back reporting no luck, and even suggested the trip appeared so futile and dangerous that he wondered if we shouldn't abandon the idea. Dr. Mullerried was a brave man. I knew that I must take his doubts seriously, but I couldn't give up.

"I am a long way from home," I told him finally. "I've come down here at considerable expense and difficulty, and I think I'll go ahead and try it alone if you don't care to go with me."

As I was hoping and expecting, he wouldn't let me go alone. He sent out runners to find someone with a truck to take us on to Xiquipilco. Late that evening, a man appeared who said he had a truck to take us, and Dr. Mullerried engaged him for five o'clock the next morning. Almost immediately, another man arrived who also had a truck, and Dr. Mullerried hired him for five o'clock the next morning. A third man appeared, and the same scene was repeated.

"What in the world are you doing?" I asked. "There are only two of us to go."

"This is the way we do in Mexico," he answered. "We will do well to get one truck. Worry about getting one, if anything."

So we got up the next morning at five o'clock. A truck appeared at six o'clock, and we went. The road was terrible. Wrecks along the way testified to many failures of vehicles to negotiate it. We made the drive of twelve-and-a-half miles by 9 a.m.

At the village, there was nobody in sight except one man, wearing a white suit and a big, broad sombrero, standing alongside the church. Dr. Mullerried explained our purpose to him, and we learned that he was *El Presidente*—mayor of the village.

The mayor listened with a puzzled manner, looked us over rather carefully, and finally invited us to breakfast. While we breakfasted, the mayor left briefly and came back carrying a twenty-pound meteorite. Was this what we wanted? Dr. Mullerreid and I could barely contain our excitement.

"Do you think you can get more of this?"

"I think I can get you a ton."

Our host sent out a runner, and while we walked about, waiting for word to get around, I myself picked up a little three-pound meteorite in one of the fields. When we came back to the town square, there were men standing about with baskets and bags and handfuls of meteorites. For the next couple of hours, we bought meteorites until we ran out of money, and the Indians were still holding up meteorites to sell, crowding about like a bunch of ants around a bit of syrup. We bought 700 pounds of meteorites.

In search of meteorites in the village of Xiquipilco, Mexico, 1929. Dr. Mullerried, who accompanied Harvey Nininger to Xiquipilco, is on the right.

One of our last purchases was a *barreta*, a crowbar-like tool, fashioned from a meteorite. It was market day, and I had gone through all the wares displayed in the public square, looking for a meteorite shaped into some kind of tool and finding none. Finally, I visited the blacksmith shops. At the fourth and last blacksmith shop, I recognized, on a tool way back in a corner, the Widmanstätten lines that proved its meteoritic origin. The figures were warped and beaten out of shape, but they were visible.

"Where did you get that?" I asked the smith.

"I made it."

"What did you make it out of?"

"*Areolito*," he replied—the Mexican word for meteorite.

Our visit was a glimpse into a distant and primitive past. The tremendous meteorite shower that occurred at Xiquipilco a hundred thousand to a million years ago marks that area as one of the most interesting in the world for any student of meteorites. It is the only place known, where from time past remembering, men had forged all of their crude knives, plows, hammers, and other iron utensils out of iron from another world.

I was accumulating considerable information and a good supply of Mexican meteorites. While examining the National Museum collection, I recognized that a small meteorite labeled as Xiquipilco was a misidentified specimen and was able to determine that it came from Chihuahua City. Through correspondence after my return home, I was able to obtain the main mass of that fall, over 100 pounds. Besides the Xiquipilco material and the Chihuahua City iron, I acquired specimens of Zacatecas, Chupaderos, Bacubirito, Tlacotepec, Puenta del Zacate, and Rancho de la Presa. The total was more than 900 pounds.

After making arrangements for shipping the meteorites back to

The *barreta* (crowbar-like tool) made from a meteorite, purchased from a blacksmith in Xiquipilco, Mexico in 1929.

McPherson, Alex and I prepared to return to the United States by train.

Although neither of us had mentioned it until we were safely housed in the YMCA upon our arrival in Mexico City, our brushes with banditry on the way down from Laredo had shaken us both, and we had promptly decided to dispose of our hodgepodge automobile. Regret was mixed with relief, for it was like parting with a fond, if somewhat crazy, friend. It had served us well.

We parted with another "friend" at the same time. As we had rattled our way through rural Mexico we often had seen boys and girls carrying crudely constructed cages containing wild birds. Their plumage was usually bedraggled and we suspected they often went hungry, but both children and adults seemed fond of such captives. Birds were always offered for sale in the small village markets.

One day, Alex returned from a market tour with a large Harris's hawk tethered by a strong cord fastened to one of its feet. A boy had offered it for a pittance, and Alex thought it might be fun to take the creature along with us. The hawk would perch on some part of our vehicle—most often a fender—and scan the passing landscape, or perch on a shoulder or knee. At markets, we purchased *carne para gato* (meat for the cat) for our feathered pet.

When we placed our jalopy in a Mexico City garage for sale, our pet went with it. The building was tall enough to accommodate two or three stories but had only empty space between the roof and ground floor. We gave the bird to the keeper of the garage, who promptly removed the cord from its foot. Away toward the roof flew the hawk, perching at a point high above us. That was the last we saw of him, but we felt sure he would be looked after.

Sale of the jalopy served as a sort of injection for our flattish billfolds. We applied at once for passes on the National Railroad and received the permits shortly before the end of our stay three months later.

Combined freight and passenger trains ran three times per week. We chose the Sud Pacific line, which would take us along the West Coast and through entirely different terrain than the high plateau country through which we had driven. Our passes permitted stopovers anywhere along the line.

Acquaintances advised us to refrain from discussing political matters while traveling by railroad. When, after a stopover at Guadalajara, we re-boarded the single passenger car at the end of a long freight train, we chose, because of the large amount of hand baggage we were carrying, the double seat at the end of the coach. When we had settled ourselves and our cumbersome luggage we looked up to see a very large, uniformed soldier-policeman just across the aisle and one seat forward, facing us.

Soldiers were present on all Mexican trains in those days as a precaution against sabotage, so we were not particularly concerned, and we were soon to leave this train anyway for a planned two-day stop to photograph tropical birds. But when we caught the next train, taking seats in the same position as before, there again was our officer, just as on the previous train.

At Mazatlán, we stopped again for most of two days, and our next train came through at 3 a.m. The bus picked us up at our little hotel, and we were relieved that our officer was not present, but then our bus stopped at the town's big hotel, and on he climbed, and when we boarded the train, as before, he took the same post facing us.

We were becoming worried that he had been delegated to see us to the end of our journey, and we only hoped there was nothing more serious in his orders. I attempted a conversation in my meager Spanish and he laughed with me a bit at my blunders, translating a few words of English into Spanish for me, and he seemed a little less formidable thereafter.

We next left the train at San Blas, a village in very bleak

desert-looking country, and here we hoped we might lose our escort, but two days later, he was already in his place when we climbed aboard, and we took our accustomed seats facing him. We felt certain we had committed no errors, but still we were apprehensive as we boarded the train after our last stopover at Hermosillo for the final leg of our journey to the border. This time, the officer failed to join us. We were back on American soil in late December.

2

Back in McPherson, shortly after the Christmas holidays, Alex brought me a meteorite about the size of a potato. He told me the twelve-ounce stone had been picked up by a neighbor, a youth who had heard me lecture about meteorites at the Paradise school more than four years before, at the very beginning of our house-car trip. A sixth-grader then, the boy had found the stone about four years later while in the field husking corn. He thought it "looked like what the little professor was talking about," he told Alex, and had brought it in with the corn. Then he got a little embarrassed at the thought that his family might laugh at him for such a notion and he tossed the meteorite into the cow lot. When Alex was at home between the end of our Mexico trip and resumption of classes, the boy described the stone to him, and they went out to the cow lot and retrieved it.

I have made many meteorite discoveries, but none has ever meant as much to me as that little rusty fragment, picked up by a boy who had "listened" when I talked about meteorites. It proved my plan workable. It had been nearly six years since the Coldwater, Kansas, fireball had inspired my program of lectures in the rural communities. The necessary element was time, a period for ripening.

I made a return lecture in the Paradise community, describing the find that had been made and suggesting there probably were other meteorites about. At the close of my talk, during the question-and-answer period, I noticed a young fellow leave by the back door. When he returned shortly afterwards, he carried a seventeen-pound meteorite that he had found the previous summer but poked under the granary so the family wouldn't ridicule him.

Within a few months, a half-dozen meteorites with a combined weight of 140 pounds were recovered in the community and given the name of "Covert" for the nearest post office. All of these had been lying around for years and were only reported after it became known that a local boy had received money for his "rock." One stone had served as a doorstop. Another, weighing thirteen pounds, held down the lid of a pickle jar for twenty-five years. A thirty-one pounder had been used to plug a rat hole in a cellar. Then the house had been moved away and the cellar caved in. Alex helped sift through the rubble, and the meteorite was uncovered.

The two Coldwater finds were the direct result of my contacts personally and through the press while making a fireball survey. The Covert meteorites were recovered as a consequence of a lecture campaign which depended on no phenomenal event to excite the people but assumed that meteorites were widely distributed all over the world from falls of ancient and recent times. Education of the public in the importance of meteorite recognition and provision of a monetary incentive were features common to both efforts. The success that attended the first survey supported my faith in a general field program, finally crystallized by the delayed reaction to the Paradise school lecture.

The Covert meteorites assured me that my theory would be fruitful, and I was determined to resign my teaching position as soon as our finances were in better shape and devote my time to meteorites. I proposed to lecture, hunt meteorites, and supply specimens to museums.

Since my guiding theorem was that the farmers and ranchmen

of America knew far more about what is in and on the soil than any group of scientists and thus were the logical source for a greater supply of meteorites, my ways and means of extracting the information they could give me had to be as effective as possible. Effective lecturing about meteorites would answer the problems of ignorance and disinterest, and spreading knowledge and interest throughout a community could lead to one find which might lead to another. The reward given for discovery would dispel any attitudes of ridicule or disbelief.

It became my standard practice to pass actual meteorites among the members of an audience so that all might examine them closely and hold and heft them in their own hands. It also was my practice to use laymen's language.

Experience showed that a strictly informative lecture could be made interesting enough to hold the attention of any group from fourth grade on up through high school and college. Gatherings of farmers and ranchmen, or members of service or civic clubs, would also listen raptly throughout a forty-five-minute presentation. Sometimes I would speak to groups of laborers coming off their shifts at mines or on road construction jobs. I would address farm workers fresh from threshing, cotton picking, hoeing the row crops. More than once, I walked out into the middle of a field where workers were hoeing cotton to show them a meteorite and urge them to keep their eyes open for something like it.

I was convinced there was going to exist a science of meteorites. Its foundation would be the field program which I believed I was initiating through my project of lecturing, distributing literature, and supplying articles to local newspapers, particularly in the Kansas farmlands. I counted on the purchases I was making to stimulate the farmers to look for additional meteorites and was certain that once my program was well underway, there would be demand for all the meteorites available.

172   IN SEARCH OF FALLING STARS

For all intents and purposes, the science of meteorites (had there been one) was dead in America by the mid-1920s, save for the activities of two men—Dr. George P. Merrill of the United States National Museum of the Smithsonian Institution in Washington, D.C., and Dr. Oliver C. Farrington of the Field Museum of Natural History in Chicago. It is to these two men that I am most indebted, despite the fact that both had passed their prime years of activity in the field by the time I entered it. Both of these men carried the responsibility of being the head of the department of geology at their museum. No institution in America had shown respect to the subject of meteorites by assigning a man officially to the field of meteorite studies.

It seemed a long jump for me in 1923 into the far corner of the scientific arena where geology and astronomy met, or rather where they came together. Neither geologists nor astronomers seemed to recognize meteorites as being worthy of full-time attention. A score of mineralogists, chemists, and geologists had made valuable contributions to the study of meteorites in the latter part of the nineteenth century, and a few in the earlier part of the twentieth, but none seemingly had visualized a real science of meteorites—that is, meteoritics.

Dr. Merrill was the first scientist I made contact with after finding my first meteorite. I sent my first sample to the United States National Museum. Dr. Merrill had a keen interest in acquiring new meteorites for the national collection, and during the remaining years of his life, he obtained portions of each fall that I discovered. The US National Museum was operating the only meteorite saw in the country at that time. When I acquired several fine specimens of the Brenham pallasite in the early 1920s, Merrill was glad to slice them for me in return for receiving half of each. It seemed a high price to pay, but it enabled me to acquire a number of saleable specimens without laying out any cash beyond the original purchase price, which had been relatively small.

Dr. Merrill did not seem to relish the idea of new men taking up the study of meteorites. I cannot say that he ever encouraged me to undertake an investigation. He had been in charge of geology and meteorites in the great National Museum since 1897 and had worked in the department there for sixteen years before that. As I came to know him better, I decided that he was interested in seeing that the national collection had portions of more meteorite falls than any other museum in the world, but that he was not interested in ascertaining, by field work, how much of any fall had reached the earth. When attempting to estimate the total amount of matter reaching the earth as meteorites, Merrill consulted the catalogs of meteorite collections in the great museums of the world.

On the other hand, Dr. Farrington was not only a curator of meteorites, but also a great student of their structure, composition, distribution, origin, the phenomena of their fall, their size, their shape, and everything else about them. It was the reading and re-reading of Farrington's *Catalogue*, together with my own field experiences, that led me to develop the conclusions regarding the frequency and distribution of meteorites upon which my hopes for a collecting career were based. My ideas seemed radical to Dr. Farrington, although he took an interest in discussing any question I asked.

I visited Farrington many times, and on the occasion of my final visit, I outlined to him my concept of what should be undertaken as a program of meteoritical research and the kind of institution that was needed to accomplish the plan. The old gentleman, then in very poor health, responded that he wished he were younger; he would like to help toward the realization of my dream. With his passing, the young science of meteoritics lost a great leader.

While the Natural History Trek students were exploring the nation's capital in 1928, I had called on Dr. Merrill at the US National Museum to outline my ideas for a field program for the recovery of meteorites. I had hoped that the National Museum would recognize the virtue of such an undertaking and perhaps support it with the understanding that all of the meteorites discovered thereby would belong to the great national collection.

During our rather lengthy conversation, Dr. Merrill's answers were all negative and finally, after my last plea, he spoke with emphasis. "Young man, if we gave you all the money your program required and you spent the rest of your life doing what you propose, you might find *one* meteorite."

To this, I replied, "Do you know what I'm going to do about it?" Smiling, he said, "No. What?"

I'm going back home, and in some way—I don't know just how—I'll raise the money and go ahead with the program, and next time you see me, I'll be selling you a meteorite."

He laughed, and we shook hands. The next time I saw him, I was selling him two meteorites! If he remembered, he never said so.

Some years later, I again visited Dr. Merrill in Washington. I reminded him of the Rosebud stone, a very interesting meteorite in Texas which I suggested he might be able to acquire by exchange or by purchase and which would be a wonderful addition to the National Museum's collection. He said he didn't believe he would be interested. "I think meteorites have already given me just about all they have to offer by way of information."

The story of the Rosebud meteorite is indicative of the confusion and disinterest that plagued me in my pursuit. Sometime prior to our 1925–1926 house-car trip, I had given a lecture at Friends University in Wichita, Kansas, following which Dr. H. E. Crow of the biology department told of having been shown an "iron meteorite" some years before near Rosebud, Texas. It was said to weigh about 125 pounds and to be in the possession of a certain Captain Jack Waters, a Civil War veteran. Professor Crow suggested that I should at some time look into the matter.

We listed Rosebud, Texas, as one of our stops in planning the itinerary for our house-car trip. When we reached Rosebud in November 1925, we found that the captain had died, but through his relatives and the local press, we learned that the meteorite had been sent to the state university several years before. According to these informants, the captain had requested its return, but the university reported that the stone had been lost.

Later that year, we were in Austin, and I visited the university. During a conversation with Dr. J. A. Udden of the bureau of economic geology, he mentioned that the university had found one meteorite during his time there, and he showed me the Tulia stone which had been plowed up the year before. He said that he had also tested a small sample of a stone that was said to have come from Rosebud, but he seemed to know nothing of the mass from which it had been detached. The sample contained nickel, but he knew nothing as to the whereabouts of the parent stone. He had been told that it had been returned to the owner.

As I was leaving him, Dr. Udden said he wished I would go over to the geology building, on a different part of the campus, and look at a rock which he believed was in the office of Dr. Frederic W. Simonds, head of the department of geology. He said it was not considered to be a meteorite but that he, Dr. Udden, had some misgivings about the identification.

A day or so later, I went to Dr. Simonds' office. No one was there, but the door was open, and I immediately saw a most beautiful meteorite lying in the far corner of the room under a long table. It was covered with dust, but I could have little doubt as to its meteoritic character because of the abundant and wonderfully oriented pittings that covered the entire visible surface. I hurried out and down the hall in search of the professor. A short way down the hall, I met

176   IN SEARCH OF FALLING STARS

a man who introduced himself as Professor Whitney. He informed me that Dr. Simonds was not on duty that morning but that he would be glad to help me as he was next in command. I repeated the request of Dr. Udden.

"Oh, yes. That stone! Well, come in. I'll be glad to show you."

As he dragged the stone out from under the table, he explained that someone had thought it might be a meteorite, but that the entire staff had held a consultation and decided it was just a basaltic boulder that had lain exposed on a hillside and had acquired those pits by sandblasting.

Dr. Whitney placed the specimen on the table, and I began going over it with my ten-power pocket lens, almost bursting with excitement. My memory was parading before me all of the pictures I had studied of the various great meteorites, and this stone under my glass was certainly of first rank, if not the very finest specimen of all.

The professor leaned over my shoulder.

"Well, what do you think of it?"

"Dr. Whitney," I exclaimed, "it not merely is a meteorite; it probably is one of the finest specimens known!"

"You think so?"

"Yes, I'm sure, but I should like the opportunity to examine a bit of the interior."

"Well, I'll get you a sample," he said, and with that, he struck a shattering blow with a heavy hammer, sending chips flying about.

"Oh, please, don't mar the specimen," I cried. "It's too beautiful. I had in mind that it should be cut."

I picked up one chip about the size of my thumb and two or three smaller fragments. When I ground and polished them later, one of these smaller bits revealed beautiful chondrules and nickel-iron grains. Another I dispatched to Dr. Merrill of the United States National Museum in Washington.

He sent a brief reply, "I find no evidence of meteoric origin," returning the specimen as I had requested.

At once, I sent it again, begging him to grind and polish the inner face. A two-word reply came back: "Unquestionably meteoric."

My request for the privilege of describing this meteorite went unheeded, and when I returned to the university two years later, I was refused permission to photograph it. Yet nothing seemed to have been done in the way of publication.

On my second visit, I asked about the place of discovery of the beautiful meteorite, alias the sandblasted basaltic boulder. If Dr. Simonds remembered, he chose not to tell. He made an apparent effort to recall, then said, "possibly Glen Rose."

Thus, in 1933, when my first book about meteorites, *Our Stone-Pelted Planet*, was published, I listed the Texas stone as *Glen Rose*, for it still had not been officially described.

I did not know during my 1925 visit to the University of Texas that this fine specimen was the Rosebud meteorite that had been described to me in Kansas. I thought it might be, but that probably it was not, for Dr. Crow had spoken of an "iron," and Dr. Udden, when he mentioned examining a small sample of a stone from Rosebud, certainly gave no hint that it might have come from the mass he sent me to Dr. Simonds' office to see.

Up to this time, I had accepted the report, relayed to me by the Waters family and the Rosebud editor, that the specimen taken to the university by the captain had been lost. Having noted the seemingly total lack of scientific interest in the subject at the university, I continued to search for the Rosebud "iron." I corresponded with and visited relatives of Captain Waters in Alice, Texas. Hearing that I had offered in Rosebud to buy the specimen if it could be found, they joined other members of the family in a spirited search for it. The captain's relatives approached the geology department of the university in an attempt to recover "their specimen," but the university spokesmen, they reported, insisted that the captain's stone had been returned.

In any event, the Rosebud specimen was not described in the literature until fourteen years had passed since my identification of it as a meteorite. There was no mention of my name in the tardy description of this fine oriented stone which I had been first to recognize and which I had later been denied permission to photograph.

***

At that time, the mid-1920s, all talk of rockets for exploration was regarded as the wildest of fiction, but I believed that man was destined to explore more and more of the universe by all possible means. Since meteorites constituted the only material source of information then available, they must be the most important inorganic objects on earth, fraught with information about the cosmos that could be obtained in no other way, and about the earth itself, which doubtless they had been bombarding throughout all of geologic time. It seemed plain fact that meteorites could never become less important; certainly they must grow more so. Response to my lectures seemed to forecast a great increase in the number of collectors of meteorites among individuals as well as institutions.

There was just one jarring aspect of all my theorizing: Scientists themselves were remaining aloof from meteorites. Dr. Merrill's frank statement that there was not much more to be learned from meteorites apparently was dismaying only to me, his attitude being no surprise to anyone else.

No one seemed to share my view that any meteorite of any size must be more important than any other material on the planet. Since my area of interest had no scientific standing, there was only one way for me to go, and that was up. I was able to turn to my advantage this near-total disinterest that surrounded the subject of meteorites at the time my own fascination with it was becoming all-consuming.

I learned that less than ten percent of the colleges and universities, and practically none of the high schools and academies or small museums, had a single meteorite specimen, to say nothing of a collection. It was my opinion that meteorites were far more important than most minerals, fossils, and other natural history specimens that were featured in those same institutions.

I also discovered what I considered to be a very short-sighted practice among the few museums, universities, and colleges that had taken an interest in collecting and studying meteorites. In general, when the discovery of a new meteorite was announced, each of these institutions would purchase a specimen from the mineral dealer who had acquired it or make an exchange for one if the new discovery had been purchased by an institution. If subsequently another individual meteorite turned up in the same general area, and it resembled the previous find, it was considered to belong to the same fall and was not in demand unless one of the few collecting institutions had failed to obtain a sample of the first one. And when a large number belonging to the same fall were found within a short time, there would be a ready market for them until each of the collectors was supplied, and then any remaining specimens would become a drag on the market.

I learned that quite a number of meteorites from the showers of Brenham, Ness County, Richardton, and Canyon Diablo were still in the hands of farmers who had been unable to sell them. These persons would be glad to market stones that for them had only passing interest other than the money they could bring.

I visited some of the nearby localities of falls. I could have purchased, had I the funds, several specimens of the Brenham pallasite from farmers who had found them years before. The same situation proved to be true with the Ness County meteorites. I was convinced

180 IN SEARCH OF FALLING STARS

that the same thing would probably apply to other falls which I had no opportunity to investigate.

Gradually, as I was able, I purchased those unwanted specimens that farmers had on hand. Before 1925, I had acquired two Brenham pallasites of 117 pounds and 85 pounds, plus a number of smaller specimens. I also obtained surplus meteorites from the Ness County and Admire (Lyon County), Kansas, falls.

By this time, my dream of an active science of meteorites included incorporation of specialized courses in all university curricula. I knew better than to publicize this whole dream, but I did do a lot of thinking and planning. For the present, my job was to get set up so that the limited market for my lectures and specimens would support the beginnings of my program. Then I would feel safe in resigning my position as teacher.

The harvest of Mexican meteorites constituted a good stock of marketable specimens, and a start in the right direction, but was not enough to allow my resignation from McPherson.

***

Just before daylight on February 17, 1930, the fall of a meteorite in the vicinity of Paragould, Arkansas, brought fear and wonderment to those in the area. The press carried various announcements concerning the great fireball, which was visible over many thousands of square miles covering portions of several states. Though the spectacle occurred at an early morning hour when most people were still asleep—about four o'clock—the fireball's dazzling light and accompanying detonations brought many from their beds. Testimony of a number of them showed that the ages-old expectation of mankind's demise by fire was still healthy in many minds.

A few hours after the heavenly display, Raymond Parkinson, a farmer who had been awakened by the light and the explosion, went into his field for his horses and came upon a freshly made hole. Investigation revealed an eighty-five-pound fragment of stony meteorite lodged in the pit.

Parkinson sent me a sample, suggesting I might be interested in the purchase of the stone. Since I had no Friday afternoon classes, Addie and I promptly arranged to leave the children with her sister next door and set out immediately on the 700-mile drive. Though roads were none too good, we took turns at the wheel, made only necessary station stops, lunched on the way, and were at Paragould in twenty-four hours.

We found Parkinson in low spirits. He told us he had lent his meteorite for display in the high school, and that the school principal and science teacher had sold it for $300 to Stuart H. Perry of Adrian, Michigan, shipping it out by express in a typewriter case.

Making the best of a bad situation, we gathered all available data and visited the site of the find. Several of the many witnesses to this great heavenly event were very acute in their observations. There were discrepancies in these accounts as to the time between appearance of the great light and hearing of the detonations, whether there were one, two, or three explosions and regarding the size, being four, six, or nine times the diameter of the moon. There was agreement that the light had gone out with the explosion, and evident consensus as to the impressiveness of the experience.

One elderly gentleman who had gone out into the yard at 4 a.m. was standing where he had a clear view of the fireball passing behind the roof of the two-story house almost directly over his head. He pointed out its course with unusual exactness because he knew where he had expected it to hit the roof.

We examined the hole from which Mr. Parkinson had removed the eighty-five-pound stone. It was on a slant of approximately 40

182   IN SEARCH OF FALLING STARS

degrees from vertical, and its axis lay in an exact northeast–southwest direction. He had preserved the footwall of the hole, which was slick from the plunge of the stone into muddy ground, because he thought it might be important.

Two small stones had been picked up in a small clearing in the nearby timber, one of which had made a vertical, clean-cut hole about four inches deep, in the bottom of which the little stone was visible.

Using a map of the locality, I plotted the course of the meteorite, certain from the nature of the reports by witnesses that the eighty-five-pound specimen must be only part of a larger mass. I drew a line on the map, designating for search a strip not less than a mile wide and on a line from the Parkinson hole a distance of six or eight miles.

Because the unsurfaced rural roads had been rendered virtually impassable by heavy rains, and because I needed to return to my teaching, Addie and I went back to McPherson, again driving straight through and arriving in time for my Monday morning classes. Soon, word came that an 800-pound stone had been recovered from a depth of eight feet just three miles from where Parkinson had found his, and right on the line I had drawn.

Before leaving Paragould, I had engaged an attorney to purchase the Parkinson stone if and when it should be regained, and to act as my agent in the purchase of any other material that might be found. A long and nerve-wracking series of negotiations began. The owner was asking for bids on the stone and was receiving offers from another source. I wrote to officials of the institution that was bidding, informing them that I had directed the search to the area where the meteorite had been found and that I had been largely responsible for the recovery of the meteorite on which they were bidding. The bidding ceased, but not before the price had gone to $3,100, more than I could afford to pay and keep the specimen. I instructed my attorney to buy, realizing I would have to borrow to make the purchase, and that I would have to sell the meteorite subsequently. I finally paid $3,600 for the Paragould meteorite.

Two of my McPherson friends had come forward when the negotiations first began, each offering to lend me $1,000 without interest. The vice president of the McPherson Citizens Bank informed me there was no need to bother my friends, that the bank would lend me the full amount. He handed me a blank note to sign and said that when I learned the exact amount needed, I should wire the bank, and the proper amount would be filled in. Meanwhile, I was to write a check upon my personal account. The meteorite was placed on display in the bank's Main Street window until it was sold a few months later.

The 800-pound Paragould meteorite was found lying in a crater some eight feet wide and just a few inches more than eight feet deep. This hole was only about thirty feet from the pasture gate of a man named Fletcher, no more than 300 yards from his house. Fletcher had looked from his window at the time of the fall, while his frightened wife knelt praying. He knew of all the excitement of the farm and town folk. He must have known of the find of the Parkinson specimen. And yet, when he noticed the yawning hole by his gate, he blamed it on the digging of dogs. It was his neighbor, W. H. Hodges, who chanced to pass through Fletcher's gate and became curious about the big hole. He brought Fletcher to the scene, and they took a slender pole and sank it through the water and mud in the hole and struck the hard mass of the meteorite in the depths.

That same evening, a small hole was dug down through the mud to the stone inside. The next morning, the meteorite was extracted, requiring three hours of work by five men with a team of horses.

While we were in Paragould making settlement for the big stone, Parkinson arrived in town. News of the generous price being paid was just more than he could bear. He strode resolutely to the high school and called on the principal, upon whom he placed blame for the loss of his meteorite, led him into the school yard and administered physical retribution. Then he marched his man to the police station, told his story, paid a $2.50 fine to sympathetic authorities and sent the educator back to his school. Parkinson ultimately was awarded the $300 that had been paid for his meteorite.

At that time, the Paragould meteorite was the largest known to have been seen to fall, and was also the largest known intact stony meteorite in the world.

Many years after the Paragould fall, I was told a tale about Paragould that is an example of the way the accounts of meteorites, particularly relative to their size and to the size of prices paid for them, sometimes become distorted. The fellow recounted that he had seen the fall of the Paragould meteorite, and that the stone was so hot, it couldn't be touched for two weeks. The fall had "lighted up the whole place like daylight," he reported. He described the meteorite as weighing a hundred tons. "Big as half a room." The buyer, he said, had resold the meteorite for $400,000.

The actual dimensions of the Paragould meteorite were twenty-nine by twenty-eight by twenty-seven inches. And the price I received for it was $6,200.

The Paragould meteorite had profound effects on our lives. I have never ceased to regret parting with it, but I had paid a price too high, and was forced to give up either the specimen or my dream of making meteorites a new vocation. And Paragould, with the $2,000 profit it brought, was the way to my dream.

Addie and I were feeling our oats. In two months, we had been reimbursed for all our time and effort, with profit besides. Neither of us could think that it might be many years before another such windfall. I had been waiting for money enough to finance a changeover from a steady, salaried job to an endeavor both irregular and unpredictable in nature. Now, on the strength of Paragould and the stock of meteorites I had built up, we seemed near enough to solvency for me to dare resign my teaching position.

I felt dissatisfaction with the situation in which I found myself, giving half of my time to the job I was being paid to perform and the other half to an idea apart from it. I was anxious to go new ways. I wrote out my resignation. We would sever the umbilical cord of salary. From this time on, meteorites would be my career. At this moment, I held no professorship, no institutional connection, no visible means of support other than an idea.

I could not expect that meteorites promised an easy living, but what I wanted was a career that would be interesting. One that would make possible the adding of some bit, small or great, to the fund of human knowledge, and at the same time contribute to the pleasure and interest of living for many people.

"Do something that needs doing," I had often told my students, pointing out the yawning breaches in man's knowledge of zoology, botany, bacteriology, entomology, geology, and astronomy. And I had been attracted to so many of these gaps that my difficulty had always been the selection of one to the exclusion of others, aware that knowledge of these various fields had already grown so extensive that one must specialize in order to be able to add anything of importance. For teachers, which most of my students aspired to become, there was always the opportunity to help others by opening their eyes to the beauty, immensity, and complexity of the world around them. But for myself, I felt that I had found a new field, where I could both widen the vision of many people and at the same time contribute to the sum total of human knowledge.

Newton had given us a world of order for one of chaos; Copernicus and Galileo had provided a solar system in trade for the old geocentric universe; Darwin had substituted an evolving world of life for a static one; Hubble had given us our multi-galactic universe for the solar one. And now, we were in the process of discovering our planet to be a youthful, growing orb, playing hide-and-seek with a multitude of attacking asteroids endeavoring to knock it off its axis. There indeed must be more things in heaven and on earth than were dreamed of in our philosophy, and surely, meteorites must offer something to our store of knowledge.

1

The Path Behind

What's past is prologue.

—Shakespeare, *The Tempest*

In the little village of Weston, Connecticut, in the year 1807, Judge Wheeler stepped from his house early one December morning and was surprised to see a great ball of fire come riding out of the northern horizon. Dazzling, brilliant, sparkling, it nearly blinded him. He watched as it passed behind thin clouds and advanced to a point almost overhead, when it flashed three times and disappeared.

The judge was a learned, thoughtful man, and he was much interested in this marvelous spectacle that was different from anything he had ever seen before. He was very puzzled. It was like a "shooting star," only greater in magnitude. While he stood pondering, a great noise broke upon his ear.

It was like thunder, except that it grew louder and louder, grumbling and roaring in a most frightening way, almost deafening. While

188   IN SEARCH OF FALLING STARS

this sound still rumbled, there came a second noise, a sizzling, whizzing buzz, as though something were falling. A little stone struck a nearby building and rolled away into the grass.

The judge, much perturbed, speculated whether there could be a connection among these three phenomena. It seemed impossible that a stone could have fallen out of the sky. Everyone knew there were no stones in the sky. How could there be? What would hold them up there?

Hosts of questions and a great puzzlement arose in Judge Wheeler's mind. He queried his neighbors. Some of them had also heard a sound, seen a light. A few had picked up odd-looking stones from their yards, wondering from where they had come. It was decided that some of the wise professors from neighboring Yale University should be asked to investigate. They came with misgivings, skeptical about the story that stones had fallen from the sky, fully prepared to dispel the peculiar and unwarranted assumption.

But as they interviewed Judge Wheeler and others, they were impressed with the sincerity and awesomeness of their accounts. Finally, after prolonged investigation, the Yale professors concluded it must be true that stones had actually fallen out of the sky. They knew that these stones were different from anything on earth that they knew of, and they had witnessed extraction of some of them from holes in nearby yards and fields.

As had happened other times in the history of science, the impossible was found to have occurred. Stones had fallen out of the sky. The story, hard as it was to believe, spread. Eventually, it reached the White House in Washington. The man who occupied the President's chair was himself a scientist as well as a statesman. When Thomas Jefferson heard this strange tale, he declared it could not be true. Stones do not fall from the sky. At first, the President treated the matter as something of a joke. But his informants insisted the stones had been seen to fall, and had been retrieved, that Yale scientists had investigated the report and vouched for its truth.

Thereupon, Jefferson, Southern gentleman, President of the new United States, whose words on science and on matters of state went far, spoke with great dignity: "Gentlemen, I would rather believe that those two Yankee professors would lie than believe that stones would fall from heaven."

And that was supposed to settle the question. But eventually there had to be a fairer settlement, and the time arrived rather swiftly. During the next quarter of a century, a number of similar falls of stones were observed throughout the United States. The scientific world became convinced that Jefferson was wrong, stones actually do fall from the sky. The name given them was "meteorites" because they seemed to be associated with the light phenomena that long had been called "meteors."

***

Similar failure or refusal to recognize meteorites as coming from "the sky," had been characteristic among scientists of Europe also, although acceptance of the fact in the Old World had come a little earlier than in America.

In the middle of the eighteenth century, an old French priest came to the Paris Museum carrying a stone which he said had been seen to fall out of the sky near his home.

"No, Father Nollet, you are mistaken. Perhaps you saw a flash of lightning which entered the ground near you and wrought a change in the soil to produce this stone. You yourself may readily see that stones *couldn't* fall out of the sky, for there are none up there to fall."

And if any further evidence were needed to disprove the priest's claim, said the great scientists, it was readily available in the fields of logic, history, philosophy, and religion. Certainly, if stones were

190 IN SEARCH OF FALLING STARS

to be expected to fall from heaven, then instruction to that effect would be found in the Holy Scriptures.

And so it was in France in 1751 that the question of "sky-stones" was settled, but just as in the young United States in the next century, it did not stay settled. A reliable astronomer, Jerome de la Lande, described a fall of stones near Bresse in 1753 and received no better answer than had the French priest. In 1769, another priest, Father Bachelay, presented to the Royal Academy of Science in Paris a piece of a stone he testified he had seen fall from the sky in Lucé. The Academy was impressed to the extent that a commission was named to investigate the matter, the panel being headed by no less than the great Lavoisier. Over that respected scientist's signature, the commission reported that the stones in question were only terrestrial stones that had been struck by lightning which fused their surfaces. A bit of this fall has been preserved in the Vienna Museum.

An Italian chemist, D. Troili, wrote a lengthy description of a fall at Albareto, Italy, in 1766. His was a worthy account, but it fell on ears as deaf as those of the French scientists.

But then, as now, the "settled questions," the beliefs or un-beliefs of men did not disturb the ordered course of nature. There were other reported falls, in 1790 in France, in 1794 in Italy, in 1795 in England. There were numbers of witnesses. Stones were produced to prove their assertions. Finally it began to seem that where there was so much smoke perhaps there was at least a little fire, and scientists began to divide into two camps, one group contending the stones had indeed fallen from the sky, the other still holding that terrestrial explanations could account for all cases.

Then, on April 26, 1803, at midday, like a determined champion of forensics making one final effort to dispose of all negative arguments, Dame Nature hurled pell-mell into northwestern France, only about seventy-five miles west of Paris, near the little town of L'Aigle, no less than 3,000 stones in a single shower. Hundreds of peasants witnessed the event, some narrowly escaping injury or death, and many were more than a little frightened.

The time had come for another investigation. A venerable commission undertook the task of studying the hundreds of reports, examining hundreds of the 2,000 to 3,000 stones actually recovered. Scores of affidavits were sworn. At last, the hard-headed scientists were converted, and a decision in favor of the affirmative was rendered. Thenceforth it was permissible for a stone to fall from the sky into the soil of France without fear of witnesses facing embarrassment or self-doubts as to its celestial origin.

Thus, the matter was settled finally among the leading French scientists, but the question was far from being settled in the minds of ordinary people. The conclusion of the French commission was so far from being common knowledge that it was not known to the Yale investigators of the Weston, Connecticut, fall four years later. It took a succession of several witnessed falls, over a score of years, for general scientific acceptance of the non-terrestrial origin of meteorites, and throughout the nineteenth century, and even into the twentieth, there have been scientific men who have doubted that meteorites actually fall, that it might not be just superstition.

$$***$$

One can understand why the fall of meteorites was a difficult fact to comprehend and was so slow to be accepted by the world of science. The arrival of a meteorite might be witnessed by one person in a million. The living people of perhaps one village in ten thousand might experience such an event, with no more than one person in ten of that hamlet actually viewing it. Generations would pass without the occurrence being repeated. Meteorite falls are so infrequent that probably no community has ever experienced two falls in one generation or perhaps in ten generations. Not only are the events rare; they are unpredictable and of a very brief duration. The majority of people in the immediate vicinity of a fall will not be actual witnesses, but will only know of the happening by hearsay, or by tremors, sounds, and momentary light flashes which merely produce puzzlement or even hysteria.

The celestial origin of meteorites was accepted by early people as a manifestation of the gods ages before it was accommodated by the practitioners of the natural sciences. Throughout his existence on this planet, man has encountered meteorites. Time and again in ancient records are references to stones from heaven or from the gods. A meteorite that fell in Phrygia about 3,000 years ago gave rise to a religious cult that exercised widespread influence throughout the then-known world. The principal shrine of the cult was in Pessinus in Galatia where a "stone from heaven" was kept in a cave under Mt. Dindymus. This was Cybele, "Mother of the Gods and of Men." The cult found favor among all Mediterranean nations. Cybele's supposed power was regarded so highly that when Hannibal had threatened Rome for fourteen years, successfully resisting the Roman armies' attempts to drive him out, the emperor in desperation consulted the Cybelline oracle and was told his only hope was to bring the great Phrygian stone to Rome. In the year 204 B.C., a special ship was built and an expedition set out to fetch the sacred object. The Phrygian stone was brought to Rome with great pomp and splendor. The revitalization of the army was immediate, and in a matter of months, Hannibal and his troops were driven away. A temple for Cybele was erected and the worship of her cult continued in the western world for 500 years, posing one of the chief deterrents to the spread of Christianity.

One of the great yearly meteor displays, the Perseids, bears a second legendary name out of ancient history, "The Tears of St.

Lawrence," in awe of the heavenly spectacle of August 10, 258 A.D., that followed the saint's death on a gridiron under the rule of Valerian.

The revered black stone in the Temple of Mecca, the Kaaba, is reputed to be from heaven, and whether a meteorite or not, its influence in Mohammedan worship is due to this belief. Plutarch told of the fall of a meteorite in the year 405 B.C., and Pliny wrote 500 years later of seeing this great stone, still held in sacred regard. The great temple at Ephesus was built for the worship of the "image that fell down from Jupiter," told of in Acts 19 of the Bible.

Factual records were intermingled with all sorts of legends, superstitions, and fairy tales. Then came the stirrings of the new sciences. The telescope was invented, the microscope. Everything began to be viewed in the light of demonstrable fact or mathematical theory. Legends and accepted notions regarding the material universe underwent critical examination and questioning. Many old assumptions could be readily evaluated, hypotheses verified or exploded by checking and rechecking of observed facts.

While science began its advances, religious views changed. The mythical cults dissolved into the mists of time, and Christianity entered into battles of the True Word and Heresy. Reports of stones from the sky were classified as heresy, which in the days of stocks and witch burnings was a punishable crime. Of course there were instances where those of a scientific turn of mind came face-to-face with undeniable evidence, and, being convinced of the soundness of the case or at least forced into skepticism regarding the dictum of the "authorities," endeavored to bring the idea up for re-examination. This required courage and sometimes resulted in shadows being cast on respectable men. No one can ever know how many authentic meteorites were disposed of during those centuries, never to be recovered.

To break through centuries of hardened prejudice required wisdom, recognized scholastic standing, and tremendous nerve. De la Lande

194   IN SEARCH OF FALLING STARS

failed, Father Bachelay also. Troili correctly deduced that the Albareto stone came from space, but his conclusion fell on deaf ears; other scientists rejected his worthy description and his reputation was imperiled. E. F. Chladni of Germany has generally been credited with being the first to properly evaluate the arrival of meteorites from space. As early as 1794, he published an able argument. Chladni gathered together a scattered lot of outlaw facts, sorted out pertinent reports and observations, made critical examinations of several "mysterious" and "miraculous" stones, and hypothesized that these certain objects were not of terrestrial origin. Although his thesis was widely questioned and bitterly disputed by those who offered some earthly explanation such as vitrification by lightning, it was only a matter of a few years before the great L'Aigle shower of stones furnished cosmic verification of his theory.

But any accounting of men who have made great contributions to the science of meteoritics must not skip over the work a generation earlier of Troili, Bachelay, and their contemporaries, who defied an established order of thought to present credible accounts of phenomena universally adjudged impossible.

***

Tales and worshipping of stones or irons fallen from heaven "in old time" were prevalent among the American Indians. Superstition, legend, miracles, and facts were all mingled together, as they were in the Old World, in a more or less meaningless hodgepodge to be dealt with only by priest, oracle, or medicine man. A number of meteorites have been found associated with Indian relics in various parts of the United States.

Among the many reports of possible meteorites that came to us over the years, there were some that, because of distances involved, we could not afford to look into immediately. We would answer the letters or make notes of oral reports, then file them away until several could be investigated on a single trip. One such report lay in our files for years. It was a message from George E. Dawson, writing from Phoenix, Arizona, that he had a 135-pound Canyon Diablo meteorite that he wished to sell. Canyon Diablos—the iron meteorites found at the Arizona crater—were the most plentiful iron meteorites on the market, and I needed no more at that time, there being little demand for them during the 1930s. I simply wrote to Dawson that I would stop to see his specimen when I was in the area.

It was a number of years before I hunted him up when passing through Phoenix. Yes, he still had the meteorite, still would like to sell it, but it was not in his possession at the moment. Query followed query about my connections and my reliability before I was finally told that I might examine the meteorite. Eventually the reason for Dawson's reluctance became evident. He had learned of the Arizona law which forbids digging into ruins—and this iron had been found in an old Indian grave. When I pointed out that the law had been adopted some seven years after he had made his discovery, and that he had committed no wrong under it, Dawson led me to the rear of a curio store in downtown Phoenix where, hidden under some unimportant old relics, was the large iron meteorite he had been calling "Canyon Diablo" to escape any possible trouble with the law.

He drove me to an ancient ruin on a mesa top a few miles east of Camp Verde where in 1915 he had come upon a stone cyst—a little pocket in the earth walled and covered over with flat rocks—in the corner of a decayed dwelling. The little cubicle appeared to be a typical child burial cyst, but instead of a mummy, the pride and joy of all pot-hunters, it had disappointed him by giving up, respectfully wrapped in feather cloth, a 135-pound metallic meteorite.

Thus did one of the first positive evidences of the American

Indian's regard for things from heaven come to light.

Pottery associated with the burial showed an age of 800 or 900 years. Removal and study of the structure of a small sample of the iron indicated Dawson was very probably right in attributing his feather-wrapped meteorite to Canyon Diablo. Except for this small section, the meteorite remains intact, priceless because of its history as an object cherished by an ancient tribe of the human race. We gave it the name *Camp Verde*, for the locality of the grave in which it was found.

Only seventy miles north of the spot where Dawson found his iron meteorite in its Indian grave, Mr. A. J. Townsend opened another such child burial cyst to find not a mummy, but a pile of green-stained rocks which he judged to be copper ore. A University of Arizona scientist, however, identified the fragments as remains of a disintegrated meteorite which evidently had been seen to come from the sky and therefore was extended the respect of an honorable burial.

The little Pojoaque pallasite was found in a pottery jar in an Indian burial ground in Santa Fe County, New Mexico, in 1931. It bore evidence of having been carried in a medicine pouch, its surface indicating it had been subject to much wear against soft materials.

In 1950, while attending a mineral exhibit at Bozeman, Montana, we secured a meteorite from C. F. Miller that he had dug from an Indian grave in 1936.

Persistent field work accumulated further evidence that ancient Americans collected meteorites. The Horse Creek iron and the Springfield stone were both found in 1937 on Indian campsites in southeastern Colorado. The Elkhart stone (1936) was found on a campsite in southwest Kansas. The Alamosa (1937), Lost Lake (1934), and Newsom (1939) stones all were found by hunters of Indian artifacts in the eastern edge of the San Luis Valley of Colorado; Cotesfield (1928) and Briscoe County (1940) were found on campsites in Nebraska and Texas respectively. Muroc and Muroc Dry

Lake of California (1936) were found in an area where Indian artifacts had been found. And the great Navajo iron of Apache County, Arizona, found in 1921, had been covered with boulders and bore several grooves that appeared to have been cut by stone implements.

Many other examples show ancient man's regard for meteorites. Certain tribes made regular pilgrimages to the meteorites of Red River, Texas; Willamette, Oregon; and Iron Creek, Canada. The Chilcoot meteorite was kept in the custody of an Alaskan chief. The Wichita iron of Texas likewise was held in high regard.

Meteorites associated with prehistoric Indian mounds of the Ohio and Mississippi valleys were found. The Iron Creek meteorite was revered by Cree and Blackfeet Indians as the "Manitou Stone" where it lay on a hilltop in the province of Alberta, Canada. The great Casas Grandes iron, 3,407 pounds, was found buried, wrapped like a mummy, in ruins of the Montezuma Indians in Chihuahua, Mexico. The three great irons brought by Admiral Peary to New York from Greenland were objects of reverence among the aborigines of the island.

If the ancients had not seen these particular meteorites fall, they had seen others and had concluded correctly that these were of similar origin, passing the story down through generations. This reverential treatment by Indian tribes of North America was in the same pattern of awe shown by other early peoples in many places scattered across the globe. In India, the meteorite falls at Durala in 1815, Nedagolla, in 1870, and Sabetmahet in 1885 were all worshiped to some extent, and the fall of Saonlod in 1867 was so feared that the natives pounded all the stones to powder.

The great reluctance to believe a fact of science was reserved for scientists, even for the first half of the nineteenth century in Puritan America.

8

Peaks and Valleys

> *Don't keep forever on the public road, going only where others have gone and following one after the other like a flock of sheep. Leave the beaten track occasionally and drive into the woods. Every time you do so you will be certain to find something that you have never seen before.*
>
> —Alexander Graham Bell

It is doubtful that any of our friends or relatives looked upon our 1930 move to Denver with sentiments other than pity. Our total assets were a 1929 Chevrolet and about $2,000 in cash, which would dwindle considerably by the time we were settled. We had a verbal agreement with Ward's Natural Science Establishment in Rochester, New York, to buy our spare meteorites for up to $4,000 a year.

We loaded our household goods into the back of the farm truck of a brother-in-law of Addie's and headed for Colorado. In late October

1930, we moved into a row house near downtown Denver. Our children were 11, 9, and 5 years old.

When I had turned in my resignation in the spring, only the last session of the Rocky Mountain Summer School remained before severing all ties with McPherson College, where I had studied as an undergraduate and had taught for ten years. We were prepared for the cut-off of all visible income other than that from the sales of meteorite specimens and lecture fees, but then an unexpected stroke of good fortune came. I was offered an opportunity to join the staff of the Colorado Museum of Natural History (later to become the Denver Museum of Natural History and now known as the Denver Museum of Nature and Science) as curator of meteorites. We would, after all, have a regular income—an honorarium of $50 a month under an arrangement requiring only that my collection be placed on exhibit and that I make my headquarters in the museum.

We had set up a meteorite-cutting laboratory in Palmer Lake, about forty miles south of Denver, in the little general store run by Ray Niswanger, a close friend during the nine years of operation of the summer school and a leader in the community. He had suggested that we set up a shop in an empty room of his store, having heard me say that I needed a saw to meet the orders of collectors. Winters at Palmer Lake's altitude of 7,280 feet were long and severe, and he assured me the work would help fill empty hours.

Ray volunteered to run the saw for a very nominal charge per hour, proposing that payment for his work be postponed "until you can pay me out of the income from the business beyond your living expenses."

Ours was the first commercial cutting laboratory for meteorites in the United States. The only other such equipment in this country was that at the United States National Museum. Often we obtained new specimens as payment in trade for the cutting. Slices of meteorites cut in our laboratory went to every great museum in the world.

To cut these exceedingly tough hunks of metal, we used an abrasive saw, a band of soft steel without teeth, run on a regular bandsaw frame such as is used in a carpenter shop, but operated at a much slower speed.

The bands of soft steel were made to order. They were two or three inches wide and were used until they were worn down to a width of about an inch. As the band revolved, a trickle of water carrying carborundum, the cutting agent, was directed against the biting edge where it entered the meteorite, mounted firmly on the saw table.

It was intriguing to watch that smooth, harmless looking band eat away at a stubborn iron meteorite. The piece would be imbedded in plaster of Paris to steady it on the saw table, and there it would rest for hour after hour or day after day, while the saw slowly gnawed through it.

The saw stood taller than a man, the band measuring eighteen and a half feet and running on two thirty-six-inch wheels. It had one stationary table on which was mounted a track which carried a movable table operated by a screw with a sprocket wheel. On this moving platform was another smaller table that moved at right angles to the larger movable one and on which the meteorites were mounted in plaster.

The cutting required an average of about an hour per square inch, with the mechanic standing by to adjust speed and pressure to accommodate any unusually hard obstructions. The largest slice we cut and exhibited, a beautiful nickel-iron that measured seventeen-and-a-half by thirteen inches, required about nine weeks of sawing.

By the close of the final session of the Rocky Mountain Summer School in 1930, the lab for cutting and polishing had been set up and paid for, and we began cutting the Mexican meteorites at once. The saw ran steadily in Ray's little store in Palmer Lake, providing specimens that sold readily. Soon I had two new meteorites from

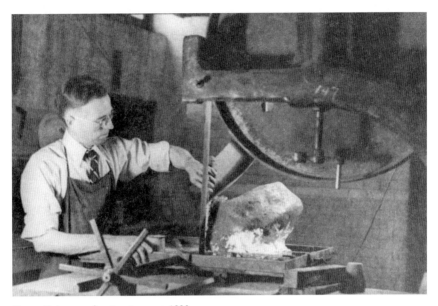

Harvey Nininger and meteorite saw, c. 1930.

Nebraska and one from Colorado. Ward's Natural Science Establishment was selling my material on consignment about as fast as I could supply it. Things in Denver seemed to have started off in our favor.

Alex Richards and I had been deep in the interior of Mexico in October 1929, when news flashed of the greatest stock market crash in history. When Alex mentioned the headlines, I remember thinking, "That will not affect me." To me, it seemed that since I had never dabbled in stocks, I would not be injured, a measure of how little attention I had paid to the financial structure of our country. I was so immersed in meteorites, that I paid little heed to talk of depression. The disastrous effects of the financial crash did not reach the western states to any noticeable extent until early 1931. By then, I had given up my $2,200 salary and the annual summer school income of about $1,000 and moved to Denver.

In the fall of 1931, we moved into a middle-aged two-story house close to the museum, undertaking a mortgage on the strength of our

202 IN SEARCH OF FALLING STARS

agreement with Ward's and stubborn optimism that the depression wouldn't reach us.

Ten years before, when I was thirty-four, doctors had told me my heart was disabled and had ordered an existence of relative inactivity—no stair climbing, no bicycling, no volleyball or tennis, no mountain climbing, no high altitudes, no this and no that. Now I had survived ten of those "careful" years, which had featured three long trips and some rather strenuous activities, and I scuttled some of my caution by adopting a bicycle for my habitual transportation to the museum, about a mile from home.

But indeed the depression was going to affect us. The purchase agreement with Ward's, our only real "cushion" of income, was terminated after about a year. The dealer could no longer accept our specimens, for no one could buy them. Our situation was precarious.

When our agreement with Ward's was canceled, I made direct contact with those institutions that had collections of meteorites and offered such specimens as I could at special prices. Results were encouraging but by no means sufficient. These direct dealings with institutions and private collectors fell off more slowly, but nevertheless, the decline was sorely felt. As the sales of specimens slowed, I began to offer my lectures more and more to colleges, universities, museums, and scientific organizations which still had lecture funds.

When our financial plight had about reached bottom, I received an invitation to lecture before the Cosmos Club in Washington, D.C. The results of this lecture were most heartening. At the close of the meeting, a committee was called together by the Director of the United States National Museum to consider what could be done to promote my work, and the result was an agreement on the part of the Smithsonian to purchase $2,000 worth of meteorites or more each year if I could furnish them at prices comparable to those charged for meteorites by regular dealers. This seemed to place a floor under our feet, and I returned home with renewed assurance.

We had not enjoyed this feeling of security long before a change in personnel in the National Museum wrecked the plan, for after I had planned a field program sufficient to assure that I could fill the orders the new curator wrote informing me that the museum simply could not buy meteorites as had been promised because all of our allotment had already been spent for minerals.

About this time, I was offered a university position that involved teaching almost entirely. I reasoned that acceptance would put an end to my investigations. Turning down the offer, with economic conditions as they were, was much like refusing a life line while aboard a drifting raft, but my life objective had become so fixed that I made the rejection without a qualm and returned to the daily struggle that was required by my chosen work. I felt that such a teaching commitment would kill my field program, because classes would take up so much of my time that very little if any field work could be accomplished. I chose to continue the risks of making my program self-supporting.

Even in McPherson, the problem of financing an expensive hobby had made our mode of living somewhat more austere than that of the average professor's family. After I left the college, it was evident that our growing collection of meteorites would absorb our entire effort and would constitute our entire estate. When the depression years reduced our market possibilities, we gave up on owning a house and moved into a rental. We borrowed to the limit against life insurance policies. There were periods when the purchase of a pair of children's shoes was an occasion for a family conference, and when the replacement of Dad's suit was solved via patches on the old one. The menu was dictated by economy rather than taste.

If there was ever a single month during that time when payment of our bills did not demand the best of us, Addie and I have never been able to recall it. What we managed to do was to tie our working arrangement with the museum, our cutting and polishing operations, the lecture schedules, and the investigative trips together with just enough in sales of specimens and fees from lectures to finance another trip, to permit scheduling more speeches, to recover some new specimen to cut and sell.

As the nation sank deeper and deeper into the trough of the depression, there was less and less money for either lectures or specimens, but even during the worst years, there remained a few good customers. Teachers were still being paid and some who had heard my lectures became determined to acquire a meteorite specimen or even a small collection. Some institutions could purchase meteorites for their departments of geology, astronomy, or earth sciences. Museum curators and even a few businessmen were among our customers.

We functioned first as *The Nininger Laboratory*, and in 1937 took the name *The American Meteorite Laboratory*.

My curatorship at the museum gave me a base of operations. There was an understanding that mutual arrangements could be made for the museum to participate in field projects from time to time on a fifty-fifty basis, but there was no obligation on either party to accept such a cooperative proposal. Our contract consisted of an exchange of letters and verbal understandings. The arrangement lasted fifteen years and ended amicably. During those years, my field program brought to light nearly 200 meteorite falls, aggregating about 1,500 meteorites, not counting several thousand from each of several craters.

There were long periods when all visible evidence spelled failure. There were many times when the larder approached emptiness, and the bottom of the barrel was scraped so often that we could almost see through it. It was a very difficult thing to make a living for a family and at the same time hold on to my objective of seeing a science of meteoritics developed, let alone my companion dream of sponsorship by an institute of meteoritical research. Our practice was never to part with the bulk of a new meteorite. We would remove one or more slices and let them go to the few institutions and collectors who furnished a ready market, then store the remainder.

The collection seemed to grow with amazing rapidity. The cataloging alone was a large chore, and correspondence was heavy. Addie, near the beginning, decided to handle the office detail involved with our meteorite search. Certainly there was no money for secretarial help. The whole family cooperated. The children did not demand luxury; no good would have been served to ask. Our one luxury was a good car, a serviceable one to carry the load of our field work.

Asked how we managed to finance our program, my usual answer was that I hardly knew myself. We always seemed so busy doing it that we never could find time to figure out how it was being done. We never stopped to worry about the overall problem of finance. We simply solved the immediate problems month by month, kept our credit good at the bank, paid off notes, signed new notes, worked harder, and forged ahead.

Always in the background of our thinking was the realization that we were steadily amassing a stock of meteorites which, if worse came to worst, we could depend on to see us through. I knew that the value of meteorites had through a hundred years proven far more stable than any currency or coin of the realm. Their status was something like that of diamonds—they were virtually indestructible, with proper care, and of everlasting value, despite the fact that for many years interest in them had faded. I held to the belief that man's next step in exploration would be into space. Certainly material from space must carry a premium.

***

At about the time the Smithsonian cut off the short-lived

arrangement to buy specimens from us in some quantity, thus just about sinking our little boat, rescue came our way in the guise of a man and a White Truck.

The house into which we first moved in Denver was only a few blocks away from Dean Gillespie's White Truck distributing agency for the Southwest. Gillespie later served a term in Congress, and his business enterprises eventually reached into other fields, but when I knew him first, he was guiding his truck agency through the hazards of the depression.

Dean was the one man who most consistently offered me a way through financial thickets during the years from 1931 to 1946. I met him soon after we moved to Denver, having been told by a mutual friend that Dean was an amateur meteorite collector. After hunting up his office, I walked in, gestured toward a big meteorite lying on his desk, said I guessed this must be the place I was looking for and gave my name.

Gillespie leaped to his feet. "We are two people of the same mind," he told me, and went on to describe his interest and his specimens.

On better acquaintance, having learned I habitually operated on a shoestring budget, he made a proposal that proved to be of mutual benefit and that served to keep me going financially in a number of instances over a period of some ten years.

Addie and I never knew when a fireball would appear that we would wish to chase down, or when we might hear of a meteorite somewhere that called for a trip of investigation and possible purchase, and of course, we were nearly always broke.

When such occasions arose, and I was short of cash, Dean would advance me money enough to make the trip, our agreement being that if the effort were successful, we would divide the earnings, but if nothing came of it, I had simply lost my time, and Dean had lost his money. Sometimes, if a project offered only a slender promise of returns, I would guarantee its cost by supplying meteorites already in my possession.

Dean helped me additionally in an unusual way. One day, he called me into his office.

"Can you drive a truck?"

"I never have," I answered, "but I suppose I could."

Dean explained to me that he thought he could help me out. As a distributor, he had trucks driven into Denver from the factory in Cleveland, Ohio, and he also delivered trucks to various localities in the West.

"My drivers get four dollars a day and expenses," Dean told me. "I'll pay you the same. If you want to stop on the way and get some work done for yourself, you can do that on your own time and expense."

So I became a part-time truck driver. I reached some of the lecture appointments I kept during those years via truck transport. When I reached my destination, I spruced up, changed my clothes, and headed for the auditorium. After my lecture and a night's sleep, I would return to my truck and the road, my suitcase in the cab behind me. So far as I am aware, neither my audiences nor the arrangement committees ever suspected their speaker had come from truck cab to lectern. I was not proud, but I respected institutional sensitivities. This system allowed me to visit museums and study collections en route, to make meteorite exchanges with various collectors, and on several occasions to recover new meteorites.

The bulk of our collection, except for specimens on exhibit in the museum, was carefully labeled and stored in a fire-proof walk-in vault in the basement of Dean Gillespie's business building. Dean never used this vault. The fact that I held the only key created a small friction. Our meteorite saw had been moved to my working quarters in the museum in Denver after two or three years in Palmer Lake.

When museum space became cramped, Dean made a place for the saw in the basement, where it remained in use until interruption by World War II. Dean enjoyed showing off my meteorites, or my cutting work on the great circular saw, or my fund of knowledge, to his friends and business visitors. He liked to be able to take his customers into the vault. Dean was a businessman through and through. He found that by getting me to talk to his customers, stirring their interest in something else, they could often be softened up for a successful truck sales pitch. The only areas of disagreement Dean and I had were the matter of the storeroom and the use of me as window trim. He was strong-minded when it came to running his business, but so was I where mine was concerned.

Our joint ventures for the most part worked very well. Frequently I would go to his office, sign a note for two, three, or five hundred dollars, climb into my car and take off on some project or other. Sometimes I'd come back with a good find, sometimes return empty-handed.

Our accomplishments over the years that were so financially critical could not have been possible if Dean Gillespie had not been there to call upon for a few hundred dollars when a loan was needed. He always stood firmly on business principles. He demanded and received his full share when a venture proved fruitful, but when a venture failed to yield, he was a good loser.

Dean recognized the importance of meteoritical research. At the time of his death in 1949, the Gillespie collection was one of the few really large private collections in existence.

***

After delivering a truck to El Paso, Texas, in 1931, I gambled some time and money on a flight to Chihuahua City, Mexico, seeking the "long-lost" Huizopa meteorite. This fall had been described in 1907 as "four irons" on exhibition in Chihuahua City, but since that time, the largest mass, weighing 258 pounds, had been lost from sight and record. I had sectioned and studied one of the smaller masses where it was preserved in the *Instituto Geológica* during my earlier trip to Mexico City, and I recalled the state of confusion and abundant mislabelings that typified that institution's collection at that time. Weighing the circumstances of the disappearance, I reasoned it was worthwhile to make a search.

In Chihuahua City, I had little trouble finding the *Museo Historia Natural del Estado Chihuahua*. It was within sight of the Hotel Palacio, and the sign read just as I had seen it worded in the British Museum's *Catalogue of Meteorites*. The door stood open, and a few steps back from it sat a woman doing needlework who evidently was in charge of the place. Immediately inside the door stood a tall, narrow pyramid of shelves loaded with choice, colorful mineral specimens, its base piled about with huge chunks of different ores from the various mines of this rich mining state. As I stepped on the threshold, I saw the blunt end of a large iron meteorite projecting from the pile of ore specimens.

After moving the ore, I was certain that this was a specimen of the same fall I had examined in Mexico City. The custodian permitted measuring, weighing, and photographing of the meteorite and said that the Museum belonged to *El Governador*.

From the hotel lobby, I called the United States consul and he agreed to assist me. The governor was out of town, not to be back until *mañana*. The next day, the reply was the same, and the next, and so on for a week. Meanwhile, however, I was proceeding on my own, conferring with the consul and going about the city in search of a proper container for the meteorite. Wooden boxes suitable for such a purpose were almost unknown in Mexico.

Finally, the governor, Señor I. Andres Ortiz, returned. A crowd of citizens awaited him in the anteroom of his office. The consul interpreted my plan. Recognizing the mutual advantage of having the meteorite studied properly and labeled correctly, the governor promptly authorized me to take the meteorite to Denver, cut it in half, return one half properly polished, etched, and labeled to the Chihuahua Museum, and retain the other half for this service.

Huizopa meteorite prior to its shipment to Denver, Chihuahua City, Mexico, 1931.

From my half of the Huizopa meteorite, I cut several slices for museums whose budget had not felt the depression too deeply, at prices that definitely bolstered our frail budget. During the course of the year following its acquisition, the Huizopa meteorite brought in a couple of thousand dollars, a considerable return for my personal investment and far beyond the wages of my truck assignment.

2

On the afternoon of May 10, 1931, W. H. Foster was hoeing in his garden at Eaton, Colorado, when he was distracted by a humming sound, not unlike that of a stray bullet, which seemed to come from the northern sky. A half-minute or so later, the sound seemed louder, as if approaching him. Foster leaned on his hoe handle and listened. This seemed a long time for a bullet to whine. He scanned the sky but saw nothing.

As the noise grew louder, he feared being hit and took a step backward. He felt the air blast in his face as an object whizzed past and struck with a thud about seven feet south and a little west of where he stood. Looking down, he saw where the sunbaked crust of the soil had been broken up, and projecting from the moist dirt thus exposed was a small, bright bit of metal.

Foster pondered for a moment, then stooped to pick up the object, "burning" his fingers, he said later, as he did so.[5] Puzzled, he walked across the street to show it to John C. Casey, the high school superintendent. Casey was as puzzled as Foster, and he called in the science teacher, Glen Mills. The janitor came also. But none had an answer. This was no bullet, but neither did it resemble the meteorites on display in the museum in Denver. Mills was quite sure he never had heard of a copper meteorite, and the nugget looked more like molten copper than anything else.

Could it be a burned-out bearing from an airplane? None had

---

[5] The usual report by persons who have picked up meteorites soon after they have fallen is that they are "cold." There have been a few reports that metallic meteorites have been "too hot to hold," and one lad described a newly fallen iron as "too cold to touch." Meteorites have been known to fall into haystacks without igniting them. The good heat conductivity of copper could explain the warm condition of the specimen reported by Foster.

212 IN SEARCH OF FALLING STARS

heard a plane that morning, nor had Foster seen any as he scanned the sky seeking the source of the strange, whizzing sound. There was not often a plane in the air over the village of Eaton in the year 1931.

The matter remained an unsolved mystery. It was given no publicity. But Mills, the science teacher, was not satisfied. Three weeks after the event, he attended a science meeting at which I was also present and described the happening to me, requesting me to investigate Foster's coppery nugget.

I called on Mr. Foster, accompanied on the visit by J. D. Figgins, Director of the Colorado Museum of Natural History, Frank Howland, mineralogist, and H. C. Markman, geologist, the latter two both museum staff.

Foster was cordial and cooperative. He showed us the specimen and told us the same story he had recited to Casey the day of the fall. The possibility that the thing was a meteorite had been suggested to him; on his own, he had no idea what it might be. We all agreed the object differed notably from any known meteorite, and though none of us could offer any other explanation of the occurrence, it could not be considered a meteorite on the basis of the accepted criteria by which meteorites were recognized.

But I could not be satisfied to leave the matter there. An intensive study of the surface features of meteorites had convinced me that their markings constitute one of the very best identifying characteristics of meteorites. When I had examined the coppery nugget under my ten-power lens, the surface was shown to be pitted in a manner that was unknown to me in any metallic object other than a meteorite.

Since the first recognition of meteorites by science in 1803, a new variety of meteorite had been seen to fall on an average of once every twenty-two months. Several of these varieties still were known by only a single fall. Could not this also be a "new" type of meteorite?

Mr. Foster loaned the specimen to me for further study and later permitted me to purchase it for the nominal sum of $5. Never did Mr. Foster seek any publicity, either at the time of the fall or afterward. For ten years, I withheld any published statement. Instead, I studied the specimen from time to time and had a partial chemical analysis made of it. I could find no flaw in Foster's story and could not think of any other plausible explanation of the nugget's arrival, but I held up final decision that it was meteoritic in the belief that sooner or later there must come some corroborative evidence.

On February 13, 1939, a nephew of Addie's, James Harold Rothrock, who lived in our home for some time and who was an able assistant to me in shop and field, was breaking in a new employee, Mr. Hummiston, in the grinding and polishing laboratory, where several slices of the Garnett stony meteorite were being ground and polished. As the two men worked on the revolving grinding table, Harold was amazed to see a grain of metallic copper that had been exposed by Hummiston's grinding. Realizing that this was very unusual, Harold laid the slice aside. Hummiston said the grain of copper had been considerably larger when he first noticed it.

This beautiful little grain of copper, embedded alongside a small sulfide inclusion, provided some of the corroborative evidence I had been awaiting. Location of the bit of copper three centimeters below the original surface of the stone ruled out any chance of entrance from surrounding soil. The meteorite had been found in a location far removed from any copper deposit. The United States Museum verified that the grain was copper.

In the Estherville, Iowa, shower of May 10, 1879, there was precedent to hypothesize that the presence of metallic particles could mean that metallic masses of larger size might once have been imbedded in stony matrices. In the Estherville case, a large meteorite of probably hundreds of tons was composed primarily of brittle stony matter with blobs of embedded metal. The stratospheric break-up

214   IN SEARCH OF FALLING STARS

which befalls most, if not all, stony meteorites reduced the brittle matrix to dust, and a tell-tale dust cloud of tremendous size remained aloft for hours, while the more resistant metallic inclusions, set free by the shattering, traveled to earth as individual pellets. In its form and surface markings, the Eaton specimen fitted such a history as that of the pellets of the Estherville fall.

When I published my findings in May of 1943 in *Popular Astronomy*,[6] there was a cry of "Hoax!" to which my answers were that Foster was of unimpeachable character, he lacked facilities for producing such an object and, above all, he sought no publicity nor profit from the episode; also, that if another party were deemed to be the perpetrator of the hoax, there was no evidence as to how such an act might have been accomplished.

Through careful questioning and re-enactment of the scene, it was evident that Mr. Foster had heard the whining noise of the "bullet" for a minute or two minutes before its landing. From what is known of the terminal velocities of such small objects, we may assume that it traveled not faster than 300 to 500 feet per second. Allowing an average speed of 400 feet per second for a period of 90 seconds, it must have traveled 6.8 miles after he first heard it and must have been traveling at least 40 seconds before the sound first reached him. If we allow a speed of a mere 280 feet per second for those first forty seconds, then the object must have started on its whining course ten miles from the Foster garden. Certainly no individual nor known terrestrial force could have put this object into a trajectory at a height of ten miles or so and kept it whining audibly for ninety seconds.

The production of an article resembling the Eaton meteorite

---

[6]"The Eaton, Colorado, Meteorite: Introducing a New Type," *Popular Astronomy*, Vol. LI, No. 5, May, 1943, pp. 273-280.

would be quite simple chemically. But the faking of flight markings—the peculiar pittings and flowings of metal produced by the passage of a meteorite through the atmosphere—would be a different matter.

Other nearly nickel-free meteorites have been seen to fall, and had they not been witnessed they almost certainly would not have been recognized as meteorites since they did not fit the accepted pattern. In 1915, Farrington listed four meteorites in which nickel, the identifying metal of meteorites, had not been found. Since then, both the Norton, Kansas, and Peña Blanca Spring, Texas, falls had produced meteorites that were so relatively nickel-free that ten- or twenty-gram random samples might show none at all.

I think that meteoritic origin for the Eaton nugget cannot be ruled out on the basis of its content of copper, lead, and zinc, roughly the formula of artificial brass, for these three metals are also sometimes found associated naturally in the rocks of the earth. It is true that in earth rocks, these metals are chemically tied up with oxygen, sulfur, and other minerals, but the same thing may be said of terrestrial iron and nickel, while the unadulterated metals are found in meteorites.

As of noon, February 18, 1948, meteorites of the Cumberland Falls, Kentucky, and Peña Blanca Spring type were so rare that only about 200 pounds had been recovered in nearly one and a half centuries. Then, five hours later, more than a ton of such material had planted itself in the fields of Norton County, Kansas, and Furnas County, Nebraska. In one afternoon, this fall had quantitatively revolutionized what science knew to be the "overall composition" of recovered stony meteorites. The Eaton specimen remains unique, but a single fall could place it among the most common of meteoritic types.

Then again, the line between authenticity and honest, but mistaken, reporting is sometimes thin. The Eaton case is typical of situations where the judgment and acuteness of observation of the

216   IN SEARCH OF FALLING STARS

witness is crucial. Did the witness see what he believed he saw? Did he pick up the object that fell, or an object occupying the space where he believed he saw something fall? Were the thing he heard and the thing he picked up one and the same?

At some time or another, perhaps, a complete chemical analysis or the new spectroscopic, isotopic tests that measure exposure to cosmic bombardment or a new, verified, witnessed fall of a copper meteorite may provide a final determination of the origin of the Eaton specimen.[7]

***

In the summer of 1931, Addie and I traveled with the family through the broad wheat fields of Saskatchewan on our way to visit my sister and her family in a northern settlement of that Canadian province.

There was little vegetation other than wheat, and there were quite a few rock piles tossed up along the fences. This seemed a likely area for meteorites. I stopped at the offices of the *Saskatoon Star* to request that a news story be run that might alert interest and result in a find.

The editor of the *Star* suggested that I write the piece myself. I addressed farmers, suggesting they might very well have hauled meteorites along with the country rock dumped beside their fences. I explained how to recognize meteorites, stressed the importance of recovering these stones from the sky, and, as always, directed that they chip off but a small corner of any likely specimen, thus avoiding damage to the whole.

---

[7] In the more than 90 years since the Eaton specimen was recovered, other scientists have tried to determine whether or not it is a meteorite. Their work has provided new and more detailed chemical analyses, but it is still not clear whether the Eaton specimen is meteoritic or not. *Ed.*

PEAKS AND VALLEYS

A few weeks after we had returned to Denver, a sample weighing about an ounce, and plainly meteoritic, came in the mail from a farmer near Springwater, Saskatchewan. The finder wrote that he had found the parent mass, weighing about forty-four pounds, many years before, and that he had sent a sample at that time to the provincial assayer, who had reported that the piece appeared to be artificial iron that probably had leaked out of a furnace. The assayer added that the thing had no value.

I requested the farmer, a Mr. Ward, to send me the main mass, which he did. When he received my check in return, he wrote that he had found more of the meteorite, and he kept on searching until he had found about a dozen individuals weighing about 200 pounds in total.

Some months later, another sample arrived from Saskatchewan, a tiny bit of meteorite about the size of a kernel of corn, sent in by Mr. A. D. Ebner, a farmer near Bruno.

Ebner reported that he had been hauling rocks from his field and when he started to put one chunk into his stone-boat, he found it was so heavy it required both hands to lift it. He gave it a second look, remembered the article in the newspaper, and decided to send in a sample.

The Bruno specimen proved to be a very well-preserved, beautifully oriented meteorite, one of the finest examples of metallic meteorites known, so unusual that its picture has been carried in many publications. Only the one piece was ever found. Mr. Ebner had treated it with great care. Had he acted with the roughness and disregard with which many meteorites have been treated, hammering and chiseling or sawing, much of the value of the beautiful Bruno meteorite would have been lost.

***

With the hope of obtaining more of the Springwater fall and with the thought that there might be a recognizable crater associated with it, Addie and I went again to Saskatchewan in 1950. In our 1931 correspondence with Mr. Ward, he had told us a really tantalizing story of the "big one that got away." Before he had learned the true value of the dark, heavy rocks, he had thrown away the largest one he had ever found.

"We were filling up a well about seventy or eighty feet deep," he related. "I was hauling rocks to fill it up, and this particular rock was so heavy we used a team to pull it onto the stone-boat, and then when I got to the well it was all I could do to dump it. I tore my hands on it rolling it into the well."

When we called at the old Ward farm in 1950, we found it being operated by a Mr. Staples, who took us out to search for the old well, but it had been filled in very efficiently and the field had been cultivated over it for thirty years; we couldn't tell exactly where it was. We went over the area with a detecting instrument in case the meteorite was in the top part of the fill, but we had no luck. Then we decided to go over the rest of the field with the instrument, and here we met a problem. I hadn't gone fifty steps until my earphone gave a beautiful bark, and we began to dig frantically. We dug down a couple of feet and brought out a beautiful granite boulder.

"Granite shouldn't give out a sound like that," I protested, but I passed the instrument over it and got the same bark. Examining the granite closely we saw that it seemed to have little particles of magnetite; I learned later that this is a very common constituent of certain granites. In that glacial region, they seemed to have every kind of granite—white granite, pink granite, blue granite, gray granite, black granite, brown granite. As we went about the field, we dug up about fifty granite boulders. One variety—neither the blackest nor the heaviest—would sound persistently in the earphones just like a meteorite.

As a substitute for exploring with the detector, we decided to scout the pastures still free of cultivation in hope of finding a mass still projecting from the ground, or perhaps a depression that might represent a small crater, partly filled in. The land was too rough and rocky for an automobile. The farmer lent us a team and a two-wheeled, rubber-tired cart used as a handy get-around over the farm. It had two automobile wheels supporting a bed about six feet long and two-and-a-half feet wide, with a seat set on it for the driver. Addie drove the two huge draft horses, and I stood hanging onto a support like some old-time charioteer, gazing out over the countryside while we swerved around or bumped over hummocks, sloughs, boulders, and holes.

Draft horses and wagon used by Harvey and Addie Nininger in their 1950 search of pasturelands where the Springwater, Saskatchewan, Canada, pallasite was found.

Except for obtaining a twenty-pound specimen that Mr. Staples already had, we didn't gain anything from that second trip beyond a fair certainty that the wheat fields and pasture land concealed no meteorite crater.

220   IN SEARCH OF FALLING STARS

The Springwater specimens belong to the rather rare class of pallasite meteorites and are among the most beautiful we ever collected. Almost thirty years after I acquired the first Springwater specimen, two young scientists of the Fermi Institute in Chicago discovered in it a new mineral, one that was previously unknown either in the earth or in meteorites.

***

On our way home from Canada in 1931, as we passed through Michigan, we stopped to visit Stuart H. Perry, builder of a notable private collection of meteorites. I asked him about the Beardsley, Kansas, fall in particular, which had occurred October 15, 1929, while Alex Richards and I were in Mexico City.

Mrs. Ray Gaines of Beardsley had been awakened in the night, probably by the light of the fireball. She had heard a noise, leaned her head out of the open window to listen, and had heard two thuds in the yard outside. The next morning, she told her husband, and they hunted around and found two small stones weighing about a pound each. She wrote to me, but by the time I had returned home and read the correspondence, the meteorites had been sold to Perry. Perry told me only the two stones had been recovered. He had tried to locate more through correspondence, but without success. I asked if he had objections to my looking further into the fall, and he said he had none.

We stopped at Beardsley, a little village in the extreme northwest corner of Kansas. The main street boasted a general store, a bank, and a service station, where a group of men visited in the shade of the garage against the late summer heat.

Carrying a small meteorite in my hand, I went over to the group and introduced myself. I reminded the men of the fall two years before and talked casually and very briefly, then started to the store across the road, since they showed little interest and seemed even irritated at my interruption of their chat. Half way to the store, however, one of the men caught up with me and told me in a low voice and with some embarrassment that he had "one of those" at home. We drove the two-and-a-half miles to his house, and he hunted around until he located a walnut-sized piece of what was evidently part of a larger meteorite that had been destroyed by hammering. He told me he had found several stones out in the field that he never had brought in, some larger than his fist.

"Didn't you know that Gaines had sold stones like this?" I asked him.

"Yes."

"Why didn't you send yours in, too?"

"I don't know."

Several of his children were playing in his yard. Their ragged clothing, the house furnishings, and the general run-down condition of the place spelled poverty. Why didn't this man, and others like him whom I met in similar circumstances at other times and places, gather in and sell those stones of whose value they were aware? Suspicion? Superstition? I ran into this taciturnity, this reluctance, often. Perhaps it was a general hesitancy to "meddle with things from heaven."

The man in Beardsley agreed to look for more meteorites. I bought the little piece he had and arranged with the local bank to advance payment for others that might be brought in. During the next sixty days, six stones, weighing from several ounces to more than twenty pounds, came in from Beardsley. Some had been in the possession of the finders since October 1929. Others were found during the autumn plowing which was in progress during the time of the several visits I made to the town during August and September of 1931. During the next two years, a total of sixty specimens were recovered from this fall.

The man who had plowed up the largest meteorite said that he had found another of about the same size or a little larger. He had carried it on his plow to the edge of the field and had thrown it over the fence onto the roadside. After learning of its value, he went to retrieve it, but highway maintenance workers had already cut away the bank to build a big grade and had evidently buried the stone.

3

The market for meteorite specimens, always extremely limited, had waned to such an extent by 1930 that the only two dealers, Foote and Ward, were trying to liquidate their stock by marking prices far below normal. Their supply consisted mostly of leftovers from old finds, examples of which long since had been purchased by the few institutions that still bought meteorites. Thus I was able to take slices or fragments of my recent finds and trade them advantageously for much greater amounts of the stock the dealers had on hand. Before making such trades, I would make such cash sales as I could to institutions and collectors for ready money to live on. I retained the bulk of every find, cataloging these main masses as the Nininger Collection.

When Addie and I began visiting Denver and Colorado Springs with our summer school students in 1922, window displays of minerals were common. Offices and homes of geologists and assayers contained handsome display cases full of specimens. By the time we moved to Denver in 1930, however, well-cared-for collections had become very scarce. Nevertheless, I made it a point to look over all the rock and mineral collections I could locate in assay offices, windows of barber shops, or small stores—anywhere there was a display of stones. Down on Larimer Street in lower downtown, in a dingy window, there was such a collection of minerals. The tenant was a bachelor of middle age who in response to my inquiry pulled out a drawer, reached in and brought forth a four-ounce stony meteorite, about the size of a golf ball but irregular in shape.

"We found this more than thirty years ago. We were stacking hay on a ranch near Doyleville, in western Colorado, and I was helping though I was only a youngster. One day, when we got out to the field, there was a black rock maybe the size of a brick lying next to the hay stack. We couldn't account for it."

His father had taken it to Gunnison to a collector who identified it as a meteorite and to whom it was given. The small piece shown to me had been detached from the larger stone.

I bought the little stone and later visited Doyleville and Gunnison. When the collector died, his minerals had been given to the state college at Gunnison. My inquiry about the "Jennings Collection" at the state college in Gunnison, Colorado, resulted in my being shown two or three cases of moderate size containing quite a number of small and poorly displayed specimens. The original mineral collection was said to have lined three walls of a room some sixteen by thirty-two feet, shelved to the height of a man's reach. It was reported that all of the large specimens had been hauled to the dump for lack of space.

I knew of several colleges which formerly proudly displayed fine mineral collections, the work of dedicated professors, that no longer spared space for such exhibits. On one occasion a janitor led me up dusty stairways, through narrow, dark halls and over a partially floored attic to the resting place of what once had been reputed to be one of the finest mineral collections in Pennsylvania. With the aid of a flashlight, I scanned a huge pile of more than a ton of dumped mineral specimens. Labels were scattered about and many fine specimens had been separated from their wrappings. Cabinet drawers lay empty around the distressing heap of specimens. The collection had been cleared out to make room for "more practical things."

224 IN SEARCH OF FALLING STARS

Like treatment was given to many other notable mineral collections during the early decades of the century. Meteorite collections were downgraded the same way. Harvard's mineralogical museum retired a large part of its exhibit in the late 1920s. The American Museum of Natural History retired all but a few specimens of outstanding size. The national collection in Mexico City, with its matchless quintet of super-ton specimens, lapsed into complete disuse and suffered severe losses. The California Academy of Sciences possessed a considerable collection, but no one attending it had been able to correctly label some of the specimens after the disruption of the 1906 San Francisco earthquake and fire. During the 1930s, I recognized one of the collection's principal specimens as marked incorrectly and found the almost complete mass of the Oroville iron lacking a label.

Harvard still possessed a magnificent collection but was not engaged actively in the study of meteorites. Yale had a large collection, but Dr. Ford told me that it had received no serious attention since about 1890. Amherst College possessed one of the half-dozen finest collections of the time, but no attention had been devoted to it since the passing of Professor C. U. Shepard in 1888. Dr. Winchell of the University of Minnesota at one time had been active in collecting but had allowed other matters to absorb his attention. Dr. Hobbs of the University of Michigan had a keen interest in meteorites but most of his time was also consumed by other concerns.

Western Reserve University, Columbia University, Wesleyan University, the University of Minnesota, the University of Iowa, the University of Nebraska, the State Museum of North Carolina, Tufts College, Drake University, the University of Wisconsin, Rutgers University, and a few other institutions of higher learning had collections worthy of mention, but none were receiving attention.

For many years, the meteorite collection at the Field Museum of

Natural History in Chicago was the largest in the world. Oliver Farrington was always glad to assist me in obtaining the kinds of samples I needed for my lectures, and to this end, we made exchanges of portions of my finds on a basis that seemed mutually advantageous. It still ranks among the best in the world, but after Farrington's death in 1933, the growth of the collection slowed. The Field Museum disposed of a large collection of meteorite casts, the chief remaining source of information as to the surface features of the original specimens, which had been cut into slices. I acquired from the Field Museum a ton of large irons from the Arizona crater in exchange for a part of a small new find which I valued at about $500.

The market for Canyon Diablo meteorites at Winslow and Flagstaff, Arizona, had completely died out in the 1930s. On one occasion I purchased several hundred pounds of small Canyon Diablo specimens for Dean Gillespie at fifty cents a pound. Large masses were cheaper.

During this period, a 400-pound specimen lay in the open between the curb and sidewalk in front of a dwelling in Winslow, Arizona. Under a sidewalk grating downtown lay three great masses, subject to regular deposits of dust and trash. Officials of the bank under whose property the meteorites lay were not very clear about their ownership. They were rumored to weigh a ton, a half-ton, and 500 pounds. Eventually, I learned this was an exaggeration. But the fact of their neglect is indicative of the casualness with which the subject of meteorites was approached. After persistent inquiry, I learned that these meteorites belonged to the Barringer estate, and I was given permission to exhibit them in the natural history museum in Denver. We weighed them and found their true weights to be 1,406 pounds, 570 pounds, and 160 pounds, respectively.

Farrington had reported that the Academy of Science at St. Louis possessed specimens of more than forty meteorite falls, but a few

226 IN SEARCH OF FALLING STARS

years later, the collection was stored, and when I went to see it in the late 1920s, I was told that it had largely disappeared. I was informed that another institution's collection of some ninety-odd falls had met with much the same fate.

There had been a good deal of interest in meteorites from the first acceptance of their space origin until the turn of the 20th century. In the early years, there had been voluminous writings. The prolific writers of the nineteenth century had been C. U. Shepard (1804–1886) and J. L. Smith (1818–1883).

The next great names in the field of meteorites were Merrill and Farrington. In addition to these two, Dr. Charles Palache was a third scientist who was helpful to me in the first years of my interest. He was in charge of the important collection of meteorites in the mineralogical museum of Harvard University during the 1920s and early 1930s. Meteorites were never Palache's chief interest or responsibility, but he contributed importantly to the literature, and I turned to him on various occasions for information and for support by way of purchase of my specimens.

In 1926, Dr. Palache cataloged the Harvard collection at 351 falls. The list grew apace during the next decade, and portions of many of my early finds were added to it. Later, much of this collection was retired for lack of space. Activity in meteorite study at Harvard slowed down and stagnated for some years.

Dr. Charles P. Olivier included a very enlightening chapter on meteorites in his masterful treatise, *Meteors*, in 1924, but his chief interest was in the phenomena of meteors, their orbits and their relation to comets. He privately confided to me, to my great disappointment, that when a meteorite landed on the earth, he lost interest in it.

During the middle and late 1920s, a number of young men began to demonstrate interest in meteors and meteorites: Dr. C. C. Wylie of the University of Iowa; Dr. F. C. Leonard of the University of

California at Los Angeles; O. E. Monnig of Texas; and Stuart H. Perry of Michigan.

But it seemed that quantitatively, meteorites were considered to have no meaning. The meteorite market was based upon the belief that after a meteorite had been described, its value lay in its being listed in the catalogs of those institutions which took pride in the number of falls represented in their collections, or in its exhibit value if it possessed features of sufficient interest to warrant display.

The fading of scientific interest in meteorites, the deterioration of collections, and the general lack of recognition of the quantitative significance of meteorites seemed to be related. The great collections of meteorites that existed all had been accumulated gradually through acquisition of accidental finds over long periods of time. Yale's collection had begun with the stones of the historic Weston, Connecticut, fall of 1807. Amherst's began soon after. The United States National Museum collection was benefited by gifts from various individuals. Ward's Natural Science Establishment was the most available market for any meteorite found, but the bulk of Ward's material was obtained through purchases of collections that themselves had been accumulated slowly. Harvard, American Museum, and the Field Museum had all built their collections through purchases of old, established accumulations.

The Nininger Collection was unique in that it was being established mainly by means of a planned search for specimens which would be added or which would provide means of exchange for others to be acquired.

In about 1930, when I was beginning to build the base for my collection, Ward's was offering the remnant of the Rochester University collection at ridiculously low prices, due to the general disinterest in meteorites. The fate of this Rochester collection provides a good indication of the failure of the study of meteorites to catch on. Rochester University—which happened to be located in the same city as Ward's Natural Science Establishment—had not really taken the study of meteorites seriously. By exchange of a portion of one of my early finds and by borrowing some money, I was able to acquire several very fine meteorites of the Rochester group at a small fraction of the prices at which specimens of the same falls had sold previously. I had taken pains to learn that only four and in some cases five institutions in the United States had samples of these falls in their collections. No one seemed to think there would be any need for more of these meteorites. The meteorites had been described, and the specimens were sufficient for exhibit. What more did anyone want? In this single instance, I acquired approximately forty-two pounds of the Gilgoin, Australia, stone, thirty-one pounds of the Estacado, Texas, stone, and thirty-three pounds of the McKinney, Texas, stone. At the same time, I obtained good-sized specimens of several irons.

Because I was able to make fast progress in building a substantial collection, because depressed interest in meteorites had led to depressed prices, and because the interest created by my lectures and writings had created a new crop of collectors, my program of search became reasonably self-supporting even before it had yielded any significant number of new discoveries. Thus, I became a supplier to collectors as well as a searcher for meteorites.

9

On Various Trails

> *Nothing in the world can take the place of persistence.*
> *Talent will not: nothing is more common*
> *than unsuccessful men with talent.*
> *Genius will not: unrewarded genius is almost a proverb.*
> *Education will not: the world is full of educated derelicts.*
> *Persistence and determination alone are omnipotent.*
> *The slogan "Press On" has solved and always will solve*
> *the problems of the human race.*
>
> —Calvin Coolidge

In Santa Fe, New Mexico, while on a truck delivery during the spring of 1932, I stopped for a brief visit to the anthropological museum there, intending to inquire about a small meteorite that recently had been found in a pottery vessel in an ancient ruin. I parked my truck in the lot and went in.

As I was passing among the exhibits, quite unexpectedly I met

230 IN SEARCH OF FALLING STARS

Dr. Alexander Wetmore, director of the United States National Museum, who happened to be in Santa Fe on a one-day visit. On a bench outside the building, Dr. Wetmore and I sat and visited, turning our talk to meteorites. When he asked how our work was going, I grinned and waved at my truck parked nearby. On the spot, he offered a tentative arrangement for some field work the following summer in the northwest part of the country on behalf of the National Museum.

We made vacation arrangements for the children, and Addie and I undertook several projects, including an investigation into the Port Orford, Oregon, meteorite story.

The Port Orford case is an enigma. This famous pallasite has never been rediscovered since it was first seen in 1859 by Dr. John Evans, a government geologist. Evans reported that he had detached a small sample from a parent mass estimated at 22,000 pounds, but he died before mapping the site. We, like other searchers, found no trace of it.

Some scientists became convinced the great mass never existed, that the sample collected represented only a small mass, and that Dr. Evans had become confused in his memory as to where he had obtained it. To me, it appears unlikely that a geologist could detach a piece of pallasite from any parent mass without holding in his mind a very reliable picture of the source of the fragment. Field experience has prepared me to understand just how plausible it is that such a great meteorite could exist just as described and yet remain unfound. Indeed, I should be greatly surprised if it is ever rediscovered, though I feel very sure it exists.

A miner from the area reported seeing the meteorite as a boy of fourteen in 1882 and finding it again in 1900 some thirty to forty miles southeast of Port Orford. In the late 1930s, he still talked of trying again to locate it.

The legal aspects involved in the ownership of large meteorites often tend to keep them from coming to light. It has been rumored that the Port Orford meteorite may have been found again, but the finder feared to announce his discovery lest the specimen be taken away from him.

***

Meteorites are where you find them. People often suggest searching for meteorites in areas where rocks are abundant, but usually I shun such places, because in an area where rocks lie everywhere, hunters will become confused and will tire of looking at every stone. If there is only an occasional rock, a searcher is much more apt to take a second glance to see if it might be a meteorite.

A daylight meteorite fell near Archie, Missouri, on August 10, 1932, while Addie and I were in the northwest. Some time after our return home, I wrote to the National Museum proposing that we cooperate in field work in the Archie area. I suggested that the museum advance cash to cover my expenses and in return receive specimens to cover its outlay, the findings of the survey to be written as a joint report. The museum offered instead to buy $200 worth of meteorites, if I could find them. This plan made it necessary for me to work on borrowed money, but it also left me free to do as I chose with my time.

During the couple of weeks that I spent at Archie, some fifty miles south of Kansas City, interviewing witnesses of the afternoon fireball, I drove regularly to and from the town of Harrisonville about ten miles north of the site of the fall and noticed that about midway, there lay a group of nicely cleared farms in the usually timbered country, with no rocks anywhere in sight.

One day, finding myself with some spare time, I decided to do some work in this rockless area, on my assumption that meteorites fall one place as well as another. I interviewed farmers, carrying with me some small meteorites for illustration, and suggested the likelihood that there could be some meteorites in the community, adding that a fair price would be offered for any that might be found. After spending two or three hours with a dozen different farmers, I drove back to Archie.

Two days later, the wife of the first farmer I had talked with came to see me, bringing a meteorite.

"Is this what you were talking about?"

"It sure is. Where did you get it?"

"Right after you left our house, my husband told me he knew he had plowed one up sometime, but couldn't remember where, so he wouldn't talk."

I recalled how the man had stood perfectly silent while I talked with his wife. I could not tell if he was resentful of my interruption of his chores, or if he was just thinking. He had been struggling with his memory. The next morning before he rose, he remembered he had plowed up the meteorite in the potato patch and had tossed it over against the hedge.

As soon as the neighbors heard of the payment I made for that first meteorite, they started looking around. In the course of a few years, forty stones were found, all proved by their distinctive structure to belong to the same fall. A fireball was known to have been witnessed throughout Missouri and eastern Kansas in about 1915. As nearly as could be reckoned from available accounts, the Harrisonville stones probably originated with that fireball.

This experience reinforced my belief that meteorites are present, though not abundant, everywhere, and that they can be located in some quantity if people can simply be trained to recognize them. A little explanation and the exhibit of a few samples to a number of persons will arouse some interest, but payment for the first meteorite located in a community very often will lead to the recovery of others—maybe just a few in a ten- or twelve-square-mile area, maybe one or more on every farm around.

Witnesses of the Archie, Missouri, fireball reported that it had disappeared suddenly, whereupon a cloud of dust had formed, about twice the diameter of the full moon. Calculations showed the dust cloud was about twelve miles above the ground, the average height at which a meteorite ceases to burn during its descent. A stony meteorite usually breaks up at about that height, the majority of the mass falling as dust particles, but with chunks of meteoritic material sometimes surviving to reach the earth.

One of the most useful accounts of the Archie fall had come from a high school senior who at first was reluctant to speak out. I presented a lecture in the high school, including information about the recent fall, and at the close of my talk interviewed all who had volunteered first-hand information. As I was leaving the building, two boys came to me.

"Bill has some very interesting facts to report, *if* you can get him to talk," they told me.

I sought out Bill and asked him to come with me to my car, where I inquired as to whether he had seen or heard of the phenomena associated with the fall. When he replied affirmatively, I asked why he had not spoken out. He explained that he knew some of the "wise guys" would ridicule him and make a public spectacle. Now he was willing to tell what had happened. He said he had been down along the creek, about three-quarters of a mile west of his home, hunting squirrels among some rather large trees, when suddenly, he had heard what he had at first thought to be the firing of a shotgun, except that it had seemed too loud. This sound soon was followed by another and then a rather disturbing roar. Then Bill heard a noise like buckshot striking the leaves and branches of the trees above him.

His thought was that someone was playing a prank, and he considered it a pretty hazardous joke.

"I was frightened and started for home, and about that time, something struck in the water of the creek a few feet away."

He said the splash was like that of a rock the size of a man's fist. A boy of his age and this place would be familiar with the splashes made by fish, frogs, and turtles as well as by stones; it was safe to assume that he had heard one of the stones that had fallen, several of which were later recovered near where he had heard these sounds.

When Bill reached home, he learned that his father had observed the fireball and the "smoke trail" of the meteorite. His father had also heard sounds like those the boy described, and he had felt a showering of small "gravel" on his hat and shirt. The youth's sister, eight years old, had come from the opposite side of the house where she had been playing, frightened and half in tears.

"Daddy, it's raining rocks out here!" she had cried.

The fact that the showering of small particles had been heard or felt by three members of this family, in two locations three-quarters of a mile apart, became more significant when it was matched with the testimony of a dozen neighbors who had been enjoying a watermelon feed under a tree at a farm a few miles away. All had heard what sounded like hail pattering on the shoulder-high corn in an adjacent field. These three occurrences of showering gravel followed shortly after the detonations from the meteor, and the locations were in a straight line extending a distance of six miles. Another report of a small stone striking the side of a house about a half mile to the south of this line also sounded entirely plausible.

This sprinkled area lay under the projected path of the fireball and was between the point above which the final burst had occurred and the area where seven stones, totaling eleven and a half pounds, were recovered.

A little calculation based upon even the meager facts that were submitted indicated that the collected stones from this fall were a mere bagatelle compared to the tonnage that must have rained down in the area. By our efforts, the seventh recovered stone was located, but we were unable to recover a single grain of fine material. Our information was obtained four months after the fall. The area was heavily vegetated and had been subjected to severe rains.

***

Several times I have "just missed" being a witness to one of the awesome sky spectacles that have sent me searching for earthly remains of "visitors from space." One instance was the Pasamonte, New Mexico, fireball of March 24, 1933, a meteor that was unusual in many respects, as were also the stones that months later were recovered along its flight path.

With my son Bob, then fourteen, I had gone to the town of Clovis, New Mexico, to obtain a meteorite that had turned up in response to some stories the local newspaper had run for me. I was awake, restless, at 5 a.m., and kept peering from the north window of our tourist camp cottage, looking at nothing. Clouds obscured the Clovis sky, but, perhaps by some vague abnormality, they conveyed a hint of the astounding phenomenon that was occurring behind their curtain. For while I stared recurrently from the window, Charlie Brown, a cowboy, at breakfast 200 miles to the north near Clayton, New Mexico, was making meteoritic and photographic history.

Charlie noted that the clock was striking five just as he sat down at the breakfast table. At that moment, the sky lit up with noontime brilliance. With astounding quickness of mind and hand, he snatched up a little pocket camera, unfolded it as he ran outside, pointed it toward a great, advancing ball of fire and snapped the shutter. When he glanced up again, the meteor was disappearing over the housetop. What he had seen, that I would have given much to see, was one of the greatest of modern meteorite falls. It was weeks later before I knew of Charlie Brown's picture.

That was the nature of most of my meteorite searches. Many were total duds. Many more brought only "delayed action" results. Others would take a sort of "one thing leads to another" course so that an initially small find, or slight lead, might snowball into something big. The last was the case with Pasamonte. By luck, Bob and I had almost been in the right place at the right time. Before we finished our investigation, we received unprecedented sketches and photographs of the accompanying luminous dust cloud, as well as the phenomenal Charlie Brown photo, and a number of specimens of a rare kind of stony meteorite. In the meantime, a number of specimens were destroyed through ignorance and out of our poor luck in not getting to the right places fast enough.

From Clovis, Bob and I drove to a farm near Melrose, twenty-five miles west, to obtain the meteorite we had come to purchase. The farmer told us he had plowed up the meteorite long before and thrown it into a ditch to save his plow. Years afterward, he needed a weight for his go-devil, an implement used to cultivate row crops, and remembered the heavy stone. He had been using the sixty-eight-pound meteorite on the go-devil ever since.

As we paid him, he said he wondered if "that thing might have come out of one of those falling stars, like the one I saw this morning."

He described the fireball he had seen. I thought perhaps the sale of his stone had exaggerated his interest in an ordinary meteor. When we got back to the little village of Melrose, we parked in front of the tiny café and opened the car door so the meteorite could be seen lying on the floor between the seats, and I started conversation with some men standing nearby. Did they know anything about meteorites?

Pretty soon one of the men spoke up: "I saw one this morning."

As others added their stories, it became evident that a spectacular fireball had crossed the northern sky. After I made some notes on their observations, Bob and I drove on west, then north, stopping wherever there was a settlement. We asked about size, appearance, and direction of travel of the fireball and drove accordingly, until eventually we had circled the end point—the point where the light had gone out—and the reports then began to describe the fireball as being in the opposite direction.

Descriptions varied. The meteor was as big as the moon. As big as a baseball. It was an advancing wedge of flame. It was a whirling ball. There were two fireballs, three, a half dozen.

In observations on direction and altitude, I have found one cannot depend on the accuracy of lay judgments beyond the limits of about thirty degrees, but I have sometimes interviewed professional geologists and physics teachers who had done no better. Even inexpert accounts can provide the basis to determine whether a meteor had passed on one side or the other of a certain point, and for calculating the limits of visibility and the vanishing point.

For two days, Bob and I searched out eyewitnesses to the Pasamonte fireball, driving ahead as indicated, and established the approximate landing area. Meanwhile, press reports placed the landing in five different states. A Denver newspaper caught up with me by telephone, queried me, and then ran a story and picture of the "meteorite," reportedly "found" in Brighton, Colorado, some 300 miles north of the area in New Mexico where the paper had chided me for still hunting it!

Over a period of four months, intermittently, I interviewed witnesses in many towns of New Mexico, Colorado, Kansas, Texas, and Oklahoma. The spectacular display was plainly visible even from points as remote from one another as Cheyenne, Wyoming,

238   IN SEARCH OF FALLING STARS

and El Paso, Texas.

Ultimately it became evident that the phenomenon was more spectacular than even the most dramatic accounts by witnesses—the fireball was cubic miles of incandescence, accompanied on its fifteen- to twenty-second flight by a discharge of material which had produced a visible column of dust a mile in diameter and lasting ninety minutes or longer. Aerial disturbances shook buildings and rattled windows as far as ninety-five miles to either side of the course. In the vicinity of the end of the visible flight, many residents insisted they had suffered throughout the entire day following the fall from a throat irritation like nothing they had experienced previously or since.

It was some months later before I was able to continue with the fireball survey. An unusual feature of the Pasamonte meteorite was that it ceased to be incandescent at an unusually high altitude of seventeen or eighteen miles, and all of the recovered fragments were deposited *behind* the end point, rather than at some distance beyond.

We had made several trips into the area over a period of months before I happened to stop at the Pasamonte ranch.

"Did you find any of that meteorite that fell near here a while back?" I asked the head man. I showed him a fresh-fallen meteorite found a few years before, retrieved before it had suffered any weathering.

"I saw it land somewhere in the vicinity," he told me. "Maybe it burned itself up." He studied the specimen I held. "I thought they were all iron."

"No," I said. "Most meteorites are stony with a little iron mixed in."

He looked at it closely. "That looks like what a herder brought in that morning. Pacheco was in camp nearly two miles from here and came running in frightened out of his wits, ready to quit his job. He said it rained rocks all around out there, and he brought one of them in. But I didn't believe him because I thought meteorites were always iron."

The boy had quit. No one knew where he was.

"What did you do with the rock he found?" I asked the ranchman.

"I don't know. We all looked at it, and some of the men pecked at it with their knives. It was kind of soft, with that black crust on the outside like the one you showed me."

He looked around and finally found the stone way back in a corner of an old cupboard, all covered with dust. It was just a little thing, but as soon as I saw it, I knew we had run down our meteorite, and we began to go around and talk to the ranch hands. Finally I located two or three fellows who told me they had seen such stones.

"We had quite a lot of them, but we broke them up."

"What did you do that for?" I asked.

"We thought they might have something in them."

They told me they had picked them up the morning the big fireball had streaked across the sky. I got them to hunt around their shack, and they found three or four stones that hadn't been cracked open.

Eventually, about a hundred fragments were recovered, scattered widely in the sparsely settled country. I kept raising the price I paid to induce people to hunt for the little stones until I was paying three dollars an ounce.

One of the finest specimens was recovered by O. P. Gard. Ten days after the fall, according to his wife's diary, Gard was driving a tractor when he noticed a little black stone lying on top of the buffalo grass. The stone didn't seem to belong there, so he veered the tractor to avoid running over it and stopped to pick it up. Gard was entirely unacquainted with meteorites, but he suspected the little black stone might have come from the big fireball he had seen ten days before. He wrapped it carefully in a piece of tissue paper, and when we encountered him much later, the specimen was still fresh and unspoiled. It weighed an ounce.

240   IN SEARCH OF FALLING STARS

It was nine months after the event before I finally held in my hand at the Pasamonte ranch a fragment of the aerolitical stone that had caused the great spectacle of light, the turbulence, the dust cloud, the fright of shaken witnesses, on March 24, 1933. The material was almost like volcanic ash beneath the outer fusion crust, so friable that in all probability no large mass had landed anywhere intact. All the recovered specimens together weighed a little less than eight pounds. Pasamonte is of the rare type of stony meteorite known as Howardite, and portions of the scant material recovered from this spectacular New Mexico fall have found their way to museums all over the world.

On my several trips into the Pasamonte area, I made it a point to instruct all of the ranch hands on recognizing both fresh and weathered meteorites in the field, for the very good reason that I was sure other falls had occurred during centuries past in the same area. On one occasion, as I drove into the ranch headquarters and turned to park in front of the little company store, I noticed a fist-size stone just inside the gate which appeared to be a fragment of an old stony meteorite. A second piece lay near by. After parking, I went back and examined the two pieces. They were both meteoritic. I took them into the ranch office and asked the manager if he knew where the fragments had come from. He told me that Fidel Lanfor had brought them in.

After talking with the various ranch hands, I drove to Lanfor's farm. There, lying in his yard, was a considerable pile of fragments like those I had picked up in the ranch yard. Fidel told me he had found a "big rock" in his field and thought it might have some diamonds or gold in it, so he took a sledge and broke it up.

"But you didn't find anything of value in it, did you?" I asked.

"No," he said.

"But you knocked about a hundred dollars out of it," I told him.

Charlie Brown, whose snapshot of the fireball in motion made scientific history, was manager of the Lyon cattle ranch, situated in a narrow valley about twenty-five miles southwest of Clayton, New Mexico.

For several weeks after the meteor, I was in the field almost constantly. On May 16, Director Figgins of the museum in Denver wrote to Charlie Brown, having received word that Brown had photographed the smoke cloud left by the March 24 meteor. Mr. Brown replied under the date of May 20 that his picture of the smoke cloud "would be of no special benefit to you as it is a very poor print," etc., but "I do have an excellent picture of the meteor in flight which I was very fortunate in getting, which shows very plainly actions of the meteor which you could not see with the eye... Some time ago, I wrote to Mr. Ninninger (sic) to call on me, that I thought I could possibly help him in determining which way this meteor went... I would be glad to give you any information I can on this meteor at any time."

The letter he mentioned never reached me. Neither Dr. Figgins nor I could understand how he possibly could have gotten a photo of the meteor in flight, but we agreed that the photograph should be examined.

When I arrived at the Lyon ranch near Mt. Dora, New Mexico, Brown was not in. I explained my errand to Mrs. Brown, adding that it was difficult to comprehend how such a feat as a photograph of a meteor in flight could have been accomplished.

If I knew Charlie, Mrs. Brown told me, I would understand. "He always does things that way, on the spur of the moment, never stops to figure or discuss; just goes ahead and does it."

Charlie Brown's photograph is dominated by the great glowing ball of the meteor, just at the point of emitting one of its periodic incandescent flares, streaming a luminous tail behind it.

Charlie told me that he habitually kept his camera on a radio cabinet near his chair at the breakfast table. He described matter-of-factly how the unearthly light had turned the wee hours of the morning into midday, and he had simply grabbed the Kodak and dashed outside—opening it as he ran—pointed it, and snapped the shutter. Then, he had wound the film and tried for another shot, but the second one was no good. He believed the light was too intense after the bursting of the big fireball.

Pasamonte, New Mexico, fireball of March 24, 1933. The first photo of a meteor in flight, captured by Charlie Brown, a New Mexican cowboy.

I had him re-enact the scene for me, timing him carefully. The entire sequence required eight-and-a-half seconds, an interval that was consistent with other reports that the fireball had been visible for between fifteen and twenty-two seconds. Consequently, for Charlie Brown, it had been no difficult feat.

As quick as Charlie was on the trigger in taking his historic snapshot, the film lay in the photo shop

Charlie Brown, who captured the historic photo of the Pasamonte, New Mexico, fireball of March 24, 1933.

for weeks. Two processors declined to print it, seeing nothing worth printing.

***

The meteorite which the farmer near Melrose had used on his go-devil, and which had taken us to the area at the time of the Pasamonte fireball, is interesting and unusual in its own right. It is a gold-bearing aerolite.

Some weeks before the March 24 fireball, I had been in western Texas on another search. Hunting had not been good, and I turned toward home, through New Mexico. As I approached Clovis, then a small village, I was impressed by the miles of almost level country, relatively free of vegetation, and with no rocks in sight. I drove on into the town, and since it was Saturday and the schools would not be open, I headed immediately for the local newspaper office. There I met Pete Anderson, editor of the Curry County News. The walls of his office displayed a number of Indian relics. After admiring his collection, I asked Anderson if he had ever run onto any meteorites during his field trips.

"I'm afraid I wouldn't know one if I saw it. I always thought I'd like to find one, though."

I pulled some samples from the little bag I carried, tossed them on his desk and began to explain the differences between meteorites and ordinary rocks. I suggested that he write a story for his paper, explaining how to recognize meteorites, telling the farmers of the area that there might be a meteorite or so among the rocks they probably had been throwing from their fields into ditches and fence corners for years, and relaying an offer from me to buy any that were found.

Pete wrote a very good story. His paper went out on Thursday, and by Saturday, Guy Groves had brought in his go-devil "weight."

Pete promptly forwarded a sample. I sent a check and arranged to pick up the stone on a future trip, the one that coincided with the Pasamonte fall.

The news that Mr. Groves had sold a rock that had been knocking around the place for years was occasion for another newspaper story, and more meteorites showed up.

When I asked my chemist, Fred G. Hawley, to analyze the Melrose stone, he discovered a trace of gold in the meteorite. Hawley was chief chemist for the Anaconda Copper Company's Inspiration mine and smelter at Miami, Arizona. I had met him during our housecar trip in 1925–26 and began at that time sending my meteorite samples to him for analysis.

Fred Hawley was extremely cautious. He wrote me in detail of the chemical test he made on a sample of the Melrose meteorite.

"If I am correct, this is surprising, for I understand that Au (gold) has rarely, if ever, been found in meteorites... I think I am right about the presence of Au; but before certifying it, I should like to make two more tests by fire and run them a little differently to separate the gold, if you will send me more material."

"If I find gold is really present, I think you should have some other chemist, experienced in this work, check my results so as to leave no doubt in the minds of other scientists regarding its correctness."

Hawley's further tests agreed with his original finding, and his determination was confirmed by assays by the American Smelting and Refining Company in Denver.

Fred Hawley need not have been so modest. When Dr. Harrison Brown carried out a program of re-checking old meteorite analyses during the 1940s, using the newest methods, he stated that those made by F. G. Hawley were among the very best.

When I published a scientific description of the Melrose stone, a problem arose. As a gold-bearing meteorite, it made headlines in

ON VARIOUS TRAILS 245

newspapers all over the world. Some newsmen made it sound as though my meteorite hunting had the objective of acquiring precious metals, and finders of meteorites began demanding high prices. Worse still, some fine specimens were broken up in efforts to recover their supposed content of gold. The $50 that I paid Mr. Groves for his meteorite was about forty times the value of all the gold it contained, but this fact seemed less newsworthy, and for years we had to contend with the misconception that meteorites were a source of precious metals.

As a result of Pete Anderson's local stories and the publicity accorded the gold-bearing Melrose stone, two falls of unknown date were discovered in Curry County some thirty miles away from Melrose, and two other very old falls were recovered about the same distance from Melrose in the opposite direction. A second stone from the Melrose fall was also recovered. Then there was no more activity in the area until 1961, when a farmer bumped into a 600-pound meteorite in the subsoil. He sent it to the Smithsonian.

2

Before any specimens had been recovered from the famous Pasamonte fall, westerners were treated to another great fireball. This was the August 1933 daylight meteor of Sioux County, in northwestern Nebraska, which created a disturbance all over that part of the state. Detonations were heard over a distance of 150 miles, and within the central fifty miles of this area, cattle stampeded, horses ran away, farmers were startled into falling from their machines in the fields, and wild pheasants were set into outcry.

Yet the stones recovered from this fall totaled only about thirty, an aggregate of six pounds, the largest of which weighed three pounds. How could such grand pyrotechnical displays occur without

246 IN SEARCH OF FALLING STARS

leaving some great memorial in the form of masses of stone or iron? The Sioux County fireball, like those of Archie, Missouri, and Pasamonte, New Mexico, added to my growing conviction that the dust clouds so often reported to accompany fireballs carry a great share of the disintegrated meteorite and that in many cases, no great mass survives.

I first heard of the Sioux County fireball when the phone rang at our home in Denver at about 10:20 on the morning of August 8, 1933. The caller was my nephew, Edgar Nininger, who was working on a farm that summer about four miles east of the city.

"I just saw a great meteor," he told me. He said he had been riding his horse from one field to another when a streak of fire had appeared in the clear, cloudless sky, descending almost vertically until it was cut off from view by an intervening haystack.

At once I drove out to check with Ed. He had marked the exact spot from which he had seen the streak of fire, and the haystack was a good landmark. I set up a forester's compass and noted direction and elevation according to his observation: about fifteen degrees east of due north, the almost vertical descent bearing just slightly westward. Ed judged the fireball had been visible for about two seconds.

I returned home. Addie had already packed our grips and a lunch for us. I obtained from the press services and newspapers the reports they had received, and we got into the car and headed north.

The nearly vertical descent, as Edgar had witnessed it, gave no indication of whether the fireball was moving toward or away from him. We never drove directly toward our objective, but to one side or the other, in order to obtain reports from observers who could give us directional information, modifying our route as we progressed. We knew this fireball had been witnessed farther north, so we drove fifty-five miles north to Greeley, where the local editor gave me the names of witnesses.

We contacted them. Each observer was requested to stand in the spot from which he had viewed the fireball and describe its appearance, whereupon I took notes and compass bearings on the vanishing point of the fireball as it was seen by each. Reports in the Greeley area were not noticeably different from Edgar's account, indicating the object must have been a considerable distance away. Also, Edgar's report that the meteor was still burning when it had disappeared behind the haystack was a good indication that it was still quite high and the landing place could be 150 to 250 miles away.

Locating the end point is an important first step in the hunt for a newly fallen meteorite. If the fireball appears to reach the horizon the meteorite probably has traveled many miles beyond the point of disappearance; if the light has been seen to go out high above the earth, it can be presumed that the surviving fragments have not traveled very much farther onward from that spot.

Paradoxically, to the observer the reverse seems to be true; the light that disappears "behind the haystack" or "just back of the barn" or "on the other side of the hill" is the one that convinces the untrained viewer that the meteorite "fell right over there."

We continued northward from Greeley. At Cheyenne, Wyoming, sixty miles farther, we interviewed observers again, and then went on to Torrington, another eighty-seven miles north. Here, the reports indicated a vanishing point considerably more to the east. Witnesses also said the fireball had disappeared while yet well up in the sky. This indicated to them that the meteorite had burned up before reaching earth; to us, it meant simply that it had quit burning.

We stayed overnight at Torrington, taking a number of bearings with the help of various witnesses, and the next morning, we drove eighty-three miles farther north, to Lusk. At Lusk, observers pointed southeast instead of northeast, indicating we had passed beyond the end point of the fireball. We turned east into Nebraska, stopping at each town until we reached Crawford, where observers pointed almost due south. A little farther, at Chadron, witnesses reported the fireball had disappeared due southwest. So we turned south and kept on in that direction until observers began to point north of west. Here, we knew, we had just about encircled the end point of the fireball. Now we were ready to plot the course of flight and to indicate the general area in which surviving fragments would have landed.

Our preliminary survey covered the whole west half of Nebraska, parts of eastern Colorado, southeastern Wyoming and northwestern Kansas. We determined that the course of the meteor was fifteen degrees west of north and that it had vanished at a point over the Niobrara River, half way between Agate and Marsland, Nebraska. Its descent was at an angle of about 30 degrees with the horizontal, and it had disappeared about eight-and-a-half miles above the ground. On the basis of these known facts, probable velocity, and other contributing factors, I had to estimate how far ahead of the point where the light had disappeared we should expect the landing place to be.

Three witnesses in Denver reported the object as "a mile or more northeast." A Colorado supreme court justice, viewing the fireball from Estes Park, was sure it had fallen in nearby Devil's Gulch. A newspaper editor in Berthoud, Colorado, relayed positive assertions that the object had fallen in a field five miles east and two miles north of that town—nowhere near the ultimate recovery site—and warned me not to bring any great number of persons to help in the search, for the owners of the field did not want "a lot of people trampling over their land." A motorist driving west from Gothenberg, Nebraska, was just as certain the meteorite had fallen in the hills near that town. And a salesman on the road fifty miles southeast of Denver—more than 200 miles from where the meteorite had actually struck—was so certain it had fallen in a field next to the highway that he made two trips back from Denver to search the pastureland adjoining the strip of road where he had been driving. Fishermen near Elk Mountain, Wyoming, spent several hours searching for the object they had "seen" fall in the hills "nearby."

What each of these witnesses possessed that was useful to the hunter of the meteorite was the observation hardest to extract from them: an accurate report of the direction of the fireball in relation to where they had been standing when it vanished. They all were so eager to help that they wanted to provide the exact location where the meteorite had landed, instead of the simple directional reports that could help to establish the fact.

By taking the varying observations and plotting them against each other, knowing the distance between the observations, it was possible to bring together the angle of descent and the horizontal path of passing and come up with a pretty good idea of the spot on earth toward which the meteor was headed. Then we marked off half[8] of the distance from the point of the fireball's disappearance to this spot to allow for further flight after burn-out, and drew an ellipse with this as the long axis.

I sat in the car on the Nebraska prairie, with map, ruler, protractor, pencil and paper, and plotted a space some fifteen miles long by about ten wide, over some part of which I figured the fragments of the meteorite were scattered. Then we went into the area enclosed by our ellipse and canvassed it house to house, talking to the residents, showing them a freshly fallen meteorite so they would know what to hunt. This survey and search were continued over the course of more than a year.

When fragments were eventually recovered, they were found

---

[8] This might be more or less than half, depending on air currents, angle of descent, height at burn-out, and other factors of the particular instance.

within the northwest quadrant of the target area I had drawn. The first fragment was picked up in a potato field, where two farm boys named Yohe had been hoeing and at the same time had been teasing the hired hand working with them. The light of the meteor had startled all three, and the detonation that followed had frightened them. That evening, the hired man complained to the parents that the boys had thrown a rock at him. The boys denied this charge, and it was suggested that perhaps it was the "thing in the sky" that had nearly struck him. The next day, working the same field, Homer Yohe picked up a stone with a shiny, glazed black surface that later was identified as a meteorite.

Again and again we returned to the ellipse we had plotted and endeavored to contact every inhabitant. One man I remember particularly well.

"No, it didn't fall here," he told me. "I saw it. It fell way south, maybe fifty miles. A team ran away with a binder down there. You're in the wrong spot."

I explained to him I had heard about the runaway team, but that it is a known fact that a big fireball will frighten horses, cattle, even human beings all along its course and far to either side of it. He was not to be convinced that the meteorite had fallen in his own locality, but finally after long protest, he agreed to watch for unusual stones as he herded his sheep. A week later, he sent me a little meteorite, one of the finest specimens I ever obtained, found right on his own farm.

The north curve of our ellipse lay in the Sand Hills, where auto travel was difficult and treacherous in 1933, and I reached that northern point only once. It was late in the day when I made my last stop and I asked if there were any other farms nearby to the north. There was just one more house, about a half mile farther into the Sand Hills, on a road that was scarcely passable for my car. My informant agreed to relay word of my search to his neighbor, but evidently he never did.

More than a year later, this unvisited neighbor picked up the largest fragment recovered from the Sioux County fall, and he took it to the stock corral on the railroad where the cattle were being loaded and passed it around for all the cowhands to see. With their knives, they chipped about half of the nice fusion crust away and one finally broke the stone in half with a hammer. It was in bad shape when it finally reached me, but obviously it had once been a fine specimen, probably equal in weight to all the other fragments recovered.

3

Problems in interpreting fireball reports had plagued us ever since the 1923 fireball. In that case, each of two farmers living 300 miles apart and in opposite directions from McPherson had both insisted the fiery object had landed in his own field, while my own observation had placed the location of the fall far from either of them. The largest distance ever recorded between fragments of the same meteorite is about twenty-eight miles, although there have been cases where news dispatches have reported a fireball has "fallen" at localities as far as 600 miles apart.

The public was nearly always willing and eager to help. There were many futile searches, but the wild goose chases were governed by the natural law of averages, not by malice. Natural mistakes were to be expected. All I really asked was that anyone volunteering information simply tell me what he knew, or thought he knew. Putting two and two together often led to humorous incidents. Measuring and utilizing the myriad untrained observations of witnesses was endlessly interesting.

Ignorance of the true behavior of meteors was widespread. People just would not be convinced that meteorites cease burning several miles above the soil. Therefore, when the fire-streak is "traced to the ground," the real fact is the object was so far away that at a height of several miles it was still passing over the observer's horizon. The only exception would be the large crater-forming meteorites, which would be destructive over an area fifty miles around the fall.

There is no mystery as to why the light of a meteor disappears at several miles above the ground. The only reason for the light is friction with the atmosphere due to the extremely high speed of the incoming meteorite. Friction causes a slowing down, and when the meteorite has been slowed down to about the speed of a high-powered rifle bullet, the friction is no longer sufficient to produce burning. The light fades out, and the meteorite continues on its way to the earth. If it is farther than a few miles from the observer, it is invisible. If a fireball vanishes at a height of ten miles, then it may be anywhere from fifteen to thirty miles from its target depending on its angle of approach.

The illusion of the distant fireball seeming to reach the ground nearby is almost universal. In the vicinity where the stones actually fall, the opposite illusion often occurs. Because the burning ceases and the fireball disappears high in the sky, observers are very apt to argue that "it was all burned up."

Discrepancies regularly accompany reports of a meteor viewed over a wide area. Suppose we take a hypothetical fireball, witnessed across an area three hundred miles wide. The site of fall will appear as the hub of a wheel, with spokes radiating out in every direction. At the end of each spoke are witnesses of a particular vicinity. From only two of these areas, those in a direct line with the object's flight, will the meteor appear as though it were moving in a straight path. From the end of the spoke following the line of approach, the meteor will be seen as a long streak ahead; from the end of the opposite spoke, far *beyond* the landing place, the meteor will not be visible; from a point closer in toward the hub, the passage will appear foreshortened, a streak of fire of much briefer passage. The wheel is in a sense lopsided, for the viewers in the area from whence the fireball came may be farther away, the spokes elongating the closer they are to the actual path of approach.

From each witness, at the end of each spoke, one gleans an additional bit of information to be used in plotting the area of fall. Then, moving inward toward the target location, the investigator will find that witnesses within perhaps thirty or fifty miles of the landing place will have heard detonations or sound disturbances. But within a still smaller area immediately around the landing place, very near to the end point, usually there are no observers at all, unless, perhaps in daylight, an individual chances to be looking at just the right spot at just the right time and sees the meteorite fall as a black-appearing object. Sounds in this immediate area are mere thuds, or patterings of falling objects.

The great rush of light and sound are experienced farther away from the end point. The horizon beyond which the meteor seems to pass out of sight may be 200 miles distant if the object is eight or ten miles high when sighted, even though it may appear that the meteorite has fallen behind a nearby obstruction such as a barn or haystack. If the light is seen to go out, or "burn up," the object has ceased burning, is less than eight miles high, and is much closer than the spectacular fireball which may be viewed by residents of a town a hundred miles away. Preliminary sifting and evaluation of mailed-in reports can indicate the area where follow-up interviews should be undertaken. Some reports are fallible, of course, because individuals are fallible. But if the best reports from all the different locations are plotted on a map, one will find a fair degree of agreement among them and can reach a very decent conclusion as to the nature of the object, the location, the height at which it was last seen, the speed at which it approached, and its direction.

On February 26, 1931, I was visiting central Kansas when I received a message from Denver that a great meteorite had fallen in daylight in western Kansas. Various press reports over the next two days estimated the location of fall in northwest Kansas near Colby, or in southwest Kansas, or in northern Colorado, or in central Colorado, or in Nebraska.

Since no accompanying noise was reported except in northwest Kansas and the adjoining state of Nebraska, I was convinced the fall had occurred in that region, though the fireball itself had been visible over a territory 400 miles wide. I traced it to this area and located the end point. To decide the angle of descent, I had either to interview persons fifty to seventy-five miles on either side of the meteor's passage, or by good fortune find an observer who was directly in line with the flight path and for whom therefore the ball of fire had appeared to stand still in the sky. Thirteen miles north and slightly west of Goodland, Kansas, southwest of Bird City, I found a family who had seen only a great ball of fire appear and disappear with practically no motion. This, then, was the target area, but no meteorites were ever found, or at least none were reported.

A man fifty miles north of Denver—possibly 235 miles from the site of the fall—insisted that this meteorite had landed within a half mile of him. No argument I could advance in correspondence sufficed to convince him that he was mistaken. Finally, recognizing it was possible, though improbable, there had been two meteors about the same time, I drove to this gentleman's home to interview him in person about what he had seen. He took me to an oil derrick where he had been working the morning he had seen the streak of fire, and he pointed out the exact spot in a neighboring field where he had "seen" this light come to the earth. When I recorded the direction on my instrument, I found it lined up perfectly with my conclusions as to the location of the fall in northwest Kansas, 235 miles away.

A second man a mile east reported that the fireball had fallen just east of him. Both these men had seen the same meteor viewed in northwest Kansas.

About seventy miles south of these two, some twenty miles south of Denver, a motorist driving southeast thought he had seen a burning airplane drop over a hill just to his left. He reported the disaster at the nearest town, Castle Rock, and townspeople spent the better part of two days searching the hills for a wreck. When I interviewed this man and had him indicate the direction from which he had seen the fireball moving toward earth, it was apparent that he, too, had been looking toward the northwestern Kansas area.

A farmer wrote to us insistently that we could collect the meteorite from a field southeast of Holly, Colorado; he had seen it strike between two cottonwood trees only a mile and a half from him. My personal judgment was that this man, too, had seen the same meteor in northwest Kansas that had fooled the men north of Denver, but on my next trip through Holly, I stopped to interview him. He led me out to his yard gate and showed me the little group of cottonwood trees one-and-a-half miles northeast, where he had seen the great fireball vanish, giving rise to a column of smoke. We sighted the instrument, and the reading indicated that again we were exactly in line with the northwest Kansas location. The "smoke" was the dust trail left by the meteorite. This dust cloud, which had been seen by all observers, was at an altitude of eight to twenty miles, and for him, its lower end was behind the horizon. He had given me an accurate report as to direction, but he was 150 miles off in his judgment. His was not an uncommon error. A professional astronomer once admitted to me that he had raced off in his car to where the dust trail of a meteor seemed to reach the ground near him. Another astronomer told me how, at the turn of the century, he had chased such a phenomenon on his bicycle. He had given up after twenty miles when witnesses at that distance told just the same story as did those at his home base.

We learned never to decide on the directional path of a fireball until we had interviewed witnesses on at least three sides of the end point or had been fortunate enough to find someone who had seen it pass directly overhead, and we also learned to establish certain units of approximate measurement to aid the descriptions by witnesses on occasions when instruments were not available. Thus, we might gauge size and distance against a hand held at arm's length, the width of the palm, the thickness of fingers, the size of a lead pencil at arm's length, or some other such homely example. Sometimes it was possible to establish very good lines by measuring the distance and height of buildings and trees standing between the viewer and the fireball.

Additional information could be gained if the meteor left a trail, or if a puff of cloud had formed. A good way to learn the elevation in degrees of the point where the dust cloud had appeared, if there were no instruments available, was to record in feet the height on some building past which the observer looked, and then measure the distance to the building from the observer. Another way was to have the witness draw a line representing the trail of dust with reference to the horizon.

In a restaurant in northwestern Nebraska, I was engaged in a discussion about meteorites with a companion when someone overheard and volunteered that he could tell me "exactly" where I could find one. As he was leaving town early one morning, he recounted, a tremendous ball of fire had come streaming in. He said he had seen it burn for some time after it struck the top of a hill less than a half mile from him. He would walk with me to the spot.

Parts of his story rang a bell. I asked him if he could recall the date of the event. His wife, who thus far had sat silently, spoke up. "I can tell you the exact date. It was the day we were going to my sister's birthday party. March 24, last year."

"Was it about five o'clock in the morning?" I asked.

She agreed that was about the right time. The phenomenon her husband had described was the great fireball of March 24, 1933, whose landing place was northeastern New Mexico. This meteor had quit burning at a height of seventeen miles above the earth. Instead of "striking" the top of the hill as the Nebraska couple had believed, the light was simply passing behind it, still at an altitude of some twenty miles from the ground.

In the case of the Pasamonte fall, I had passed through the area on my errand to gather up the Melrose meteorite just a day before the fireball, causing some local citizens to accuse me of having purposely stationed myself in the vicinity in order to be present when the thing arrived. Why else would I have been so far away from home and in this particular location?

The Pasamonte fireball produced typical examples of the fright and confusion often associated with the passage of a great meteor. Sim Cally, a veteran rancher who had spent most of his sixty-five years on the New Mexico and Texas plains, had moved with his men to a new sheep camp the evening before. As usual, he was getting up at five o'clock to start the day when a blinding light and horrible roar startled him. Their camp was in a canyon, and when he saw that ball of fire come into view over the canyon wall, he told me he thought the whole earth was burning up.

"I yelled to the boys, 'It's the end of the world! It's all over now!'"

The fireball passed about seventy-five miles from him and at a height of about twenty miles.

Another rancher 200 miles farther from the meteor's course reported that it had killed several of his cattle. In reality, their deaths were due to lightning striking the wire fence against which they had huddled during a storm.

Scores of early risers who had witnessed the Pasamonte spectacle picked up odd bits of stone which they had not noticed before and erroneously reported finding fragments of the meteorite. We learned to approximate all our judgments to take into account our dependence on the memories of untrained witnesses, unprepared for the phenomenon they had viewed and unprepared to report it.

10

Not All Knowledge is in Books

Every great scientific truth goes through three stages. First, people say it conflicts with the Bible. Next, they say it had been discovered before. Lastly, they say they always believed it.

—Louis Agassiz

A century ago, cowboys riding the naked plains of what is now Kiowa County, Kansas, came upon occasional heavy black rocks scattered over the buffalo grass. No other stones were to be seen for miles around the treeless plain, almost as level as a floor, with a fine, dark, sandy loam under the buffalo grass turf.

These unusually heavy rocks, with their odd dark color, attracted the attention of the cow hands, who took to using them for weight-lifting and shot-put demonstrations. One cowboy, suspecting these heavy stones must have value corresponding to their weight, hid a few in a badger hole. He never found any sale for them, however, and revealed their whereabouts only on his death bed.

When Frank Kimberly brought his bride, Mary, to Brenham to homestead in the raw Kansas prairies at the end of the nineteenth century, the first thing she noticed was not the buffalo grass, nor the sod house, but a rusty-looking black rock.

"Frank, do you see that rock? Do you know what it is?"

Of course, he told her, it was a rock.

"Well, it isn't any ordinary rock. It's a meteorite."

And Mary carried it up to the door and kept it. She added to her collection more and more of the "iron stones" that were common in the neighborhood, but Frank, and the neighbors, too, only laughed at her growing rock pile.

After the area was opened for homesteading, the "iron rocks" were accepted as unusual, but useful, objects native to the locality. In an area that otherwise was bare of rocks, the black stones were used as weights for rain barrel covers, corner stones for chicken houses, ballast for dugout roofs, and to hold down fence lines.

It remained for Mary Kimberly to prove they had value beyond their daily uses.

When Mary had been a little girl in Iowa, her school class was escorted by the teacher to the railroad station to view a great meteorite that was being transported to an eastern museum. Mary, a perceptive child, never forgot the experience nor the appearance of the meteorite.

Frank laughed at Mary's belief that the stones had come from the sky. He went back to his work. When he plowed, he occasionally ran into a hunk of iron rock that crumpled the share point, knocked the plow handles into his ribs, and inspired language far more color-ful than he usually employed.

As Frank plowed the meteorites out of the ground, Mary dragged them back to her pile. One day, when she brought him his lunch in the field, she saw he had dug out a particularly big one, weighing

NOT ALL KNOWLEDGE IS IN BOOKS 261

probably a hundred and fifty pounds. They loaded it onto his wagon, but her young husband, who was becoming increasingly disgusted with Mary's "crazy notion," dumped it out later instead of taking it in from the field.

Mary was not one to let go of an idea. She wrote to everybody she could think of who might have knowledge of geology, asking that they examine her collection. Finally, after a five-year campaign of letters, she reached Dr. F. W. Cragen, a geologist at Washburn College in Topeka, who agreed to look at her iron stones. He was amazed and delighted at her hoard, and on the spot paid her several hundred dollars for the better half of the approximate ton of material she had gathered by that time.

That first big sale price was enough to buy a neighboring farm, and on their new property, the Kimberlys found more meteorites to sell. Other scientists followed Dr. Cragen to the "Kansas Meteorite Farm," and the market was brisk for a number of years. Frank took up the cause enthusiastically and would go about the neighborhood offering to buy up the iron stones. He sold for $100 a meteorite he had used to plug a hole in a pig fence. He was not so discerning as Mary, however, and sometimes paid his price for a hunk of slag or a chunk of sulfide from a coal pile. He went back to his field and hunted in vain for the heavy meteorite he had dumped in dudgeon from his wagon. Years later, when he was an old man and Mary an old lady, he told me how he had hunted again and again for that stone, but had never found it. He would laugh at himself, and boast of Mary's cleverness, and she would sit by, proud and glowing.

Frank had one story of his own acumen that he loved to tell. He said that one of the first of the large meteorites to be found in the area had been taken to Greensburg by a lawyer, a Mr. Davis, and placed on the sidewalk just outside his law office door as a curiosity. This was prior to recognition in the community of the true nature of

262 IN SEARCH OF FALLING STARS

the heavy stones.

Before Mrs. Kimberly had made her first sale, she and Frank had tried to persuade Davis to give Mary that "rock" to add to her collection. He even refused $10 for it. The specimen weighed more than 200 pounds, and after Mary had sold half her hoard to Dr. Cragen while she and Frank still were keeping the sale secret, Frank took $200 in five-dollar bills and went to town, determined to buy the big stone.

"Mr. Davis, I want to buy that rock. What'll you take for it?"

Davis struck his courtroom pose, cleared his throat, and replied in stentorian tones. "Now, Frank, I don't care particularly about selling that rock. I sort of like it out there, but if you are determined to buy the thing, it's going to cost you $50."

Frank reached for his roll of bills and peeled off ten of them, which he placed in the lawyer's hand.

Davis couldn't help seeing the bank roll. "Damn you! You were prepared to pay me a couple of hundred for it. What's gotten into you, anyway?"

Frank proudly loaded the meteorite into his wagon and drove home. Later he sold it for more than $200.

I developed a great respect and fondness for the Kimberlys, and when writing my first book, *Our Stone-Pelted Planet*, in 1931 and 1932, I paid several visits to their home in Haviland, taking notes for a chapter about them and the "Kansas Meteorite Farm," which naturally was a story mostly about Mary Kimberly. When I told her of this, she was delighted and proud, and she looked forward to seeing a copy of the book, which was being published by Houghton-Mifflin.

Publishing the book took longer than expected. Meanwhile, Mary Kimberly, then in her eighties, was failing noticeably. In early 1933, I received one or more cards in her shaky handwriting inquiring as to when she could see "that book." I assured her the publishers had promised its early arrival. In April of that year, I made a lecture

NOT ALL KNOWLEDGE IS IN BOOKS

263

trip to the east, and before I left, Addie and I agreed that if the book arrived before my return, she should send one immediately to Mrs. Kimberly with a note that I would autograph it later.

Addie rushed a copy in the mail to Mrs. Kimberly, but it was too late. She received in return a letter from Mary's daughter with the news that her mother had passed away six days earlier.

In our meteorite laboratory in Denver, I cut many fine specimens from this Kansas fall, given the name "Brenham," after the nearest post office. It belongs to the unusual pallasite class and is very beautiful, containing greenish crystals of olivine surrounded by bright nickel-iron. The olivine was nearly always fractured and discolored by iron oxide due to long exposure to weather after arriving on earth, but in one instance, my cut revealed a perfectly beautiful crystal, transparent but of a greenish-golden hue, which I determined to remove and have mounted in a ring for my wife. Addie had never worn a ring, since before we left Kansas we conformed mostly to the modes of behavior favored by the Church of the Brethren, including the ban on jewelry.

I thought: How romantic, to present a ring with a gem from *out of this world!* I took the olivine to a jeweler and cautioned him to be most careful with it and directed him to bring me something really beautiful. For a long time, he kept telling me that he hadn't gotten around to it yet, and when I finally went to urge him to get it ready for Addie's birthday, he told me that in attempting to mount it, he had broken it. I've never found another perfect crystal.

The Kimberlys had marketed a ton and a half of meteorites by about the turn of the century, and then could find no more buyers. For twenty years, nobody seemed to be interested in what they had found. In 1923, when I had first gone to their farm, they still had a number of small meteorites and two large ones that they had not been able to market. I managed to buy them with borrowed money and encouraged the couple to search for more.

These specimens from the "meteorite farm" made a substantial addition to my young collection and were a substantial help to my thin bankroll when I made resales of parts of them. In 1927, I purchased a mass weighing 465 pounds, turned up by a plow boy. Ultimately, I added a half-ton of Brenham to my collection before the supply seemed to be exhausted.

Soon after the publication of my book, I gave a lecture in Hutchinson, Kansas, taking along copies of the volume for sale. After my talk, a member of the audience, H. O. Stockwell, whom I had met briefly through my brother John some years prior, came forward and purchased a copy, asking me to autograph it.

Much later, he recounted to me that he had gone home that night, read the book, and became instantly excited about hunting meteorites, but because of lack of finances to pursue the matter, he'd laid the book and the interest aside. Some ten years later, in 1947, when he was in better financial shape, he'd read the book again. Then he'd set out to equip himself with a metal detector and had driven to Kiowa County, where he'd spent the day combing a field on the old Kimberly place.

Weary and with eyes burning, he decided to quit. As he walked toward his car still carrying his machine, he got a clear signal, started digging and unearthed a 750-pound meteorite. That enthralled him. Days and weeks on end he pushed his wheelbarrow-mounted detector until he had searched the neighborhood and recovered a total of one and a half tons of the Brenham fall. More than three and a half tons is known to have been collected, and more was probably picked up but never reported. Undoubtedly, some remains unfound.

One day, Stockwell read of a meteorite find made in Wisconsin almost a hundred years before in which only two small irons had been picked up only a few rods apart. Playing a hunch, he loaded his machine on top of his car and drove the 1,300 miles to the village of Trenton. There he arranged with the landowner to search and began sounding the soil. Almost immediately, he received a signal indicating the presence of metal and dug out a 450-pound iron. After a little more work, he recovered a second mass of about the same size. With these two great specimens, he headed back to Kansas.

Various kinds of detecting devices have been used in scouting for meteorites, particularly since the war with the easy availability of military mine detectors, to varying degrees of success in the hands of various people. Stockwell, who combined his layman's interest with a practical wizardry in the handling of electrical and electronic appliances, must certainly be among the most successful dowsers for meteorites.

$$***$$

The first meteorite crater recognized by white men in the United States was the great crater in Arizona near Winslow. People began to suspect the pit was of meteoritic origin in 1891, but it was not given serious consideration as such until 1903. Even then, only a few scientists, led by a mining engineer, Daniel Moreau Barringer, were willing to advocate seriously that meteorite craters existed. The scientific world in general rejected the idea, especially after the Barringer group had carried out an extensive and expensive program of exploration in the crater without discovering the great meteorite mass for which they were searching.

At the time of my burgeoning interest in meteorites in 1923, the general opinion among geologists and astronomers was that the Arizona crater had been produced by a gaseous volcanic explosion, a hypothesis advanced by Dr. G. K. Gilbert of the United States Geological Survey in studies of 1892 and 1896.

In addition to the Barringer group, some others clung to the impact explanation, and probably a larger number than those who had spoken up, because doing so would have meant labeling oneself as a radical.

Dr. Merrill had spent some time at the crater and leaned strongly to the meteoritic hypothesis, but his mathematical reasoning and advice convinced him that a meteorite of sufficient size to produce the crater would of necessity have exploded on impact. Merrill sought in vain for evidence of such an explosion in the form of "volatilization products" and stated in 1908 that their absence constituted "the greatest difficulty in accepting the meteoric hypothesis." Merrill had not moved from this position as late as 1928, nor, so far as I know, at the time of his death in 1929.

Farrington wrote in 1915: "Complete proof of the (meteorite) hypothesis would be obtained by finding within the crater above the undisturbed sandstone a meteoric mass or many of them which would together approximate the size mentioned (500 feet in diameter)."[9] So far as I know, Farrington had not changed his opinion ten years later.

Personally, I shared Barringer's belief that the crater was meteoritic and that the mass of the meteorite lay somewhere in the crater. I did not then believe that a large mass would have to explode on impact.

I reasoned that meteorites of all sizes up to those that were described by astronomers as asteroids must have bombarded the earth occasionally during its long history and that therefore other meteorite craters would be found. In fact, I saw no reason why meteorite craters should not be a prominent feature of the earth's topography except

---

[9] Gilbert in 1891 had hypothesized that *if* the crater had been produced by a meteorite, that body should have been of the order of 500 feet in diameter.

NOT ALL KNOWLEDGE IS IN BOOKS

for their concealment or destruction by the forces of erosion and weathering. As I traversed the Kansas plains in search of meteorites during the late 1920s and early 1930s, I sought constantly for evidence of such features as I suspected an ancient crater would present.

In early 1929, visiting Frank and Mary Kimberly at their Haviland farm, I inquired as usual whether they had found any more meteorites. They had not, but Frank, as he was used to doing, began to reminisce. He was telling about one of the occasions when his plow had struck a meteorite and how profitable the damage to his implement had turned out to be, when Mary interrupted.

"Which one was that?"

"Don't you remember that sixty-eight-pounder that I hit in that old waller?"

I broke in. "Do you mean to tell me that you found one of those meteorites in a buffalo wallow?"

At Frank's affirmative answer, I inquired further. Mary said she had picked up "about a bushel of those real small ones" around the wallow, but not in the hole. I asked Frank to show me the wallow. It was on the same quarter as the farmhouse, and we went right out to it. I carried along a shovel.

Frank showed me a shallow depression about forty feet across. The feature that attracted my attention was a rim around the edge, rather conspicuous against the otherwise flat landscape. Winter wheat covered both the depressed floor and the elevated rim, but the rimmed depression was plainly evident.

"Can you show me just about where you hit that meteorite?"

Frank pointed to a spot inside the eastern rim. With his permission, I dug a short trench along the inner side of the southeastern rim. When I reached a level below plow depth, I began encountering rusty, brown, potato-sized nodules. Breaking one revealed rounded olivine crystals exactly like those found in the pallasite meteorites that had

268 IN SEARCH OF FALLING STARS

been found in this and neighboring fields for over forty years.

Except for the olivine crystals they contained, these rounded, rust-colored lumps bore no resemblance to the meteorites that had proven such a boon to the Kimberlys some thirty years before. Frank recognized the significance of the olivine inclusions. He and Mary had found some oxidized specimens on their farm. They had never been able to find a market for them, however, so Frank attached no importance to my find in the wallow.

I was elated at what I had found, but I tried not to show too much interest, thinking it best not to give Frank any reason to get commercial ideas concerning the possible contents of this little depression. He would recognize that the "burnt" specimens, as he called them, had "no value," but I didn't know what he might think of the possibilities in the depths of the bowl if I gave him the idea of a meteorite crater.

I inquired whether he would be willing for me to excavate the wallow, and Frank indicated that if anything was to be done, he would do it himself. He said they had been trying to fill in the hole during the years of its cultivation, that it "used to be a lot deeper," and that in fact it had been quite a water hole for the cattle when the field was a pasture. It had held water for long periods after rains. If he had thought there were meteorites in the hole, he would have dug into it himself.

I explained to Frank that more important than any material that might be found in the depression would be the information concerning the manner in which it had been produced, but I did not press the matter of further excavation for fear that he might undertake it himself and ruin an unprecedented opportunity for securing data on the question of meteorite craters. I conceived that these oxidized forms, with their shell of oxides containing remnants of a true pallasite, might be scientifically more valuable than unoxidized specimens. Besides, I wanted to ascertain the form of the buried crater. If Frank

NOT ALL KNOWLEDGE IS IN BOOKS 269

were to do the excavating, I knew he would go after a "big one" and ignore everything else. I secured his promise that I should be present when and if he did any excavating.

Since I had been his only buyer of meteorites since about 1900, Frank Kimberly could see no great urgency about tearing up a bit of his wheat crop merely to satisfy his or my curiosity.

I returned home convinced that I had discovered a meteorite crater. F. G. Hawley verified the meteoritic character of a specimen I sent him for analysis. I wrote a paper for the *McPherson College Bulletin* in which I briefly described the finding of this strange type of oxidized meteorite, which I named *meteorode*, under and inside the rim of the depression in the field, thus placing it on record but withholding the use of the term meteorite crater, until I had opportunity to make a proper excavation.

The specimens I had found in the crater looked so much like the ordinary nodular masses known as "iron concretions" that I thought it would be interesting to send some to professional geologists for identification. Accordingly, I sent samples to four leading universities where the subject of geology was prominent in the curriculum, simply requesting that they be identified for me. The answers that came back were unanimous that my specimens were "ordinary concretions."

In 1932, I visited the Odessa, Texas, crater. The discovery of this crater had been announced by A. B. Bibbins in 1926, but its meteoritic origin was held in even greater doubt than that of the Arizona crater.

When I saw what others had described as a "peculiar hole" on the high plains within a mile-and-a-half of US Route 80, one of our great national highways, I was thrilled. The great saucer-like depression was almost 600 feet in diameter, with a depth of about eighteen feet and a rim rising some four feet above the level of the plain. I came away from this "questionable" Texas crater with a 300-gram meteorite, a number of smaller ones, and not the slightest doubt as to the crater's true origin.

In 1933, I presented a paper at the annual meeting of the American Association for the Advancement of Science, southwestern division, under the title, "Meteor Craters vs. Steam Blowouts." At the same meeting, a paper written by L. J. Spencer of London describing the Henbury meteorite craters of Australia was read.

By this time, it had been four years since my Kansas discovery. I felt I could wait no longer. I hurried home to the Colorado Museum of Natural History and made such a plea to Director Figgins that he agreed to put up the expense money if I could get permission from the landowners and would give my time.

Excavation of the Haviland crater—the first such excavation in the world—perhaps was the most significant event in the history of the Kansas Meteorite Farm.

My son, Bob, then fourteen, was with me. Frank and Mary Kimberly were dead, but the younger generations of Kimberlys, both adults and youngsters, showed an active interest in plans to dig out the old "wallow."

In 1933, powered machinery was not used as much on Kansas farms as it is today. We hired two teams of horses and two old-time road scrapers.

We made several careful cuts, excavating and examining the soil between each operation. The excavated crater was in the form of an elongate bowl. The first cut yielded many meteorites ranging in size from that of grape seeds to as much as fifteen pounds. Each was surrounded by a layer of rust-colored sand or soil about a quarter- to a half-inch thick.

Our method of operation was to remove the crater fill with team and scraper until we began to see the rust stains that marked the meteorite-bearing zone. Then the team stood by, and with hand shovels we dug meteorites in about the same way as we might dig potatoes on the farm. For the most part, the specimens ranged in size about the same as would a good crop of potatoes, with occasional larger ones. At the mid-point of our second cut, we uncovered the top of a rounded mass larger than a basketball and right beside it, two others almost as large. Several smaller knobs were interspersed in such manner that we supposed they all were merely prominences on a mass some three feet across. Darkness overtook us before we finished uncovering this

Excavation of the Haviland, Kansas, crater, 1933.

The excavation "crew" at the Haviland, Kansas, crater, 1933.

object, so we went to bed contemplating the big day to follow when we would feast our eyes on a mass of a ton or more.

All of the specimens we were finding in the crater were completely oxidized. Moreover, some of them were quite moist and had to be handled with great care to prevent their falling to pieces. Consequently, we found it necessary to clean away all dirt from around them before attempting to move them. In the morning, we renewed this process with all the patience we could muster. This operation gradually disclosed the disappointment that we were not uncovering one large mass, but several smaller ones nested together. The largest weighed eighty-five pounds, another was half as large, and a third was slightly smaller. There were several lesser ones. Our "ton" had gone the way of many a "big meteorite" hope. When the job was finished, we had about 1,200 pounds. Half of this weight was in specimens ranging from ten to eighty-five pounds.

We later searched the Haviland crater area for magnetic nickel-iron particles scattered by the crater-forming explosion. Bob Nininger

Bob and Harvey Nininger standing beside "meteorodes" recovered from the Haviland, Kansas, crater, 1933. Members of the Kimberly family are in the background.

and Alex Richards dug 1,400 post holes, the removed soil of which was screened and magnetically combed for nickel-iron particles. A considerable scattering of such particles was found. In 1937, using such detecting devices as were then available, we attempted subsoil search for additional meteorites but had no luck. The first successful work of this nature in that area was Stockwell's, ten years later.

The Niningers at the Haviland, Kansas, crater, 1938. From left: Addie, Harvey, Doris, Margaret.

***

Excavation of the Haviland crater in Kansas increased my interest in the "peculiar hole" at Odessa, Texas. Of all known meteoritic craters on the surface of the earth, the Odessa crater seemed best adapted for excavation. The Kansas crater—thirty-six by fifty-five feet—was tiny by comparison. The Odessa crater—almost 600 feet in diameter—covered between six and seven acres. Surely the meteorite mass or masses which splashed into the earth to form it must have weighed hundreds and possibly many thousands of tons. I estimated the original depth of the hole at about 180 feet, most of which had been filled in with sediment, supporting a surface not unlike the greasewood-dotted plain on the outside.

On an earlier visit in 1932, I had equipped myself with an electromagnet with a cord attached to a battery. With the battery as a hub, I proceeded to comb the surface in a large circle. The electromagnet was mounted on a long handle, so the procedure was not unlike raking leaves from a lawn. For a couple of days, I stuck to my self-appointed task, shedding off my magnetic gatherings at intervals into a box. In this way, I collected about 1,500 fragments. My harvest aggregated about eight pounds of material, with a couple of sizable chunks picked up on the surface. So far as I knew, this was the first electromagnetic rake used in meteorite collecting.

In the fall of 1935, I received financial help from Dean Gillespie and, with Bob accompanying me, went again to Texas. We were armed this time with an entirely new device, a magnetic balance invented by G. L. Barnett of Oklahoma. He had been working for a number of years to perfect his instrument with two principal objectives in mind: to visit old camps on Spanish trails and old battlefields to hunt for historic relics, and to attempt to locate some of the traditional buried treasures of the Southwest.

"If your instrument operates on the principle of magnetics," I told Mr. Barnett, "it ought to be at its best on meteorites—if it works." Mr. Barnett demonstrated its virtues on a stony meteorite, and I borrowed the machine.

There was nothing in sight on the surface when Bob and I arrived at the Odessa crater, and six-plus acres is a lot of area to work over with any kind of instrument, but I was conscious of the old thrill of anticipation when we set up Mr. Barnett's balance. Would it work? So much might depend on the answer to that question.

Noon had almost come by the time we had set it up and connected it to the battery.

"We won't eat until we find a meteorite," I told Bob.

With a grin, he adjusted the headphones. Then he began walking about, carrying the awkward gadget with him, while I kept the long cord free from surface obstructions. We had worked for only a few minutes when Bob began moving the machine carefully back and forth over a certain spot. Just as I walked toward him, he uttered a disgusted groan and kicked a small tangle of rusty wire away. Ten minutes went by. We were approaching the fifteen-minute mark when Bob stopped.

"Dad, I've got something," he said tensely.

I freed the trailing cord from rocks and bushes and joined him, armed with a probing iron. We moved about cautiously until the instrument was in balance, locating the right spot, and I dug in with the probe. The first prod struck a one-pound meteorite just under the surface.

We ate lunch in elation and haste, eager to continue finding meteorites. That afternoon we located seventeen meteorites in all, then

Bob Nininger using a magnetic balance, an early type of metal detector, at the Odessa Crater, Texas, 1935.

we ate another lunch and made camp in the crater, under the sky.

Long after Bob was asleep, I gazed up at the clear heavens. Ever and anon a meteor would shoot across the starlit vault. Lying there I found myself picturing the great fall that had made the Odessa crater: A huge mass of iron from outer space, streaking across the sky, seen over a distance of a thousand miles or more. A blinding light, going out as suddenly as it had appeared, thunderous detonations, heat destroying all vegetation within a diameter of fifty or perhaps a hundred miles, and the mighty splash as the mass hit the ground and ploughed its way far beneath the surface of the plain, sending thousands of tons of rocks and dirt and fragments of itself upwards and outwards to form a rim that had survived through millennia.

I slept at last but was up at daylight to walk around. I picked up two or three meteorite fragments weighing an ounce or so and came at last to the northeast section of the rim, where I happened to notice a chip lying partly buried beneath a greasewood bush. I attempted to gather it in, but it wouldn't be picked up. Even with the toe of my boot I could not budge it; finally I dug out a specimen weighing eight and a half pounds, the largest found at the Odessa crater up to that time.

We breakfasted and put the instrument to work again. By this time I regarded Mr. Barnett's invention with profound respect, for each time Bob indicated a spot to dig, the probe found a meteorite without any fumbling around. Every hole yielded its specimen. We dug out ten more for a total of twenty-seven meteorites aggregating thirty-four pounds, located by Barnett's magnetic balance. All were on the rim of the crater, the deepest about seven inches beneath the surface.

This experience was as successful as any I have ever had with any mechanical detector. It was as close to Stockwell's kind of luck as I ever came, lacking his wizardry with instruments.

I came away from the Odessa crater with three convictions:

First, that there must be within the crater tangible masses of metallic meteorites beyond the reach of Barnett's instrument.

Second, I would recommend that the federal government excavate the crater according to the means I had utilized at Haviland, and that a shaft should be sunk for underground exploration. The aim should be to preserve the surface appearance as nearly intact as possible, all excavated dirt being conveyed away from the crater itself.

Third, I was convinced that the crater was of such interest and scientific importance as to deserve the creation of a park by the State of Texas to include its environs, with provision for a museum to offer explanatory exhibits to the public.

My first conviction, that a mass of meteoritic material lay within the depth of the crater, dissolved in later years as I learned more about the nature of impact explosions. My dreams of a careful and thorough excavation and of preservation of the unique geological feature for future generations dissolved also. My recommendations to appropriate officials met with little response and no action, while the crater was, to my belief, desecrated by the nature of crude surgery performed upon it a few years later. They dug holes here and there, leaving the excavated rubble right where it was handiest to throw it. There was no trenching to reveal the structure of the crater. Finally, a ten-foot shaft 160 feet deep was sunk at the center of the crater and the rubble piled around it. O. E. Monnig wrote me after he looked at it and said that from now on it is not Odessa Crater, but "Oh, desecration!"

2

In 1915, a young Texan sent a small stony meteorite to the National Museum of the Smithsonian Institution. The Texas specimen was received with pleasure, for it represented a new find. Until then, only twenty-two stony meteorites that had not been seen to fall had been placed on record over the entire United States. The young man was sent a check and asked to search for additional meteorites. In the course of two years, he supplied about a dozen stones, averaging about five pounds each. Then it was concluded that the collecting job was complete. A scientific paper was published describing the find, and the books on it were closed.

In 1928, I asked Dr. Merrill about this find at Plainview, Texas, which he had described in 1917. Dignified and scholarly, he gave me the facts, talking to me as a country school teacher might speak to a first-grader. Then I asked him if there were plans to explore further, and he assured me the job was finished and he had judged it good. When I asked if there would be any objection to my trying my hand at a search in the Plainview area, I was assured there would be none, but he offered no encouragement.

For years, I was unable to initiate such a search. I was always so broke that I was forced to plan my expeditions to kill two birds with one stone, or even three or four—that is, to search for several meteorites at a cost of time and gasoline for one trip. I did make several indirect efforts, working through residents of the area or visitors to Plainview. In 1931 and 1932, I visited the county surveyor at Plainview, and he assured me both times that if there were any meteorites lying about in his county, he would have found them, for he had lived there forty-two years and "had been on every square foot of the county."

Finally, in December of 1933, my brother John and I were returning home from a long, fruitless, and tiring trip into Mexico to investigate a "fifty-ton meteorite" that had turned out to be an outcropping of iron ore, and a half-dozen other stories that had sounded as good and had ended as poorly. We neared Plainview about an hour before sundown.

"We will stay here tonight," I told John, "and before we go to bed, we will go out into the country and get a meteorite. We'll sleep better."

John looked at me with real alarm. Was I feeling all right? What was I talking about? Did I know of any meteorite? Had I seen it? Hadn't we just wild-goose-chased a half dozen reports in Mexico? This, the first trip he had taken with me, had proved to be anything but pleasant, as well as luckless. The food had been terrible, the water worse. Fleas, ticks, ants, scorpions, and snakes had helped to make our days and nights in Mexico miserable. We had found absolutely nothing, and here we were, three days before Christmas, returning home nearly broke.

But I did have something of a plan to account for my optimism. There was pretty good reason to think it might work, and anyway, I told myself, it was better to think up than to think down. This Plainview prospect had yielded nothing yet, but I was confident it would if I made a house-to-house canvass.

As we approached the outskirts of Plainview, we pulled into the first cabin camp we saw. We paid for our four walls, with iron bedstead and mattress, and drove into the countryside.

I started rapping on farmers' doors, catching the men just back from the fields at dusk, showing them my sample of a stony meteorite, and explaining that I was in the market for similar stones. The first surprised farmer told me he had seen that kind of rock many times and that it was about the only kind of rock around the area, but that he had never picked up and saved any. I gave him my card and told him I would be back.

At the next place, I got about the same results. At the third farm, the family was new. The farmer remembered seeing nothing like my sample. At the fourth house, the large family and hired help were at the supper table, eating by lamplight. The patriarch of the group greeted me coolly and with some puzzlement. When I felt that impatience was turning into hostility, I managed to interject that I would pay for stones like my sample at the rate of one dollar per pound.

280 IN SEARCH OF FALLING STARS

The father jumped to his feet, grabbed the lantern and rushed out into the yard, followed by an eight-year-old daughter. In a couple of minutes, they were back with an eight-pound meteorite.

"Is this what you are looking for? We've been using it to hold down a chicken coop against the wind."

When I handed him a check, he said he believed he could find more specimens around but would have to wait for daylight.

Out at the car, I laid the meteorite in my brother's lap. "Let's go back to town and go to bed," I told him.

The next day, we continued our survey and by evening had acquired twenty-six meteorites totaling 152 pounds, paid for with checks I could cover with a note before they reached Denver.

One elderly gentleman listened with considerable curiosity as I explained my mission and displayed my sample, then something seemed to click in his memory.

"Let's go out and look around the barn," he suggested.

From one place to another he searched, finding nothing. Finally, he went to the cow shed and from a crossbeam took a small fist-sized meteorite.

"I've been keeping this around to take after my bull with," he told me. "My bull has a pretty mean disposition, but when he goes on a rampage I only need to hit him once with this, somewhere around the head, and he quiets right down."

The old farmer had to find a new method of bull-subduing or a new instrument, for he sold the meteorite to me.

At a well-cultivated farm, I received a somewhat stern reception from a generously proportioned woman of middle age who appeared at the back screen door. Her hand fixed firmly on the latch, she demanded, "What do you want?"

As affably as possible, I described my purpose. Telling her I was hunting for rocks "like this," I showed her my specimen, said I had reason to think there might be similar stones on her farm, and concluded with the pronouncement that I would pay a "dollar a pound" for them. My hostess's stare had grown progressively colder and threatening until those last magic words. Then her face had melted into a thin smile as she seemed to gaze past my head in an effort to recollect something.

Finally, she asked, "How long you going to be here?"

"Just long enough to find out if you know of any stones like this," I told her.

"Well, there's some of that kind around here some place. You go look around the garage, and I'll go down in the basement."

When I came back from my search, she was waiting on the porch with a meteorite in each hand. Each bore a coating of salt a quarter-inch thick.

"Is this what you want?"

I nodded.

"Well, we've been using these as weights in the pork barrel. That's why they're covered with salt. Wait here. I think there's more."

She brought three more stones. When I paid her for the lot, she stared at the check a moment and shook her head.

"You know," she said wonderingly, "that's more than I got off the farm this year."

Crops had been an utter failure in that dust-blown depression year of 1933.

John and I drove into one farmyard that seemed to be unoccupied and decided to have a look around. I walked out to the windmill and the adjacent milk house. In the dry milk trough lay a four-pound meteorite that probably had been used as a weight on a milk crock cover. Three more stones had been thrown at the foot of an old apple tree in the small grown-over orchard. Another lay by one of the windmill legs. At the next house, where we were able to purchase several meteorites, we asked about the vacant place.

"That belongs to Dean Huff, but he lives in town. There's nobody living there now."

I said that we had walked about the place and had picked up some meteorites, and now we wished to settle for them. It was the farmer's recommendation that we just take them along and not bother, but he told us how to reach the farm's owner.

That evening, I rang Dean Huff's doorbell. He looked rather skeptical as I stood there with an armful of rocks. When he asked what he could do for me, I replied that I thought we could do things for each other. I had gathered these rocks on his farm, I explained, and had come to settle for them.

"Well, I reckon you are welcome to any rock you find on my place."

When I told him I had paid his neighbors for similar stones, he invited me into the house.

"These are meteorites," I told him as I laid down my burden. "Legally, of course, these five stones belong to you, but I don't imagine you have much use for them. They contain nothing of value except information, but since meteorites are the only objects from beyond the earth that man can touch, it is important that they be studied. That is my reason for wishing to buy them."

"What do you pay for them?"

I told him I was paying a dollar a pound to induce the farmers to watch for and preserve meteorites. He asked a series of questions.

I wrote him a check for $28. We had been talking at the foot of a long flight of stairs leading to the upper floor of the rather pretentious, old-fashioned, high-ceilinged house. I had assumed we were alone, but no sooner was the check handed over than there came a feminine voice from up the stairwell.

"Now I know where we'll get the money for that Christmas dress!"

Mrs. Huff came bounding down the stairs, young and beautiful. I'm sure we helped her get her wish in time for Christmas.

With two days between us and Christmas Day, John and I headed northward, feeling a touch of Santa Claus glow after our weeks of bad luck. On the way, we stopped at the Pasamonte ranch and obtained the first specimen from that fireball chase, then nine months old.

The brief stopover at Plainview was the beginning of a search campaign that yielded 900 meteorites by 1948 in an area about sixteen miles long by four miles wide. There are doubtless other meteorites to be found there yet.

We always carefully re-checked each specimen recovered on a field trip. We were aware of the possibility of two falls occurring in the same area over a period of a thousand years and of the fact that even stony meteorites of the sturdier types may survive a thousand years or longer. There was always a chance that two strewn fields might overlap.

Because we wanted to determine the bounds of the actual Plainview field, we would sometimes extend our canvass to farmhouses just outside the area known to have yielded meteorites. On one such survey, I had driven along the western edge of the producing area and then turned east on a line about two miles south of the recognized field. I found nothing and was speeding up on my way to Hale Center, a village about three miles south of the proven field, when I noticed a large rock garden in front of a house set back several rods from the road. I stopped, backed the car, parked beside the road, scanned the rock garden, and then rapped on the house door.

A pleasant lady asked my errand. I apologized for bothering her and picked up a dark-brown six- or seven-pound stone.

"Why that's the ugliest one in the whole lot!"

284 IN SEARCH OF FALLING STARS

"Yes, I know, can you tell me where it was picked up?"

She replied promptly that she had found it over near Abernathy, a village about eighteen miles away.

When I explained that it was a meteorite and therefore was important for scientific purposes, she wanted to give it to me, but I insisted on her accepting the dollar a pound that I had paid for others in the area. Leaving her puzzled but pleased, I went on my way.

At least three and probably four different falls were recognized in the crop of meteorites that we harvested in the Plainview area that had been "finished" decades before.

3

The discovery of new meteorites carried a bit of glamour in the minds of many persons. Offers to participate in my quests came rather frequently from friends who had been keeping an eye on my adventures. Actually, of course, my efforts often seemed more successful as viewed from the sidelines than as seen through the eyes of a man who had money invested in them. Though often a project fell flat, the usual outcome was that my backer or "cooperator" took the disappointment philosophically and waited for another opportunity to come along, which might be luckier for both of us.

Meteorite finds have been rarer than discoveries of comets in the sky. Consequently, any new meteorite find was an event considered newsworthy, but the many fruitless efforts that plagued our days were not so well known. We never made any effort to keep them secret, but the newsmen did not see them as worthy of type space, nor was I anxious to have the failures publicized.

I always warned "investors" that there might be no visible results from our effort. Generally the proposition was that they would furnish the cash outlay, and I would give my time and knowledge, any

NOT ALL KNOWLEDGE IS IN BOOKS

discovery to be shared fifty-fifty.

The Colorado Museum of Natural History acquired quite a number of important additions to the museum collection through their support of a number of field efforts. A number of individuals shared projects. There were others besides those already mentioned. They included Arthur Thompson, a good friend who was keenly interested in my program and who had supported minor excursions at different times, sometimes accompanying me; W. F. Wasson, a Denver attorney; and Frank Clay Cross.

Frank Cross introduced himself to me in the 1930s as a Denver writer in search of a story. His first story about meteorites was followed by others. He wrote for several of the nation's leading magazines, and his stories sold so well that I once told him he was making a better living out of me than I was out of meteorites! Finally, he started issuing what he hoped would become an established periodical which he entitled *The Explorer's Club*. This little bulletin featured various aspects of natural history with special emphasis on meteorites. A small fee was charged for membership in the "club," which covered a subscription to the bulletin. The periodical contained a set of instructions about the recognition of meteorites in the field and new tips on places for search, which brought to light several new meteorites. This venture might have proved a real success, but Frank was unable to finance it for very long.

The little Enon, Ohio, stony meteorite might have been lost to science but for one of Frank Cross's articles. An Ohio farmer had found the stone, about a pound and a half in weight, in 1883. Because it was unusually heavy for its size, he had saved it among his various curiosities for the next fifty-five years, carrying it with him to new residencies in Ohio, Wyoming, Washington, and Colorado. He had never submitted it to anyone for identification until he had read one of Frank's stories.

286 IN SEARCH OF FALLING STARS

It was a decidedly mixed blessing that newsmen saw in meteorites a source of eye-catching news. Publicity is now regarded as a necessary part of launching any new project in science, but forty years ago, publicity was frowned upon by scientists, and any scientist whose name appeared often in headlines was regarded as something less than truly scientific: he was deemed a publicity-seeker. I recognized the peril in press reports of my escapades "chasing meteors," but how else could I gather the reports that were absolutely necessary to the work I had undertaken? I was really after meteorites, not headlines.

Eventually, I came to agree with David Dietz of *The Cleveland Plain Dealer* that the trouble lay as much with scientists as with newsmen, and that the two were necessary to each other. David Dietz was one of the leaders in establishing a corps of reliable science writers, and he was among those helpful to me and interested in my program. The late Ernie Pyle featured my work in his columns. The broad readership of his writings inspired a good deal of correspondence and sending of samples. The local Denver writers who were most interested and helpful included Alexis McKinney, Roscoe Fleming, and Gene Lindberg of *The Denver Post*; the late Lee Casey of *The Rocky Mountain News*; Frank Conly, a freelance magazine writer, and, of course, Frank Cross.

There may have been some thorns in the publicity bouquet. When he had seen an article about my work, my friend and fellow professor, E. L. Craik of McPherson, who had been with me to witness the 1923 fireball, wrote: "Congratulations! I see you made the same front page as John Dillinger." At that time, Dillinger was Public Enemy Number One. But I could not avoid some feelings of pride years later when public attention came from such publications as *Literary Digest*, *Life*, and *The Saturday Evening Post*.

***

One day in Denver in 1934, I was called to the telephone.

"Hello? Is this Professor Nininger? I hope you'll pardon me for bothering you. I'm Herbert Fales. I'm out at the airport and—"

"What did you say the name is?"

"Fales, Herbert G. Fales, from New York. You don't know me, but Lewin Barringer suggested I should get in touch with you. I fly my own plane, and last winter while I was flying over the Carolinas on the way to Florida, I saw extensive areas of elliptical scars. When I told Barringer about them, he said you are devoting your life to meteorites and I should talk with you. I wonder if you would like to have me come out next summer and fly over the plains in search of some meteorite craters."

"Wait a minute," I said. "Do you know who you're talking to? I hope you're serious. You're liable to get yourself into a job for the rest of your life."

And that is the way I first heard of Herbert G. Fales, Vice President of International Nickel Company, who has had an important part in whatever contribution I have made to the science of meteoritics.

When I learned that Fales was with International Nickel, I told him I believed the Sudbury Basin in Ontario was caused by meteorite impact, and I wanted to obtain aerial photos or fly over the Basin with him. That was in the summer of 1934, and for years, geologists did not accept the impact theory.

Subsequently, Robert S. Dietz observed that some rocks around meteorite craters have been distorted by pressure. He called the resulting conical structures "shattercones." In 1961, he visited Sudbury Basin and found many shattercones. Later, Bevan French made a thorough geological study of Sudbury Basin where he found shock effect in the minerals there. Geologists now accept the meteorite theory and say that the Sudbury Basin is an astrobleme, an ancient impact scar produced by a large meteorite.

Fales did come out to Denver the next summer, accompanied by his bride of a few months, who proved to be a very good sport as her husband piloted us about, dipping, circling and dodging to inspect many depressions in the hope some of them would display tell-tale features that would justify further ground investigation. The late Otto Roach of Denver, one of the West's finest photographers, accompanied us, proficiently recording our study on film.

The first reconnoitering trip occupied three days and embraced a journey of 2,000 miles over parts of Colorado, New Mexico, and Texas. The plains country, largely uncultivated, seemed to offer the greatest opportunities for discovery of significant features. We thought that we might recognize craters and scars that had retained most of their original characteristics despite the passage of centuries, undisturbed except by surface erosion. We photographed numbers of buffalo wallows and prairie lakes.

Over Texas, we flew back and forth for some time before we identified the Odessa crater, which is located in the midst of an expansive plain dotted by hundreds of playa lakes that are dry most of the time. Herbert was alone in the cockpit and the other three of us, in the cabin, were busily scanning the landscape, when finally I spotted the crater and penciled a bulletin that was passed to the pilot. Our Lockheed Vega banked sharply and went into a complicated pattern of circling, ascending, swinging, and dipping over and back, round and round while Roach snapped his shutter repeatedly. Otto Roach secured what is to my mind the only truly serviceable photograph of the Odessa crater ever obtained.

Two years later, Herbert Fales flew again to Denver, and again I went with him on a reconnaissance flight, this time with Dr. Alfred M. Bailey, newly appointed director of the Colorado Museum of Natural History, and a friend, Chester Lee. We flew across southern Colorado and northern New Mexico to the Arizona crater. We circled this great bowl several times, then landed in Winslow, Arizona, to make a ground visit.

Aerial view of the Arizona Meteorite Crater, looking south, c. 1937.

We spent that night at *La Posada*, the Fred Harvey hotel in Winslow. The heat was terrible and our discomfort was not mitigated by the constant puffing back and forth beneath our window of an old coal-burning switch engine. At breakfast, when I asked Herbert how he had slept, he replied that he thought he had not slept at all.

Herbert Fales' interest in meteorites was not an evanescent thing, but grew from his training and experience in metallurgy. He set out to secure representatives of as many of the different varieties of meteorites as possible and built an exceptionally fine collection.

Through the years, on occasions when times were unusually tough for Addie and me, I would look through our stock, pick out an especially desirable specimen and write to Fales, offering it for his collection. Invariably, he would buy it. He was one of three or four eager collectors whom I always could count upon for the price of a specimen in an emergency. Each of these collectors presently owns certain meteorites which I never would have let go except that necessity demanded it. I could always console myself that these specimens would be in good hands.

In the middle 1930s, I first met Oscar E. Monnig of Fort Worth, Texas. We had corresponded as early as the late 1920s. Oscar asked me if the "Nininger method" of discovering meteorites was secret. He reminded me that he had been interested in meteorites almost as long as I had and yet, he said, he had "never found a meteorite larger than could be accommodated in an ordinary desk drawer." We worked cooperatively on numerous occasions. As the years passed, Oscar's collection filled many a desk drawer, or comparable receptacle, and several specimens had to be furnished larger quarters.

When he was quite a young man, Oscar Monnig initiated a little bulletin, *The Texas Observer*, recording observations in astronomy and meteorites. He was soon on the way to becoming an avid collector of meteorites. At the same time that he achieved business success in a very busy life as president of a large dry goods company, he contributed notably to meteoritics.

4

Notes from a night's camp on a stream in western Nebraska, 1937:

I weary of folks sometimes. Mostly I'm a sociable person, but still there are those rare hours when I want to be alone, not indoors—happiness never finds me alone inside a house.

This evening, I stopped beside a clear stream on the plains, one of those few survivors of the days before the soil from cultivated fields

NOT ALL KNOWLEDGE IS IN BOOKS

burdened the drainage courses.

The little creek lies between steep terraces a quarter-mile apart. The terraces are dented like crimped pie crust. I have built my fire in one of these terrace dents. Some killdeer are feeding and screaming in the valley... and white-tailed jack rabbits lope in and out of the little meadow. I have been eyeing a coyote hole in the opposite terrace, wondering if some action may start as the evening comes on... Meadowlark songs fill the air, accompanied now and then by the mournful note of the dove or the chir-r-r-e-e-e of the redwing. The sun is setting now and the flickers are ceasing their diligent search for ants and seeking the nearby cottonwoods as a great blue heron wings his way up the stream for his evening of fishing. He squawks unmusically as he passes.

...I poke the coals, add a few sticks to the fire...reflect on the tasty meal of sauerkraut and wieners which I've just finished. Another heron squawks his greeting from high overhead where I see him still reflecting the rays of a sun which for us has set.

What a pleasant ending for a busy day among folks. Three lectures since breakfast this morning, continual conversation between programs. What a great boon, this hour alone. Normalcy has been restored. Two more herons go fishing but no longer any sunlight even for herons... The coals at my feet are gloriously red.

***

The success of our search for meteorites always depended upon the effectiveness of our message to school children and to farmers. But besides the search for meteorites, there was the challenge of presenting to scientists facts and opinions concerning the importance of meteorites in the history and development of our planet.

Over the years, I delivered hundreds of lectures across the nation in colleges and universities, in addition to the talks without number given before classes and assemblies in elementary and secondary schools, and the other hundreds of addresses before miscellaneous groups and institutions.

Although I have no complete file of this lecture activity, I have been able to reconstruct a list of 182 colleges and universities where I spoke. Included on the list are many of the great educational institutions of our land. At some of these, I spoke several times. I spoke on street corners, in country schools, in the Carnegie Music Hall of Pittsburgh.

Many lecture engagements were made possible by the arrangement for dual-purpose travel I had worked out with Dean Gillespie for the pickup and delivery of trucks. There were occasions when my only opportunity to attend a scientific meeting depended upon this combination of labor and culture, and by notifying schools and institutions of my coming visit to their city, I often was able to pick up a number of "extra" lecture engagements, with honorariums of $25 or $50. Besides such miscellaneous appointments, there were more formally scheduled lecture tours, arranged in advance to cover a large area and a fairly long period. On some occasions after the children had reached college age, or were otherwise looked after, Addie would accompany me, and we would travel to the east or west coast with a trailer adapted to use as mobile home and laboratory. But sometimes I drove on these tours alone, or traveled by train.

On one occasion, after I had lectured before a Harvard geological group, then stood through an hour of questions, I took hurried refuge in my hotel room, getting into bed at once with symptoms of a heavy cold coming on. Almost immediately there came a knock at the door, and I arose to admit a member of the geology staff who had come to deliver mail that had arrived in care of the department. He apologetically asked some more questions and then had a final comment before departing thirty minutes later.

"I have taught in Harvard twenty-five years," he told me, "and I never thought there was as much to be learned from meteorites as I learned tonight."

Such praise helped to counterbalance the judgments of those who so often viewed me as a "mere collector" of meteorites. It was always a source of some chagrin to me to be introduced, as I was frequently, as "the man who has found more meteorites than any other man in history."

Such a statement missed the main point of my life. Collecting occupied much of my time and effort, but collecting served as a sort of platform or footing on which to stand while I sought to educate, and while I pleaded constantly for an organized program of meteoritical research.

Dr. Harlow Shapley of the Department of Astronomy at Harvard always supplied inspiration when I lectured at that great old university. The sense of humor of this famed and respected astronomer sometimes came to my rescue. On one occasion, I spoke at some length of the importance of meteoritic dust clouds, deploring the fact that nothing ever seemed to be done about them beyond pure speculation. I exhibited photographs of the dust clouds of the great meteor of March 24, 1933, and also that remarkable, almost unbelievable film record of the fireball itself caught by Charles M. Brown.

One of the young Harvard astronomers rose to describe a remarkable daylight meteor that had been witnessed by members of a Harvard solar eclipse expedition. He told how members of the expedition had just unloaded their equipment to be set up the next day, when suddenly, a huge fireball swept down the western sky, leaving in its wake a long cloud-like streak which hung in the evening sky for some twenty minutes. He described how this trail lay at first in a straight line, then became warped and twisted and widened into a bunchy sort of cloud.

Dr. Shapley spoke. "How did it sound, Nininger? A little amateurish?"

His question gave me precisely the opening I needed.

"I was just thinking, Dr. Shapley, what a pity that our astronomers are not trained to take advantage of such events the way some of our western cowboys do!"

Probably no other great fireball was ever witnessed by so many trained astronomers, surrounded by an unmatched collection of photographic equipment.

***

On my eastern lecture tours, it was my custom to detour through Rochester, New York, and look over the stock of meteorites at the Ward's Natural Science Establishment. If there was something I needed for my collection, perhaps some of my duplicate specimens could be exchanged for it, or sometimes Ward's would buy an item from me.

On one of these visits in the early 1930s, the late Dr. George L. English opened a drawer full of small, unlabeled specimens that had been recovered from the rubble of a recent fire and remained unidentified. He said that if I could identify any of these odd pieces, I was welcome to help myself.

I recognized a few common things and then came to a small, triangular corner piece of a nickel-iron meteorite showing untarnished fusion crust on one side and a polished cut surface with a structure I recognized immediately.

Holding up the thumb-sized piece for Dr. English to see, I said, "Here is one that I don't think you want to give me."

"Yes, anything you find in there is yours. Do you recognize it?"

NOT ALL KNOWLEDGE IS IN BOOKS 295

Quite sure that the corner represented a specimen that I had never felt I could afford to purchase, I suggested to Dr. English that he bring out copies of old catalogs. Meanwhile, I weighed the specimen.

"Now," I said, "look for a listing of a 26-gram piece of Braunau, a corner piece with crust."

Sure enough, there was a listing of the scrap, even to a description of its exact shape, quoting a price of $26.

Dr. English seemed quite taken aback, but he still insisted that since he had told me to take what I wanted, I should take it along; however, he wanted to know how I had recognized it.

I explained that the untarnished crust told me it was very likely a witnessed fall, since the surface had not had the opportunity to deteriorate. The etched surface showed the specimen was a hexahedrite and the only hexahedral iron of witnessed fall I knew of that had ever been put on the market was this one that had fallen through a house in Bohemia in 1847. The Braunau meteorite had narrowly missed striking two children who were sleeping in the room it had passed through. Because of this unusual circumstance, it was regarded as very precious and only a small amount had ever been put up for sale. The price was $1 a gram, a high price for a meteorite in the 1930s. I was most happy to add this piece to my collection, but I refused to accept it without making payment by way of an exchange.

On another occasion, I went to Ward's hoping to sell a five-pound stony meteorite, one of several recovered from a newly discovered fall. Dr. English explained that funds were short just then and suggested I look over their stock and find something I wanted in exchange. I looked about, but there was only one thing I needed, and it was a large slice priced at $330, entirely too much to trade for the stone I was offering. I didn't even mention it, saying there seemed to be nothing suitable for an exchange and I supposed we could not make a deal.

Dr. English leaned back in his swivel chair, smiled rather mischievously and said, "I have a piece down in the basement I'll give you for it. It's a great big thing and I don't think it is Canyon Diablo, but none of us know what it is since its label was lost years ago."

Of course he knew that I was not in the market for Canyon Diablo, the most available of all meteorites.

"Well, I'll go down and have a look at it."

We went downstairs, and there was a half of a large iron. The shape of its cut surface told me at once that it probably was the very mass from which the $330 slice I coveted had been cut, although the 200-pound hunk was so badly rusted, I could not be absolutely sure.

"I'll take a chance on it and accept your offer."

Almost before I had finished speaking, my host called to his helper at the far end of the room.

"Ambrose, crate this big chunk and ship it to Nininger's laboratory in Denver."

Evidently, he feared I might change my mind. Before they crated it, I borrowed a hack saw and cut off a very small corner which I polished and etched on the spot. When I finished, Dr. English asked me if I was satisfied. I assured him that I was.

"Do you mind telling me what it is?"

When I told him it was St. Genevieve, a prize item from St. Genevieve County, Missouri, he was clearly quite shocked. Apparently the shock deepened after the shipment was made, for he wrote me that he had been reconsidering, and he wondered if I would be willing to trade back. My answer was to the effect that I would not consider trading back, reminding him that he knew well from experience that had he asked me to identify that mass for him I would have done so gladly without charge. But instead, he had approached me in horse-trader fashion and I had accepted his proposition. However, I did cut a large slice from the mass, polished and etched it, and sent it to him with my compliments. We sold several other slices of the beautiful St. Genevieve meteorite, then the remaining mass became one of the prize specimens of our collection.

***

Frequently, the lecture trips were grueling, particularly those I made alone by train. But they extended both the collecting and the educational phases of my program, and they aided the family budget.

For weeks, I would be confined to cities. My days were spent in deep, narrow, noisy canyon streets, in offices and halls. The evenings found me looking from auditorium stages into the faces of unknown listeners. Hours in lonely hotel rooms were passed puzzling over problems of providing for the needs of a family in a highly intricate and mechanized society. I carried my ideas for a solution which at the same time could fulfill my hopes for a science of meteoritics from office to office, from lecture platform to auditorium stage, from one railway station to another, on one train and then another.

Days in cities, evenings in trains, nights in hotels. Then onto another train, dozing bumpily along through the night and at daylight peering from my berth through a smudgy window at the smoky, stack-studded industrial district of another great city. Down again onto uncaring concrete to

Harvey Nininger boarding a train for a lecture tour, c. 1937.

hurry along through cinder-tasting air under a block-long smoky shed into a cavernous waiting room, snatch breakfast at a counter, purchase ticket, transfer baggage, board one more railroad car, and settle again with a magazine.

At last, I could look out from a train window to see that the cities had vanished. In their place were tangled thickets of brush and briar harboring contented rabbits and industrious squirrels, hawks floating leisurely against an unstained sky, chickadees searching the twigs acrobatically for food. The uncrowded west and home were at the end of the rails.

***

At a little village school in Marsland, Nebraska, I addressed about eighty pupils from grades four to eight, describing what a meteorite might look like should they happen to come across one in a field of this farming country. To help the youngsters visualize such a find, I said that meteorites are about the color of rusty iron, appearing to have dents in them. They might look, I said, like an old, battered, rusty tin can, but of course, would be very heavy. A little red-headed sixth grader with a freckled face put up his hand.

"Did you say they look sort of like an old, rusty, battered tin can and are awful heavy? I'll bet Dad and I found one when we were building fence."

After school was out, the principal and I went home with this boy and found he had a ten-pound meteorite, now known in collections as the Marsland meteorite.

In meteorite-hunting, I had several methods of operation. When I tired of one because it was unproductive, I did not decide it was no good; I would simply turn to another task for awhile. If lecturing two or three times a day with no visible results was about to get me down,

NOT ALL KNOWLEDGE IS IN BOOKS 299

I'd turn to another method—visiting country editors, perhaps, or talking with farmers on the streets of a small town, or with prospectors in a bar or pool hall. Armed with a couple of small meteorites to gain attention, it seldom was necessary to say much before someone began asking questions. Within minutes, half of the men within earshot would be gathered around, and sometimes very important leads came from casual encounters.

If the program had been routine, doing the same sort of work over and over, day after day, I would have floundered in a short time, for I am not a routine man. I doubt if any man was ever less-fitted by disposition to be one. If my job required that I climb a certain hill from a given starting point to a particular destination every day, I would do it faithfully as long as I held the job; but I would probably find a dozen different routes by which to make the climb.

One day in western Kansas, my work with schools had become rather disappointing because the particular district in which I was working had adopted a rule against any teacher tolerating interruption by persons not specifically authorized by the board of education. As I drove along, I noticed a number of cars parked at a farm house and guessed correctly that this was a public sale or auction. I stopped.

Fortunately, the auction had not begun yet, but everything was ready. I approached the auction clerk and was escorted to the auctioneer. I briefly explained my purpose, promising not to use more than three minutes of their time, and showed the auctioneer and clerk a few specimens. Both officials were pleased to have a surprise to offer a crowd used to waiting through unvarying preliminaries at sale after sale, year after year. I was given five minutes to speak. The officials and the audience became so interested, that I could have talked much longer, but I kept my personal rule never to exceed the time allotted. Out of that crowd came a very good lead.

***

The viewing of a great fireball can be both earth-shaking and emotionally jarring. It is an experience long remembered, both the scene itself and the associated feelings often being retained with clarity. During my very earliest meteorite hunting in Kansas, I came upon an old man who held vivid memories of the Modoc shower in Scott County of that state in 1905.

"Tell me," I said, "just where were you the night of the great shower of stones?"

The old farmer sucked at his smokeless pipe, cupping his horny hands over it with a lighted match. He drew a few hearty puffs, spat at the discarded match stem. Then, resting his hand on the corner of the wagon box and pushing his slouch hat to the back of his head, he gazed into memory.

"I was driving along that road going north. It was this very wagon. Different team, though. I reckon it was about nine o'clock in the even, clear as a bell—not a cloud in the sky and no moon. All of a sudden, everything got light as day, and I saw a fire as big as a haystack in the western sky! The team went down on their knees. I thought the world was coming to an end. Before I had time to think about anything except the horses, the whole blazing mass busted, looked like a million stars coming right down on us. I looked round at my son, about sixteen, in the back end of the wagon box, and he was down on his knees praying. That's just how it was, and then came the noise—louder than thunder. It boomed and roared and rolled way to the west. I was so busy holding the team, I can't say how long it lasted. A few days later, the neighbors brought me some funny rocks, all blacked over, but inside, if we chipped off a corner, they were gray like cement. I reckon about a hundred were found in the neighborhood.

"Now you tell me, young feller, did those rocks come out of that fire? And where do you reckon they came from in the first place?"

NOT ALL KNOWLEDGE IS IN BOOKS

I told him we still were trying to answer that last question, and that was why I was there.

"But you can rest assured," I said, "that those stones, that same evening at sundown, were farther away from the earth than the moon is tonight."

***

During our Denver years, we participated as fully as time permitted in the life of the community. I undertook leadership of a troop of Boy Scouts and also served as a camp instructor for the YMCA summer mountain camp.

I took my Scouts on a couple of long tours. Each chipped in for his expenses, and I managed to schedule lectures along the way. In this way, the boys visited the Midwest and Chicago, and the Southwest and California, and at the same time, I was able to do some work and make some contacts at somewhat less expense than otherwise.

I always seemed to be tying shoestrings together to make do in one way or another. Once, I brought home 500 pounds of peanuts which I had been able to buy for three cents a pound in Texas. Sometimes I brought home quantities of pecans or eggs, or a carful of apples or peaches. Such provisions served a dual use: some went into our own larder, while the rest was sold to the neighbors at bargain prices, with some profit to me as middleman to mend the budget.

Sometimes, we scarcely knew which way to turn, but we had made our decision and did not waste time in regrets. Fortunately, I was so persistent by nature that I seemed almost immune to discouragement.

Bob finished high school in 1936 and spent an additional year taking extra courses and earning money toward college. As the time for college approached, we had some serious conversations about his education and how it could be financed. His mother and I agreed that usually it is best for a young person to enter a college far enough away to avoid any temptation to run home every week. We advised him that far and then left it up to him.

Selfishly, I felt that I would like to see Bob go to my alma mater, but from the standpoint of a greater benefit to him, I believed a school of greater repute would be better.

Bob was much impressed by Dr. F. B. Loomis of Amherst College, who paid us a visit, and he decided that he would like to attend Amherst. Then came the question of meeting the expense. Times were a little better, but our income still was not sufficient to lay out much for college. We had always told Bob that he should expect to work his way through college to a large extent, but that we would help some as necessary.

During a lecture tour among eastern universities and colleges, I visited with Dr. Loomis at Amherst. I asked him to tell me about Professor Hitchcock, whose portrait hung on the wall of his office. Dr. Loomis told me how in its early years, Amherst had faced bankruptcy and the trustees decided to close it down, whereupon Professor Hitchcock, who recently had discovered a great deposit of dinosaur footprints in the Connecticut Valley, begged the board to allow him to run the college for another year, bearing the expense himself. He said he would have his students help him quarry the footprints, which then could be sold for enough money for his own living expenses and to pay the faculty. Teachers were few, and salaries were small at that time. The trustees agreed, and Professor Hitchcock by this means carried the college through until other resources were obtained. As I listened to this story, a thought took shape in my mind.

"Dr. Loomis—a college with a history like that might be willing to help another teacher, pioneering in a different field, put his son through college. Would Amherst accept a collection of meteorites in lieu of regular tuition for my son, Bob?"

Dr. Loomis smiled and thought for a moment. "Sounds like a good idea. I'll speak to the president about it."

It was agreed that the college would receive Bob as a freshman on that condition. If he did well, the same arrangement would hold for the other three years. However, Bob managed to earn most of his expenses, so we did not need to supply many meteorites after the first year.

With this exchange of meteorites in trade as a precedent, we later approached two other colleges—Grinnell in Iowa and Carleton in Minnesota—with similar propositions for our two daughters, Doris and Margaret. Their tuition, room, and board through four years were mostly paid for with rather extensive collections of meteorites.

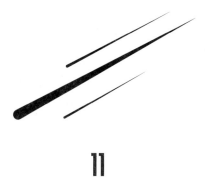

11

Not All the Big Ones Get Away

Nature is the authority.

—G. W. Stevens

In 1915, in his *Catalogue*, Dr. Oliver C. Farrington recognized 247 North American falls prior to 1909, sixty of which had been witnessed and the remainder merely found and recognized without any data as to the time of the fall. Of the sixty witnessed falls, 56 were of the stony varieties, three were irons, and one was a stony-iron. Of the 187 finds, 158 were irons, while twenty were stones and nine were stony-irons. About twenty times as many stones as irons were seen to fall, while about eight times as many irons as stones had been found.

Dr. Farrington puzzled over the fact that certain large, cultivated areas like the state of Illinois had never turned up a single meteorite, while Kansas had yielded fifteen, all but one being stones or stony-irons. Nebraska, adjoining Kansas on the north, had yielded six irons previous to 1909 and not a single stone, while to the south, the great level plains of the Texas and Oklahoma panhandles had yielded neither stones nor irons. Dr. Farrington also called attention to the marked concentration of stony meteorites of undated fall in western Kansas and in the coastal plain of Texas.

Various ingenious speculations were advanced to account for the apparent non-random geographic distribution of falls—obstruction by mountain ranges, abnormally high localized gravitation.

Dr. Farrington offered the suggestion that the soil of these areas was favorable to the finding of meteorites because of the relative scarcity of terrestrial rocks. He also pointed out that dry climate favored preservation for a longer time. At the time of his writing, only twenty stony meteorites of unwitnessed fall had been recorded for all of North America.

In 1925, I suggested what I considered to be a third reason to account for the rich yield of Kansas meteorites, an explanation I termed "the interest factor." By 1929, when the Covert meteorite was discovered as a direct result of my lecture program, I was convinced that one part of the earth's surface was just as likely to receive meteoritic falls as another, and that the education of the public as to the importance and value of meteorites would prove the largest factor in the discovery of meteorites.

It seemed to me that not only must it be due to failure of recognition that more meteorites were not found in more places, but that the same answer must also apply to the discrepancy in numbers of stones and irons. If few meteorites of any kind were recovered due to lack of knowledge about them, then it was natural to expect that still fewer of the stony variety would be found, since they differed less from ordinary terrestrial rocks.

A chief objective of our initial program was to discover stony meteorites of unwitnessed fall.

The task of teaching the people of the plains to know meteorites and distinguish them from other rocks had seemed formidable, but not nearly so formidable in our thinking as it actually proved to be.

Opportunity came to speak before scientists as well as laymen. It soon became apparent that it wasn't only the plainsmen who needed instruction. The scientists were just about as ignorant when it came to meteorites. In certain states where we conducted field work, I could not find a single man who could be relied upon to recognize a meteorite if it were brought to him.

But the interest factor operated in our favor wherever enough instruction could be given to encourage people to "keep on the lookout" for possible meteorites.

The people of Kansas have been credited with being "meteorite-minded" since 1890, but Mary Kimberly had collected heavy black "rocks" on her Kansas farm for five years before she had succeeded in attracting the attention of scientists, aided by the interest aroused by the great Farmington fall of 1890 in Washington County.

At the time of the Farmington fall, and after the recognition of Mary Kimberly's meteorites, Kansas experienced a classic example of what I have observed to be a tendency of meteorite finds to group themselves in the wake of some outstanding meteoric event or discovery. During the next eight years, another ten finds were made in as many localities, four of these meteorites having been in the hands of their finders and unreported for years.

After subsidence of the casual publicity given by Kansas newspapers to the various discoveries of the 1890s, finds in the state were scarce between 1898 and 1923, and, except for the witnessed Modoc fall of 1905, little was heard of meteorites. There was a complete gap from 1906 to 1923, except for the Cullison stone plowed up in 1911.

It is my opinion that far less than one percent of Kansans were meteorite-conscious previous to 1924, and less than ten percent of the farm population even twenty-five years later. By "meteorite conscious"

I mean a degree of awareness that would lead a finder to consider inquiring about the nature of an unfamiliar stone.

Beginning in 1924, I believe that by our efforts, Kansas did begin to become meteorite-minded. It required hundreds of free lectures to many thousands of students, farmers, and farm wives to bring about the thirty-four actually verified finds in the state which resulted from our program during the years from 1923 to 1948.

Our research program depended heavily on the interest and cooperation of schools and their pupils. Any man who offers himself as a target for queries from a group of high school or college students is bound to learn something.

Gradually, I had come to realize that the world of science was only mildly interested in that aspect of the universe that monopolized my thinking. Educators presented a slightly different problem. Once exposed to the subject of my endeavors, they became enthusiastic and would cooperate in any way possible, but there was one fly in this ointment; their enthusiasm was likely to be short-lived.

The high school principal or teacher is so laden with curricular duties and daily reports that any extra-curricular activity can enjoy no more than an evanescent welcome into the whirl of the daily grind. Had there been ways to build meteorite-hunting into a team sport, with the development of keen competition among schools, this might not have been so.

For years, I have been carrying on a sort of single-minded argument with educators concerning the manner in which introductory courses in geology are presented. In many institutions, the subject of general geology is handled primarily as preparatory to more advanced geological study rather than as a course with a character of its own, designed to help fit the average student for greater enjoyment in life. The college graduate whose transcript includes credits for geology often finds himself somewhat helpless to interpret the most elemental aspects of his geological environment. He perhaps has learned the names of a given number of rocks and minerals which made up the study collection of the department, and has learned something of the various formations, all in proper succession as listed by the stratigraphers, but his acquaintance fails to extend to the ability to identify these in the field.

It seemed inconsistent that meteorites were deemed unworthy of attention while care was taken to see that general geology students, most of whom certainly would not become paleontologists, were given sufficient instruction to enable them to recognize that a fossil is a fossil, and to identify several of the more common species. Much attention was devoted to glacial erratics. Every geology student was expected to be familiar with traces of the continental or local shifting of surface rocks. Cosmic craters, however, were seldom even mentioned in geological texts.

Despite my argument that it is just as ridiculous to graduate a geology major without equipping him to recognize a meteorite in the field as it is to turn him out without the ability to recognize that a fossil bone is of organic origin, the fact remained that ninety-five percent of the geologists I met admitted that they could not distinguish meteorites from certain common rocks, and that they did not even try. They indicated that they thought the subject was not important enough to bother with. A good percentage could with more or less certainty recognize a metallic specimen, but many did not even realize that stony meteorites existed.

***

Our field work procedures became ever more productive, and as broader areas were reached by lecture and survey, the ramifications of initial contacts began to yield results.

We were continuing our original procedure of school lectures, leaflet distribution, and press stories. We encouraged the rural population to make simple field tests to distinguish meteorites from terrestrial rocks, offering to buy meteorites as an inducement. Lectures in a given community included illustration by samples of the types of terrestrial rocks most commonly found in the area, together with warnings as to the particular kinds that in the respective locality might most likely be confused with meteorites.

This was the general search. Wherever this initial effort resulted in the discovery of an unrecorded fall, an intensive search was then conducted, involving a house-to-house canvass and in some cases, personal field work by a member of our "staff."

Our staff, for all practical purposes, consisted of myself, Addie, our son, Bob, as he grew older, Alex Richards and his wife, who took on random assignments to cover some specific area for a period of time, and incidental arrangements for individuals to perform field work when occasion demanded.

I soon learned that "meteorite-mindedness" was not sufficient in itself. About four out of five of the individuals who were alerted by the lecture and press program would let their interest die on the vine instead of acting on it. A plowman would turn up what seemed to be the kind of stone I had "lectured about that night in the schoolhouse." He would lay it out under the fence at the edge of the field, intending to send it in or to write me, but he wouldn't get around to it—too busy during the day, too tired at night. And, anyway, he had heard that wise old merchant in town say that all this talk about "meters" was bunk, for he had never heard about it in the university.

Only when we made repeated visits to the same community would most of the finds be reported. Naturally, many communities were never revisited, or not revisited often enough.

In one area where a large meteorite shower had occurred, we

collected about 300 stones aggregating about 800 pounds from approximately ten square miles of level farm land—but it took us four years to accomplish this task after our initial discovery, and that first find itself occurred only after we had made four different and unsuccessful efforts to find meteorites in that vicinity.

This sort of experience taught us very early the importance of follow-up searches. Sometimes, we found additional meteoritic matter in follow-up searches aggregating many times the amount of material gathered at the time the fall was reported and recovery efforts were made and "completed."

Two main facts were impressed on us during the early years of our survey: there was an enormous inconsistency between the amount of meteoritic matter actually collectable in a given area and the extremely small amount formerly recorded for the same area, and even in regions where the amount finally collected was surprisingly large, it could be gathered only with extreme difficulty. We became convinced that probably no search could ever be made to approach completeness. Our results indicated that enormous quantities of meteorites escape notice entirely, due to the obvious difficulties of conducting any kind of intensive search over any great amount of area where some likelihood of success has not been indicated due to a find already reported in the area.

In many cases, the intensive work done in a given area in search of meteorites of a particular fall, however, produced discovery of meteorites of unrelated falls.

Our efforts proved most successful in regions where the native sod had been broken out by the generation yet living, perhaps with the old walking plow, and in regions where the soil is relatively free from terrestrial rocks and dense growths of vegetation. Since a reasonable density of population proved to be desirable, along with the cultivation of the soil, our methods showed their greatest success in limited areas—prairie farmlands and southwest range lands.

Our strategy was devised too late to be applied effectively in many areas of the earth's surface. Enlightenment of the populace, of course, remains a key factor anywhere, and, ultimately, new technological developments of civilization will have to be utilized to make up for opportunities lost in the areas encroached upon by that civilization.

2

It seemed only necessary for a rock to look out of place for it to become a meteorite in the minds of many laymen. Absence of other similar rocks in the immediate vicinity seemed proof enough that it must be a meteorite which had fallen from the sky.

So our mail would bulge with samples sent to our home laboratory for inspection—out of 2,000 of these, only two were meteorites. The whole great conglomeration represented the best efforts of many persons to put into practical use the information I had endeavored to give them. The ratio of meteorites among specimens submitted by persons who had heard a lecture or been contacted personally was much better—one meteorite out of twenty-five or fifty submissions.

Sometimes, we were faced with real riddles of masses of iron or stone which in convincing ways showed characteristics of meteoritic origin and yet could not qualify for positive identification. Sometimes there seemed to be no logical explanation for the presence of such masses.

Again and again, from the northwest there came samples of what appeared to be slag iron but which the senders insisted could not possibly have come by human agency to where they were found. In one instance, the finder asserted that a mass of several tons was found in the mountains some sixty miles from any railroad.

About once a year, or more often, during the 1930s, a sample would arrive from the mountainous terrain bordering the Columbia River. The specific locations of these samples were always kept more or less secret, however. I suspected that they may all have come from the same mass because all of them showed the same structure, but all efforts to obtain exact field information failed, and we were never able to investigate the source. Others have reported similar experiences of failure in attempts to secure accurate information about the source location of iron samples from the northwest.

These apparently unanswerable riddles were one thing, but all across the country in the 1920s and 1930s, leading universities, colleges, and museums were exhibiting spurious specimens as meteorites or the reverse—displaying true meteorites under labels of ordinary rocks and minerals.

In 1931, I stopped at a Canadian university to view a mineral collection reputed to be one of the finest in North America. I went up and down the aisles looking over the cases of beautiful minerals that filled the room. When I reached the display of heavier iron minerals, I saw a little specimen I was sure must be an iron meteorite. It was labeled as magnetite from Leeds, Quebec. I hunted down the custodian to ask permission to examine the specimen, but was told there was no way of obtaining access to any of the cases, and that certainly there could be no error in the labeling since the curator was one of the top mineralogists of North America. Finally I found my way to the curator's office and asked an assistant if I might check the specimen in question. He was insulted on behalf of the absent curator, but he reluctantly agreed to open the case for me. With his permission, I used an emery wheel in the museum shop to grind just a little corner of the specimen, where it promptly showed bright metal instead of the black of magnetite. When we ground down a bit more, I polished it by hand and then etched it, bringing out a beautiful Widmanstätten figure.

The little Leeds iron was magnetic, it was black, or rather dark brown, and it was heavy. Magnetite has all of those qualities. And so the little Leeds iron, an immigrant to our planet, was given membership in the great family of terrestrial iron ores.

At eight different times in eight different institutions, I found meteorites masquerading under other labels. As many times or more, I discovered terrestrial rocks classified as meteorites.

The Rosebud meteorite, which had originally been pronounced to be a wind-blasted lava boulder, was covered over every square centimeter of its surface with fusion crust found on no other type of rock. It bore hundreds of characteristic pits or pizoglyphs. It showed perfect orientation of those pits and absolutely no marks of terrestrial erosion. Yet the entire geology staff of a university had given it an erroneous classification. Perhaps those men had been trained as readers and memorizers of text books, presented with solved problems. I heard a geologist deliver a popular lecture within sight of perfect examples of all types of faulting, erosion, lava flows, talus slopes, and alluvial fans, while using only book illustrations which referred to features hundreds or thousands of miles away.

In 1931, when I was passing through a university city, I telephoned the department of geology to ask if there was a collection of meteorites. I was informed they had one meteorite, which Dr. Farrington had identified for them a few months before, after it had lain in the rock collection for fifteen years as a favorite example of a glacially scratched boulder. This was the little Lafayette, Indiana, stone, one of the most perfect and beautiful specimens. All of its most important features—its orientation, fusion encrustation, crinkly glass thread lines, and readily distinguishable forward and rear surfaces—were later described in print, with illustrations, as they related to other meteorites.

In one university museum, two polished sections of a stony

meteorite were exhibited as limonite, while the third fragment of the same meteorite lay in a box of discards destined for the city dump until I rescued it.

Following a lecture in Fort Worth, Texas, in the middle 1930s, I invited questions from the audience of geological society members. After several other queries, a middle-aged geologist spoke up in a tone that reflected both impatience and skepticism:

"What evidence have you or anyone else that any of those specimens which you displayed tonight came from space?"

3

Many times, results followed a long while after the event, and a tour of lectures or explorations would be capped with success years later, or miles away, or through some indirect association.

A lecture I gave in the Sharon Springs, Kansas, school about 1933, yielded only one visible result at the time. However, it brought me into acquaintance with Mr. R. A. Dollarhide, who then was the telegraph operator at the railroad station in that town. Dollarhide was an avid collector of Indian artifacts and various kinds of rocks.

He sent in a few samples, none of which proved to be meteoritic, and I lost track of him for some time. Then he sent in a chip that he said had been taken from a mass of several hundred pounds near the village of Morland, Kansas, where he was telegrapher at that time.

This was in 1935, a year of heavy expense for us, and I was broke. But Dollarhide's sample was genuine, and his story was good. Dean Gillespie agreed to put up money for the trip and purchase of the meteorite in return for half interest.

I made the trip at once. Dollarhide told me where to find the farmer, Sam Hisey, and told me how Sam had broken his lister—a kind of plow—on the meteorite in a field he had farmed for many

NOT ALL THE BIG ONES GET AWAY

years. It was "the heaviest darn rock" he had ever seen, Hisey had said, "big as a wash tub and in a field where I've never even seen a pebble before."

We went out to the field, and what I saw there set my nerves tingling. A mass that weighed in later at more than 600 pounds had been pried out of the soil. It had been broken into three pieces, probably at the time of fall, but they could plainly be fitted back together. Morland turned out to be one of the world's eight largest aerolites.

Mr. Hisey had supposed that the stone had arrived since the last year's plowing. However, the meteorite was visibly aged and no doubt had landed long before settlement of the area by white men. It had been embedded near the top of a small hill where erosion by water and wind had gradually lowered the surface several feet, eventually exposing the stone to the plow.

Had Dollarhide not heard the story, investigated it, and sent me a small sample, Hisey would have carried out his plan to hitch a team to the troublesome stone and drop it into a deep ravine that traversed his farm.

As the story of Hisey's find got around, neighbors remembered that a fifty-pound stone of the same kind had been found by an early settler and used in the foundation of a neighboring farm building some fifty years before. Some fragments of this piece, with cement still clinging, were recovered from the building remnants. Years later, another ninety-seven- pound mass was found by pheasant hunters about three miles south of the Hisey farm. This stone lay in a deep ravine under a fence that still held an empty wire loop to which the meteorite had been tied and buried as a "deadman" to hold the fence down. Soil erosion had apparently deepened what had been a mere ditch when the fence was erected. Analyses of the three stones proved they belonged to the same fall.

I became more convinced than ever that there must be hundreds of other meteorites that had gone undiscovered because the margin of chance was not in their favor.

Our methods of field search were built on logic in a situation where the mathematical odds against us were great. We had studied our map of the Great Plains. In the southwestern corner of Kansas, there was a block of six counties from which not a single meteorite ever had been reported. This was strange in an area where soil conditions and population density were favorable to finding specimens. Alex Richards was instructed to concentrate on this area. He worked persistently, lecturing in high schools, speaking before clubs, and reaching by personal contact as many people as he could. But our small fund for the assignment soon was exhausted, with nothing to show for it but the recorded fact that hundreds of school pupils and other hundreds of parents had been acquainted with meteorites.

Again and again, as we were able, we invaded more and more of this area of some 6,000 square miles. Repeated negative results failed to dent my certainty that meteorites were distributed indiscriminately and that failure to find them reflected only the incompleteness of search.

In March 1935, Bob and I were returning from a trip into New Mexico and chose a route through southwestern Kansas in the midst of dust bowl storms that were blowing acres of topsoil from the farms. At the end of four days of a discouraging battle with dust-laden winds and sand-blown roads, we drove wearily into Hugoton, Kansas, the county seat. We stayed the night with some former students of mine, the Hubbards. These hospitable people were tireless in their efforts to make us comfortable in spite of the blowing dirt, but when we awoke in what had been a freshly cleaned room the night before, the once immaculate bed linen was the color of creamed coffee, with two incongruous white spots where our heads had rested on the pillows, and our feet plainly tracked the floor.

Hubbard urged us to arrange for a lecture in the high school, but we were anxious to be on our way. Then he mentioned that the principal was another former student of mine, so I changed my plans. I was greeted warmly at the school and invited to address a special assembly. Since there was no time for the janitor to remove the deposit of blown-in dust from the auditorium seats, the lecture was given in the library, where the 300 pupils sat on chairs or table tops or stood, all of them examining with interest the specimens I sent around for inspection.

At the close of my talk, a senior boy, John D. Lynch, Jr., came forward and told me that one of the specimens resembled a stone his father and he had plowed out with the lister some years before. The stone had been thrown under the fence and, so far as he knew, was still there.

Bob and I, with Hubbard and the boy, drove out to the farm. The stone was found, just a corner still protruding from a sand drift rapidly forming under the fence. It was plain the sixteen-pound object was but a fragment of a much larger mass, probably severed by the plow. We followed the corn rows a half-mile into the field to the spot where John Lynch recalled that the meteorite had been plowed out. There, we soon found fragments of meteorite that had been scattered over an area of about an acre in size during the years of cultivation. I began scratching with a shovel where the concentration was heaviest. Hubbard wielded a second shovel, and soon we could hear fragments grating against the implements at every stroke. All were sifted out carefully and set aside, and we proceeded cautiously with our digging, on the lookout not to disturb any piece that seemed to be anchored, for fear of destroying the parent mass which must be in an advanced stage of decay.

Hubbard struck stone that seemed to be anchored, and a moment later, I did also. Our two stones were twenty inches apart. It was like drawing straws—who would get the big one? We laid our shovels aside and scratched away with bare hands, as excited as treasure hunters. In answer to repeated queries, I predicted that we might find a stone weighing as much as 200 pounds, though I doubted it, since our two protuberances were not deeply buried.

Five of us dug—Bob and I, Hubbard, the Lynch boy, and a friend of his. We worked patiently in two rival groups, one at each knob, until we excitedly found we were all working on the same large mass! By noon, we had unearthed the second largest stony meteorite ever discovered, a weathered, brown aerolite of 749 pounds, with twenty-one pounds more of separated fragments. During several seasons of cultivation, more fragments were gathered, bringing the total to nearly 900 pounds.

We uncovered the stone and photographed it where it rested. Then it was wrapped in burlap and plaster over two thicknesses of damp newspaper. Two six-foot lengths of gas pipe were wrapped into the plaster jacket to serve as handles. When the plaster had set, six men lifted the mass into a trailer for the move to Denver. Then the wrappings were removed and the more broken portions were taken apart piece by piece, cleaned, and set together again with plaster and gum arabic. The extreme upper portion had been so crumbled in the soil that

The Hugoton, Kansas, meteorite after excavation, 1935. From left: Mr. Hubbard, John Lynch's friend, and Bob Nininger.

it could not be reassembled. A plaster base was built onto this disintegrated end to serve as a permanent support, reversing the position of the stone as found.

The Hugoton, Kansas, meteorite, wrapped in newspaper, burlap, and plaster, being loaded for transport, 1935.

Recovery of the great Hugoton stone was dependent upon a series of chances—our selection of the return route from New Mexico and our visit to the Hubbards, the last-minute decision to call at the school, the hastily arranged lecture and the presence of John Lynch, who had graduated from the school only weeks later, a thoughtful boy's memory and curiosity, and finding the nearly buried fragment before the wind-drifted sand had completely buried it. This remarkable find came at the end of four months of otherwise fruitless field work.

All in all, the 1930s—such bad years in so many ways—were good years for adding to man's knowledge and inventory of meteorites. Perhaps it was because in those dreary, worrisome days, people

320 IN SEARCH OF FALLING STARS

were glad for a new interest, especially if it offered a slight chance of obtaining some money. More likely, it was because these were the first years anyone had given all of his time searching for and studying these visitors from outer space.

In 1936, I called in Alex Richards.

"I want you to go down to Gladstone, New Mexico, set up your tent, and stay three weeks," I told him. "Keep your meteorite samples on display in that country store, and try to contact every farmer. There should be a meteorite somewhere thereabouts because none has been found nearer than about twelve miles, and quite a lot of land is under cultivation.

At the end of two weeks, Alex wrote that he had found no leads, he had seen every farmer and rancher in the vicinity, and he himself had walked fields and pastures until he was so sore and stiff, he couldn't sleep. Should he stay on another week or go to the next location?

"Stay on," I answered.

Three days after he had written that letter, a rancher who had been passing by his display and ignoring it regularly finally stopped to look.

"Is that the kind of stuff you offer to buy?" he asked. "Hell! I've got one as big as a coal bucket."

Alex promptly bought his meteorite, a 128-pound stone that had been lying up against the barn ever since it had been plowed up four years earlier. And within a week, Alex purchased four others, representing three different falls.

***

Pasamonte and Sioux County were landmarks for 1933. The two great stones, Morland and Hugoton, marked 1935. But 1937 and

1938 were our most productive years.

During 1937, thirty-one previously unknown meteorite falls were recovered by our little organization—more than had been recorded in any three-year period throughout the entire history of the world. None of these were seen to fall, so there was no exciting event to stir any segment of the population into observational activity beyond our lectures, leaflets, and newspaper stories.

When I first went through the town of Miami, Texas, in 1937, I noticed the courthouse was surrounded by fossils and went inside to ask who the collector was. I talked to the hobbyist, Judge J. A. Meade, admired his collection, and asked if during his field work he ever had happened onto a meteorite. Judge Meade said he had never seen one and wasn't sure he would recognize one if he did. I showed him some specimens and left a little sample of a stony meteorite with him.

Field work took me through Miami several times in a brief period, and each time, I stopped to visit with the judge. At last, he reported that he had located a meteorite, but he said he couldn't talk the farmer into bringing it into town. The judge said it was just like the sample I had given him, but "as big as a half-bushel measure."

Judge Meade told me how to reach the farm of Mr. Thornhill, fourteen miles out from town. When I arrived, the farmer was busy training a young stallion. There was no sign of a meteorite about the yard. When I told him of my conversation with the judge, Thornhill insisted he didn't remember any such thing. Finally, I brought a specimen from the car.

"If you can remember where it is, and if the meteorite is genuine, I'll pay you fifty dollars, sight unseen."

Immediately the farmer yelled for his hired hand.

"Flaky, show these men where Arkansas built the fence for us."

Thornhill explained that the fence builder had taken a pile of white caliche rocks—desert limestone—for deadmen, and had carried the "black rock" along for the same purpose.

There were about fifty deadmen along the fence line, all of them white but one, and nearly all of them buried at intervals along the two-mile fence wherever it crossed a ditch or a dry wash, as weights to hold the wire down to keep in the cattle. Bob was with me. We both took shovels and started digging down to the deadmen. The twenty-sixth was a 127-pound meteorite. We stowed it in the car trunk and gave a check to the farmer, and he responded with a broad smile.

"You like fried chicken? Pick out a couple from the pen."

He dressed them and packed them in ice, and directed us to stop at the next restaurant. They fried us a fine fresh chicken dinner.

***

One hot, windy summer day, I sat down at a lunch counter in Sterling, Colorado, and ordered a hamburger and coffee. I laid a small stony meteorite beside my water glass. Shortly after, a cattle truck stopped in front and the driver entered and sat down beside me. He wiped the sweat and grime from his face, rubbed the sand from his eyes, wiped his eyeglasses with a napkin, replaced them, and reached over and picked up the little meteorite. He examined it briefly and put it down.

Harvey Nininger "exhuming" a "deadman" (weight used to anchor a fence), 1937. The Miami, Texas, meteorite.

"Do you know what it is?" I asked him.

"Well, it looks like a rock."

"Yes, it is a rock, but a very special one. It fell from the sky."

"Oh? You mean it's one of those meters?"

He picked up the meteorite again, hefted it, and turned it over. "You're not kidding?"

I assured him I was not.

The driver took a swig of coffee, wiped his mustache, and examined the meteorite again.

"You know, my brother-in-law over in Nebraska may have one of those things in his yard. It's a big thing. He bumped into it when he was plowing several years ago in the field where there were no other rocks. He brought it up to the house, and it's been lying in the yard ever since."

I asked him how far we were from his brother-in-law's place. The distance, 110 miles, seemed too far to risk a wild goose chase, so I handed him a leaflet on how to recognize meteorites and asked him to send another to his brother-in-law. A few weeks later, I received a sample of a somewhat old and weathered meteorite that represented a rather rare type. When I went to Nebraska to see it, I found the farmer had about 450 pounds of meteorite in the yard, and we were able to gather another 150 pounds from the hilltop where he had plowed it up and from a fence row where he had thrown a forty-five-pound fragment. This was the Potter, Nebraska, find.

We recovered a total of sixty-five meteorites during 1937–38. Each has something of a story. But to me, the most exciting aspect of our mounting success was the fact that my contention that meteorites were an important feature of the earth, both geologically and astronomically, was being proven.

4

Among the most interesting of witnessed meteorite falls is the one that interrupted a burial service being conducted near the little town of Johnstown, Colorado, on July 6, 1924.

The explosions and puffs of gray "smoke" that accompanied the shower of stones were heard and seen by no fewer than 200 people, mourners in the church yard or workmen in the fields. A fifteen-pound specimen was dug out of the roadside just thirty feet from the church door, where it had fallen with an audible thud and buried itself twenty inches into the ground. A fifty-pound stone was dug immediately from wet ground in which a farmer had seen it bury itself to a depth of five feet. A seven-pound stone and two weighing three and a half pounds each were picked up in fields where they had been seen to strike. Numbers of pea-size to walnut-size stones were gathered between the cemetery two miles west of Johnstown, and Mead, ten miles south.

Fifteen years after this fall, in 1939, I was invited to address a group of employees of a Denver dry goods firm who wished for a speaker on some subject considerably removed from matters of merchandising. After I spoke to the group of about thirty, the chairman, Sam Dreith, came over to me.

"I've got a meteorite that I saw fall," he said.

I asked when and where, and he told me it had been about fifteen years prior, near Johnstown. He had been out raking hay with a team and dump rake when the noise of the meteor had frightened his horses. He had seen a stone drop not far away in the field, a black stone leaving a little vapor trail, but he hadn't broken his raking pattern. It was on the next day that he reached that point in the field and his rake had kicked up the stone. He had carried it home, put it in the bottom of his trunk, and had never taken it out since. When he went home the next weekend, he pulled the meteorite from the trunk and brought it to me in Denver. It was a beautifully oriented, fresh-looking stone weighing five pounds, perfectly preserved after its long years in storage.

Our field work of 1939 was crowned by the recovery of the largest meteorite that Addie and I ever retrieved personally, the one-and-one-third-ton iron meteorite of Goose Lake, California. The discovery of this great meteorite can be credited to a pair of deer hunters. Recovery of the ponderous iron from its lava mesa in uninhabited wilds can be credited to the ingenuity and years of logging experience of two old-time woodsmen, Olin E. Ake and I. C. Everley.

Deer hunting had not been good in the fall of 1938 for Joseph Secco and Clarence Schmidt. Deer were plentiful in the rugged Devil's Garden atop the Modoc lava beds of the extreme northeastern corner of California, but luck had failed the two hunters until they had just one day remaining to get their bucks. In mid-afternoon of the last day, Clarence sat down on a log at the edge of a rocky flat. Joe had started across the clearing to hunt the neighboring forest. Suddenly, Clarence heard a strange sound from across the lava field: melodic "pings" and "bongs" as of iron. He followed after his partner. A few hundred feet away, he found Joe sitting astride a great irregular brown mass of iron, pounding away at it with a chunk of basalt. The resulting sound seemed to Clarence like a cross between a mission bell and a blacksmith's anvil. Though he never in his life had seen a meteorite, he had read about them, and he told Joe he was sure this must be one.

The boys wanted to take a sample, but with no amount of pounding were they able to detach a fragment. Finally, they went back to camp for a meat saw. By the time they had cut off a tiny corner of the steely mass, their saw was ruined.

During that winter, we received a letter from Oakland, California, concerning a large "possible meteorite in northern California," and we added this one to the list of prospects to be investigated as soon as we could arrange for a trip to that state. We received a second report of a suspected large meteorite from Dr. Frederick C. Leonard of the University of California at Los Angeles. Dr. Leonard was a great enthusiast of meteorites. He talked of almost nothing else. As an astronomer, he looked at the study of meteorites as astronomical. I insisted that it was as much a geological as an astronomical subject.

The following April, when we were in Los Angeles, Dr. Leonard showed us a small sample of this "questionable" meteorite. Leonard said he was puzzled. He had thought the piece to be meteoritic but had been given a rather indefinite report by the geology department following examination of the specimen. He wanted my opinion— which was, at once and with no uncertainty, that the material was genuinely meteoritic. Right away, Leonard wanted to know what the proper procedure was to recover the mass. First, I told him, we needed to learn whether the specimen lay on public or private land. From the general locality of the reported site, it appeared the land might be government-owned, in which case a report would have to be made to the United States National Museum. Leonard asked me to proceed with a recovery effort, but stressed that he wanted very much to participate.

Addie and I drove to Oakland, where I talked with Dr. Earl G. Linsley, director of Chabot Observatory, who had written the first letter we had received about the possible meteorite. He referred us to Clarence Schmidt. We sought him out and arranged for him to go with us to the northeastern corner of the state and lead us to the meteorite. We also secured from the Forest Service a promise that a surveyor would immediately begin to determine the ownership of the land involved.

Because Schmidt had not returned to the spot, he realized it might be difficult to locate the meteorite again. He knew it lay in a clearing on top of the lava mesa out of Alturas, but this was vast and rugged territory.

Addie, Schmidt, and I started out, driving first to Alturas and then going some thirty miles farther over dirt road to the foot of the lava mesa that rises from the western shore of Goose Lake so tortuously and ruggedly as to earn its name of Devil's Garden. A logging road led by a round-about course to the "Garden," fifteen miles away, but melting snows had rendered it impassable for ordinary passenger vehicles. Schmidt thought he could lead us on foot by a much shorter course. We spent a half day tramping through dense forest, up canyons and over ridges, and finally reached the top of the mesa only to find we were in the wrong spot. Disappointed and tired, we returned to camp.

The next morning, we obtained horses and rode the longer way round. This time, Schmidt led us to the magnificent iron, lying just as he had seen it before and just as it had lain no doubt for centuries. We rode back down to civilization, telephoned Dr. Leonard, invited him to come join in the fun, and made plans to bring the meteorite down from the mountain.

Everley and Ake, with two of their neighbors, rigged up a four-ton log wagon drawn by four heavy draft horses with a block and tackle. We loaded it with tools, bedding, provisions, and cameras. Dr. Linsley and an assistant had arrived with a panel truck. Dr. Leonard, accompanied by Dr. Robert Webb, professor of geology at UCLA, was en route. Leaving word for the two Los Angeles professors that the first half of the climb could be made by auto, we set out on foot, following the wagon, hoping they would catch up with us. When we were about half way up the mesa slope, we could hear Dr. Webb calling, and I ran back down the trail to meet him and Dr. Leonard, who had driven as far as the road permitted. We all walked on, soon overtaking the wagon. We reached the mesa top and then proceeded across the two miles or so of boulder-strewn, muddy mesa, dodging ponds and fording small streams of spring run-off until we approached the meteorite where it lay on reasonably dry, boulder-paved terrain.

By this time, Frederick Leonard had shed most of his academic dignity. When we came within sight of the big iron, the pudgy little professor ran on ahead, placed his hands lovingly on the great meteorite, bent and kissed it. Then he lifted his hands skyward and turned to face us.

"This is the greatest day in meteoric astronomy!"

Now we all joined in to fondle the prize, an almost unbelievable chunk of metal that seemed to link us, there in that wild corner of the earth, with worlds beyond.

We examined and photographed the mass where it lay and then made camp.

Harvey Nininger with the Goose Lake, California, meteorite in situ, 1939.

The next morning, the men worked for hours with block and tackle, trying to load that mass of iron on the wagon. They had constructed a tripod of timbers cut from the nearby forest, and they tried to slide and haul and push the meteorite up this crude ramp to the wagon. There were several breakdowns and repairs, but finally, after persistent efforts and with infinite patience, the loading was completed.

The Goose Lake, California, meteorite ready for loading, 1939.

Some miles of almost hub-deep mud, boulders, treacherous ponds, and streams lay before us. Scouts went ahead to pick out a course and in some cases to build a makeshift road, clearing away boulders, filling in low spots with rocks and brush, and cutting down small trees.

It required a day and a half to work our way back down the mountain to the road. There the meteorite was transferred to a truck and taken to San Francisco, where it was displayed at the Golden Gate International Exposition at Treasure Island until it was eventually moved to the Smithsonian Institution in Washington. The finding-place was national forest land, so this great meteorite belongs to the nation.

The Goose Lake, California, meteorite on the wagon used to recover it, 1939. Frederick Leonard of UCLA is on the right in the suit and tie.

The Goose Lake, California, meteorite after being transferred to a pick-up truck, 1939. The American Meteorite Laboratory trailer is in the background.

NOT ALL THE BIG ONES GET AWAY 331

***

Who owns a meteorite?

In the popular mind, finder is keeper.

In 1890, Professor Horace Winchell of the University of Minnesota paid a man named Peter Hoagland $105 for a sixty-six-pound meteorite that Hoagland had seen fall as part of a shower in Winnebago County, Iowa. Hoagland had received permission to search for the stone—but from the tenant of the farm rather than the landowner. When the latter determined to assert his rights of ownership, the state supreme court ruled the meteorite was "part of the soil" and so belonged to the owner of the land.

This principle was upheld in the Oregon "iron rock" case a dozen years later. Ivan Hughes had discovered a fifteen-ton iron boulder on land belonging to the Oregon Iron and Steel Company. Hughes told himself, "finders keepers," and decided to appropriate the huge block of metal for his own. This was no small task for a woodsman in remote forest in 1902, but he believed once the mass was on his premises his ownership would be secure, and so he addressed the removal task with singular ingenuity. He fashioned a capstan with chain anchor and a braided wire rope to wind on it. Then he constructed a log chassis to which he attached sections of logs for wheels. Using improvised blocks and levers, he skillfully engineered the giant meteorite onto his crude carriage and, with windlass and horse, moved it inch by inch and foot by foot three-quarters of a mile to his own land.

The ensuing excitement over Hughes's iron rock alerted Oregon Iron and Steel, and when it was learned that the mass originally had lain on company land, a lawsuit was filed against Hughes. Again the matter reached a state supreme court, and again it was held that the meteorite constituted "part of the soil." Hughes saw no part of the

$20,500 for which the Willamette meteorite was sold, though had it not been for him, the meteorite might have remained unfound, and would not be seen by thousands of people annually in the American Museum of Natural History.

During our years of collecting, we purchased several thousand specimens without a single complaint ever being filed in court. Where the find had been made by other than the landowner, we always recommended a division of the sale price between the finder and the property-owner. Usually the owner insisted on the finder receiving the proceeds, though in a few cases, the money was split. Some sort of division of ownership between finder and landowner seemed to be an equitable arrangement.

A more sensitive problem than the matter of ownership can arise if one collector "encroaches" on an area of fall to which another makes prior claim on the basis of a program of search already undertaken. When a man or an institution through its representatives in the field discovers the remains of a dinosaur or some other important fossil, it is regularly accepted that such discovery establishes a prior right to work in the area until the entire skeleton or skeletons have been safely recovered, assuming, of course, that in any such case, proper arrangements have been made with the landowner. A meteorite fall can be studied properly only if all of the recovered material can be examined critically and compared.

The going price to owners of meteorites had been established pretty well before my time at a dollar a pound. We made this our basic figure. Deviations up and down from this figure were based upon the type of specimen, its size, its state of preservation, its special features as to form and surface markings, its internal structure and composition, as well as any facts known about the fall. For my predecessors in the collecting field, who collected only casually, the price paid to the finder usually constituted the entire cost to the buyer.

For me, however, it was often the minor part of my costs. Sometimes Alex or I or another of our "staff" would spend weeks in the field. Much time, too, was spent in the office and laboratory, answering the letters and testing the samples that came in response to our program of lectures—many of which were free—leaflets, radio, and press appeals. All of these activities were costly in both time and money. By the time a meteorite had been acquired, classified, and undergone some study, it had become an expensive item. In our most successful year, the average cost of acquiring a specimen was about $14 a pound in addition to the price paid to the farmer who owned the land where it was found. At the end of the 1930s, in terms of dollars of the 1930s, I figured that the average cost of discovering a new meteorite amounted to about $300 in addition to the purchase price. For one who was working on as frayed a shoestring as I was, the rare occasions over the years in which our prior field work was usurped and we lost out in the acquisition of the specimens recovered seemed like major catastrophes.

Selling specimens had drawbacks, too. Word came to me that a man whom I shall call Mr. Von Hess desired to purchase a meteorite which he would later present to an institution. This was sometime in the 1930s during unusually tough times, and I could not afford to pass up an opportunity to cash in on a specimen. Because I made it a point to know who had collections of meteorites, I knew that Von Hess had none, but that at one time he had owned a seventy-eight-pound iron which had a rather interesting history. It had been found in 1881 by a sheep-herder, Ignacio Martin, who had traded it to Thomas Tobens for a small pony. Thomas, believing it to be solid silver and fearing it might be stolen before he could market it, buried it in a manure pile in his barn, and there it lay for several years. Mr. Von Hess heard about it, determined that it was meteoritic, and purchased it for a scientific society.

Since I was told that he wanted "something nice" and that he was well-to-do, I looked forward to a pretty good sale. I selected a very fine specimen, beautiful more than scientifically valuable, to present for his inspection. This was an end piece from the Xiquipilco fall, with a polished face the size of a slice of bread that displayed exquisite Widmanstätten figures. The nicely sculptured natural surfaces were beautifully black. The iron weighed about six pounds.

I was cordially received by Von Hess. He leaned back in his swivel chair and admired the meteorite. He inspected the specimen carefully with a magnifying glass, inquiring as to the techniques of polishing and etching. As our conversation proceeded, Von Hess reached into his pocket, took out a penknife, opened it and began very casually to scratch an area in the center of the polished and etched surface. I almost cried out. He remarked that the narrow bands were harder than the wide bands in the pattern. Yes, I replied, I knew this, and I might have added that I could have told him as much without his damaging investigation. But I was still consoling myself, or trying to, with the thought that the specimen soon would be his and the arduous re-finishing job would be at his expense, not mine.

No such luck. He greatly appreciated the opportunity to examine the specimen, but he didn't believe he was quite ready to make a purchase. All the fortitude I could muster was required to bid him a pleasant "good-day" and take myself home, contemplating my loss of time already spent and another two days for the regrinding, polishing, etching, and administering of anti-rust treatment to the disfigured specimen. Subsequently I was less timid about cautioning prospective customers not to touch the finished face of a specimen.

***

Not long after our Goose Lake experience with our largest

meteorite, Frederick Leonard, Addie, and I had the less-spectacular, but just as significant, thrill of finding the smallest meteorites then on record.

These tiny particles, remnants of the great Holbrook, Arizona shower of 1912, reinforced my belief that a rain of small particles must be associated with every major meteorite shower, carried in the great accompanying dust clouds.

Reports of such rains of gravel had come most prominently in connection with the Estherville, Iowa, shower, and the Archie, Missouri, fall, but there was no proof in the form of recovered, finely divided particles.

On the hot evening of July 19, 1912, the Lopez family sat at dinner in the Aztec section house eight miles northeast of Holbrook on the Santa Fe railroad, when a brilliant flash of light attracted the attention of the small boy of the family. A moment later, there was a thundering sound and then a noise as of an object striking the roof. The youngster dashed outside and came back quickly, shouting that it was "raining rocks."

The father reached the door in time to see little puffs of desert sand rising all over the surrounding area, and in the sky to the west, there was a grayish cloud unlike the small cumulus clouds studding the rest of the sky.

Members of the Lopez family gathered some of the strange black stones. One was taken to Mr. Lloyd Henning, who identified it as a meteorite.

After learning that the stones were saleable, the residents of Holbrook swarmed over the area gathering the little black-coated pebbles that came to be known popularly as Aztec peas. About 14,000 of the little stony meteorites were purchased by W. M. Foote at that time. Twenty years later, I conducted a search that yielded an additional 2,000 specimens. These 16,000 stones varied in weight from a tenth of a gram to about fifteen pounds, with an unusual intermingling of the large stones among the small ones, but there was nothing small enough to have been carried in the dust cloud.

Little is known of how these clouds travel and deposit their loads of meteoritic particles, and I hoped somehow to recover material of a size that could tell something of the significance of meteor trails.

Frederick Leonard joined Addie and me at Winslow, Arizona, on April 13, 1940. We spent a couple of days at the crater, and then, on April 15, we drove to the little railroad station of Arntz, some seven miles east of Holbrook, to search for remnants of the Holbrook shower.

A forty-mile-per-hour wind and whipping sand made a visual search almost impossible, so we devoted ourselves to a magnetic search for small particles, using magnets mounted on canes. We were hampered by a heavy sprinkling of dark-colored gravel, terrestrial in nature, of which some was magnetic.

Finally we turned our attention to ant hills, where large red ants had piled up mounds of gravel and sand. Slowly and systematically, we dragged our magnets over these mounds. We collected the adhering magnetic particles and took them with us for examination under better conditions. We found among them three perfect little aerolites, measuring from 2 x 2 x 3 mm to 2.5 x 3 x 4.8 mm, and weighing collectively less than a tenth of a gram. Magnification showed each of these tiny stones to be completely or nearly completely enclosed in fusion crust.

Our find occurred nearly thirty years after the Holbrook shower. Probably a stony mass of many tons, perhaps hundreds of tons, went to pieces in the stratosphere that July afternoon of 1912, most of it being reduced to dust. An investigation made immediately after such a fall could shed important light on the nature and behavior of dust clouds. Unfortunately, such a showering is a rare occurrence, and any study may be hindered by factors of location, such as a fall in a heavily vegetated or sparsely inhabited area with few witnesses. Additionally, wind currents may drift the fine particles in any direction and over many miles.

I could see in my imagination the disintegration of such a mass, possibly the size of a moving van, as it plunges into the earth's atmosphere at a speed of 90,000 miles an hour (25 miles per second). By comparison, a high-speed bullet would be practically standing still. Striking the atmospheric blanket would be comparable to collision with a granite mountainside. Only the atmosphere would save the earth from a great hole blasted by such a meteorite in its transformation on impact into rock vapors and dust. The atmosphere over our heads is an effective shelter, equivalent to four-and-a-half feet of steel or fourteen feet of solid granite, against penetration by meteoritic visitors. Only a few are large enough to survive passage through this barrier, and these in most instances are slowed to the speed of an ordinary bullet and are generally pulverized in the process except for a few fragments that escape.

One can read the story of the stratospheric smash-up above Holbrook in the surviving fragments. Each tells its story in the black coating of slag that resulted from brief exposure to the enormous temperatures developed when the parent mass bored its way through the atmosphere as a cutting torch burns through steel.

12

Prospects and Prospectors

*Opinion and force belong to different elements.
To think that you are able by social disapproval
or other coercive means to crush a man's opinion,
is as one who fires off a blunderbuss to put out a star.*

—John Morley

During the winter of 1940 Addie and I took our house-trailer into the Southwest for some lectures and some follow-up field work. We parked for a night in the town of Wickenburg, Arizona, and Addie settled down with some reading material and her knitting.

"No telling when we shall be here again," I remarked. "I think I will scout around, see if I can locate some prospectors. We might find a new meteorite."

There was no lead in our files suggesting a meteorite at Wickenburg, but this was in our favor. Since no meteorite had ever been reported from this part of Arizona, there might be one to be discovered.

I wandered about the town hang-outs—the drugstore, a bar, a restaurant—asking about prospectors. Twice I was told of a Mr. Kellis, who no longer was prospecting but who for forty years had explored the surrounding hills. I asked directions to his house. After climbing a dozen blocks of dark streets, I reached the white house on the hill where he lived in retirement. A large pile of rocks marked the front yard. It was about nine o'clock in the evening, but my knock brought Mr. Kellis to the door, courteous, white-bearded.

When I identified myself and my mission, he was quick to tell me I was looking in the wrong area, that I should visit the meteorite crater near Winslow to find what I was seeking.

No, I told him, presently I was more interested in stony meteorites, not Canyon Diablo. I showed him a small specimen from my bag.

"You call that a meteorite? You must be mistaken. Meteorites are made of nickel and iron."

A good deal of persuasion was required to convince him the stone I showed him was a meteorite.

Finally, he shrugged. "Well, you *should* know your business, and if this is a meteorite, I may have one out here in my rock pile."

He brought a flashlight and we went out to his pyramid of rocks. There must have been two tons of stones, and among them, plain in the light, was a twenty-pound meteorite that had lain there for thirty years. Mr. Kellis recalled where he had found it, but he never found another there, nor did we. His was the first stony meteorite to be recorded in the state of Arizona, except from the widely known shower of 1912 at Holbrook, in more than thirty years.

Our luck went up and down. For every exhilarating experience of discovery, there were many disappointments. There would be months on end during which we worked just as hard, borrowing and spending, without receiving even a sliver of meteorite in compensation. After many years of chasing fireballs, I finally accepted recovery

340 IN SEARCH OF FALLING STARS

of one meteorite from a half dozen surveys as a good average.

I remember struggling for hours in the Sand Hills of western Nebraska, checking out just one among many reports in an effort to map the course of a fireball. The car sank belly-deep in blown sand on a lonely road, many miles from any source of supplies and far from hope of help. Night was approaching, and I was beginning to expect to sleep with sand for a blanket. Finally, I foraged weeds at some distance from the car to lay against the wheels and then, by jacking the wheels one by one and by various other contrivances produced enough traction to get under way just as dark closed in. No meteorite was ever recovered from that particular fall, nor from a majority of all those that were mapped over the years.

The leads that sounded best might bring the deepest disappointments. Some seemed so certain that it appeared they could not possibly fail to prove good. Every once in a while, a tale would prove genuine and traceable and so pay off for some of the false leads. Some of those old stories still haunt me—some of them sounded so good, in spite of investigations that produced nothing, that it is hard not to believe they might have had a core of substance.

On our first visit to Wickenburg in 1926, then just a village, we met an old prospector with an interesting tale. When I asked him about meteorites, he knew what I was talking about.

"I've got three thousand pounds of 'em. Gathered them myself around the crater."

Could I see them?

"No. They're buried."

Would he sell them?

"Yes, if you'll give enough. Want $3,000."

Would he sell part of them?

"No. Won't show anyone where they're buried until I can make a deal for all of them at a dollar a pound."

He was adamant. "If I don't sell them, the secret will die with me."

He said only that the cache was "not too far" from the crater. On later visits to Wickenburg, I never found the old man again.

Was his story true? The tale fit the nature of a man old in the ways of prospecting and experienced in the disappointments and waits and secrets of that manner of life. I'm sure he is no longer alive and also that he never disposed of his meteorites. I haven't the slightest idea, however, as to where they were buried.

There were so many tantalizing reports. One of the most intriguing and convincing was a tale told to me by a man in Prescott, Arizona. I had gone into a bar on one of my customary tours of community hangouts seeking people who might have stories to tell. I laid a nickel-iron meteorite on the bar beside a man who was drinking a bottle of beer. He looked it over carefully.

"Are those worth anything?"

I told him they were. He stood, looking past me for a moment, as if gathering details out of his memory. Then he pointed to the cigarette vending machine.

"I found one as big as that machine one time, but it's been fifteen years." He stroked the little specimen before him on the counter. "It was iron just like that, and had dents in it just like that, and when I hit it with my hammer, it sounded just like an anvil. I know it was the very same thing."

He had been a surveyor at the time and was running a line for the government. He told me what line it was. They came to a ravine they couldn't cross, and he had gone south about a quarter mile to where it could be negotiated. There in the ditch was this great hunk of iron.

Neither he nor I had the time nor money to make a ten-day trip to the spot at the time, and I never was able to contact him again.

Somehow I lost the notes I scribbled down, including the man's name. I had fully intended to follow up this lead, having him guide me to the spot. The territory in which he was working is seldom visited by anyone except deer hunters, and it may be a hundred years before another man sees that great iron—if it really exists.

Of all the hundreds of reports I investigated in my years of meteorite hunting, more than a hundred yielded meteorites, while the other hundreds were all duds. Few of the productive tales were more convincing than this man's report. It is my belief he had found Arizona's finest meteorite specimen. Whether it could be found again is the question.

***

The "Danforth puzzle" was another of the teasing, unproductive reports that sounded almost too good to be true, but on the other hand, also too good not to be true.

An enormous fireball was witnessed over the great Maine wilderness. The next day, two men out deer hunting were trudging through a cover of snow over a swampy, timbered area when they came upon a strange scene. The trees on all sides of them were spattered with mud, and right in front of them was a big hole in the forest floor revealing an underground lake. This hole was nearly round, about nine feet across, and its edges were clean-cut, almost as if cut with a saw. In this swamp, the forest floor was simply a tangle of roots in a foot-thick blanket of sphagnum moss that covered a former lake which still persisted under the forest cover. Such conditions were common in that part of Maine—but the hole! Such a hole was something new to those hunters, and it proved no less a puzzle to all who heard their description of it.

So ran the story that was sent to me by George Sprague, who

had read one of the magazine stories about my meteorite-hunting and had a hunch that I might have the answer. His letter intrigued me as much as the tale had interested him. I wrote back for more details. He explained that he and his brothers had visited the spot and found things almost exactly as the hunters had described. They had probed into the pool and had reached a muddy bottom only seven feet below the surface. No scattered roots were found—only mud. This had not been an explosion of gas from below; it must have been caused by an object speeding downward from above. The only possible answer they could muster was that the hole had been created by a large meteorite. If I would come, Sprague wrote, he would guide me to the spot.

Naturally, I would come. Where I would obtain money for the undertaking was one of the unknowns in the problem, but the answer was yes, I would come. Art Thompson came forward this time, offering to take the gamble and eager to go with me to Maine for this adventure. We were met at the town of Lincoln and were driven in an aging Ford to within a three-hour walk of our destination.

As we approached the spot, shrubs and small trees bowed to our passing, and we found that by standing in one spot and alternately squatting and straightening, we could soon have the trees on all sides curtseying gracefully. We were walking over the once-open small lake that gradually had succumbed to the encroachment of mosses and ferns and finally had come to support a young forest.

In the midst of this strange floating forest, we faced a clean-cut opening to the lake below, cut right through the forest floor. Although it seemed so clean-cut, close examination of the exposed root-ends showed that no cutting instrument had been used. All the roots had been broken as if by a violent blow. There was no evidence of any lumbering, wood-cutting, or other human activity in the vicinity or, according to our guide, for miles around.

A few years had elapsed since the hunters had first found this hole; consequently, all mud had been washed from the tree trunks, if they had once been spattered as reported, but there still was evidence of violence dating back to the time of the hunters' report. The top of a sapling near the hole had been cut off, and the annual rings in the healing growth corresponded to the time of the first report. Also, bark had been knocked from the trunk of another tree a few feet from the hole, yet we found no scattered roots such as one would expect to find if any had been thrown out by the force that had cut the hole.

We prodded the pool and found muddy bottom overlying gravel at a depth of seven feet. A drill bit brought up nothing but granitic gravel when it refused to go deeper. Our mine detector failed to register anything, even though Art, an adept operator, explored all within a radius of some fifty yards. A search for fragments of any odd-looking stony matter was also in vain. We returned baffled and by no means satisfied.

Thompson and I made this initial trip in the late 1930s. I was able to secure funds to employ a magnetometer operator to go with me on a second trip to the spot. Our results were no better than the first time. The same was true of other efforts over a decade. Two possible explanations came to mind: 1) A stony meteorite of an extremely fragile and friable variety may have perforated the forest floor and skidded some distance on the muddy lake bottom. Such a mass would not respond to the instruments that we used. 2) A very unsupported, but possible, hypothesis is that an ice meteorite (a variety not yet proven to exist) had cut the hole and left no other trace of itself.

It was suggested that perhaps a bomb or a box of explosives had been dropped from a military plane from a military installation not too far away, but this idea seems a bit far-fetched. Besides, the evidence appears to point to a cutting agent too large to fit such a solution. Also, our careful search for indications of anything associated with explosive devices proved fruitless.

2

Accompanying many and probably all falls of ponderable stones and irons, there is a far greater weight of finely divided material (of dust, sand, and gravel sizes) that is showered over several square miles. What we collect as meteorites represents merely a small remnant of that which entered the atmosphere. Under the stress of impacting the atmosphere, a meteorite disintegrates, and only the fragments that accidentally escape the general disintegration of the mass reach the ground as stones or irons.

Efforts have been made intermittently for more than a century to collect meteoritic or cosmic dust. Two sources—sea-bottom oozes and arctic snows—were searched for matter of such origin. I tried several methods. Finally, in September 1938, on the assumption that considerable portions of meteoritic matter would be magnetic, I utilized an alnico magnet wrapped in rubber (to prevent nickel contamination from the magnet) to collect material at the exits of roof downspouts. These proved to be a fruitful source of a heavy, slate-gray dust which gave a positive nickel reaction. Microscopically, the material was seen to be composed of irregular particles among which were multitudes of minute spherical globules, many of them highly magnetic.

To remove our experiments from city contaminants, we traveled to a remote mountain top and anchored a vessel to a mast made fast to the top of a tree at the 11,000-foot summit. This was left in place seven months. We collected and melted ice from the surface of a mountain lake, searched the rough bark of trees by means of a magnetic device, and floated a magnet from a weather observation balloon for six hours in a twenty-mile-per-hour wind over a snow-clad mountain. For seven days, air was sucked over powerful magnets by an electric fan on the top of a high tower. By these means, the same kind of material was collected, but in immeasurably small quantities.

The collections from downspouts permitted a crude measurement of the rate at which the material accumulated as each rain shower or melting snowfall washed off previous accumulations. Thus, regular collections after each shower supplied quantitative data. The rate of fall indicated was on the order of several thousand grams per square mile annually, figures in strong contrast to previous estimates of an increment of only ten to one hundred grams annually on a square mile of surface.

If such quantities of meteoritic matter are arriving on Earth, why should the soil not be very rich in nickel and why should we not be knee-deep in meteoritic dust? Nickel leaches rapidly out of oxidized meteorites. The surface of the soil is so churned and altered by the ordinary forces of weathering that the most scrutinizing search would be necessary to find even a trace of a rain of spacial matter, even though in the course of a million years, it might amount to a layer several inches thick. A layer only the thickness of ordinary paper might accumulate in a period of a thousand years.

The study of several hundred meteorite falls and spectrographic studies of meteor trains have shown that the great bulk of this cosmic dust should be nonmagnetic, but no satisfactory identification method has been devised for the nonmagnetic portion of dust.

I was convinced that the comparatively crude methods I was using pointed the way to a study that could be of real value. I suggested such a controlled research project, on a scale of scientific significance, as part of detailed plans that I formulated proposing the establishment of an institution devoted to meteoritical studies—the long-time dream which seemed then about to become a reality, but that very shortly was to fade.

Almost since we had first become acquainted through correspondence in 1930, Frederick Leonard had been pushing for the establishment of a meteoritical organization. My early reaction was that I favored the idea but was opposed to taking any official part. My entire time was being devoted to meteoritical problems, and I feared that participation in administration of a formal society would only interfere, so far as I was concerned, with that which it proposed to promote. When we met for the first time in July of 1932, Leonard pursued the matter again, and on September 2, 1932, I wrote Leonard I was willing to help him set up an organization.

An organizational meeting was announced for August 1933, at the Field Museum of Natural History in Chicago. A small but eager group attended and established the Society for Research on Meteorites (renamed The Meteoritical Society in 1946). Leonard was made first president of the group, and I became secretary-treasurer.

The organization enjoyed a rather healthy growth. At the end of the first four-year term of office, I was elevated to the presidency. By the end of my service in that post, the membership had grown to something like 160, with several countries represented.

In 1935, I taught a night course on the downtown campus of the University of Denver, the first college course in meteoritics offered by any institution. The enrollment was small, but the response was enthusiastic.

In 1941, in my report as retiring president of the Society for Research on Meteorites, I was able to say that such courses were being offered in at least two American universities, that more than a score of new institutions had been added to the rather meager list of those possessing worthy study collections of meteorites, and that for the first time, a textbook on geology had appeared wherein several pages were devoted to the subject of meteorites.

It seemed there was growing awareness of the importance of meteorites, and there were growing numbers of individuals interested in devoting time to promoting research and study of meteorites. My

348 IN SEARCH OF FALLING STARS

hopes grew that there might be a way for me to channel my own knowledge and interest in the field in some way without having to battle for mere survival. Absorbed as I was in recovery and study of meteorites, I tried ceaselessly to interest individuals and institutions in establishing the kinds of investigations and searches and research that were beyond my personal capacity to undertake.

In 1933, at the organizational meeting of the Society for Research on Meteorites, I set forth "A Suggested Program of Research on Meteorites." An outline of this address, carried in the November 1933 issue of *Popular Astronomy*[10] includes as Steps 5 and 6 the following:

5. Relation of meteorites to earth-growth, both past and present, will prove a very important and fruitful field:

(a) As evidenced in traces of huge crater-producing meteorites;

(b) As evidenced by traces of meteorites in various geological formations, when we have learned to recognize them;

(c) As evidenced by the possible causal relation between certain major topographical features of the earth and the impact of meteorites of asteroidal dimensions.

6. The possible relation of huge impacts to the glacial periods and the shifting of the poles of the earth.

In a lengthy paper presented before the American Association for the Advancement of Science in Minneapolis, June 25, 1935, I visualized a comprehensive program of recovery and study. I proposed a

---

[10] "A Suggested Program of Research on Meteorites," *Popular Astronomy*, Vol. XLI, No. 9, Nov., 1933, p. 521.

National Institute of Meteoritical Research and recommended a country-wide alerting system.

Photographs are of inestimable value in the work of locating and determining the magnitude of falls. The institution would have to maintain in operation at all times several batteries of cameras, equipped to operate automatically by photo-electric cells, trapping the record of any fireball of large magnitude which falls within its range of vision. These batteries would be operated in connection with various selected institutions spaced approximately 300 to 400 miles apart throughout the country. Each battery would be composed of cameras pointing to the various sections of the sky so that in case of fireballs of the magnitude which deliver meteorites they should be definitely recorded by two or more stations photographically.

In cases where the meteor, or fireball, leaves a persistent train, or cloud, these automatic batteries would be supplemented by operators with ordinary cameras. Photographs of these clouds are notably scarce in our records. Yet a careful check over only a part of the United States during the last several years has demonstrated that it will be possible each year to secure photographs of several persistent clouds left in the wake of falling meteorites. The significance of these clouds must be a very important matter, yet almost nothing is known about them at the present time, either as to the process of formation or the material of which they are composed. Nothing whatever is known concerning the mass of meteoritic material which they represent...

When the approximate location of a meteorite fall has been determined, there should be an immediate visit to the locality by a designated investigator. By telephone, the local newspaper, and personal canvass, he could inform residents what to look for, its scientific importance, and its preservation ... This investigator should remain in the vicinity until the collection of all the immediately available material has been completed.[11]

At the close of a lecture I gave in the Adler Planetarium in Chicago early in 1936, Dr. Forrest Ray Moulton came to the platform and introduced himself. I was a great admirer of Moulton, who was one of the giants of American science during the first half of the twentieth century. With T. C. Chamberlin, he was co-author of the planetesimal hypothesis of the origin of the earth that dominated much of the thinking in the fields of astronomy and geology for several decades after the turn of the century.

He was most complimentary, and we chatted for some time in the emptying auditorium and then went on to his home. He expressed great interest in the work I was doing and asked some questions as to how it was being financed. When I described my program of searching for meteorites, the selling of part of them, and my lecturing as a means of both aiding the search and helping to pay for it, he wondered aloud why my home city of Denver had not come forward with some manner of help. This I could not answer, but I had to admit the reason might be because I had not asked for anything.

Dr. Moulton then suggested that he would like to come out to Denver and approach such organizations as the Chamber of Commerce, the Colorado Museum of Natural History, and the University of Denver on my behalf. I was surprised and heartened by this unexpected offer of help from a man who was a member of the National Academy of Science, the American Philosophical Society, the National Research Council, and numerous other important scientific

---

[11] "Proposed National Institute of Meteoritical Research," *The Pan-American Geologist*, Vol. LXIV, Sept., 1935, pp. 107-124.

organizations, not to mention the author of several books, the recognized authority on ballistics and celestial mechanics, and one of the leading mathematicians in the nation.

In December of that year, Dr. Moulton came to Denver to undertake advance planning for the meeting of the American Association for the Advancement of Science to be held in Denver in June 1937. During this visit, he studied the collection of meteorites at the Colorado Museum of Natural History and addressed a dinner gathering of city officials and community leaders arranged by Dean Gillespie. Dr. Moulton announced to the guests and newsmen that Denver was being considered by the National Academy of Sciences as a possible headquarters for a national institute for study of meteorites. He described Denver as an ideal site for such a center and specifically praised the meteoritical research being done by me and the museum, suggesting also the possibility of a grant to me for some immediate research as a preliminary step toward large-scale operations from Denver.

Dr. Moulton not only lent his great prestige to my work and focused the hopes I had held so long for a more solid and formal program in meteoritics, he also made an immediate and positive contribution to my efforts by using his good offices to secure a grant for additional work at the Haviland crater in Kansas. It was soon apparent, however, that the formation of a center by the National Academy as he envisioned it would not come to pass.

In 1938, a committee was organized in Denver, led by Frank Cross and the Chamber of Commerce, to propose the American Foundation for Meteorite Research, which was to be supported by sustaining and contributing memberships. This proposal received printed praise and good wishes from scientific colleagues around the country, but this also faded. Then, in 1939–40, hope brightened with what looked like a very real opportunity for foundational support, but this, too, proved to be just a flash in the pan.

While I was in the east on a lecture tour in the fall of 1939, word reached me that the late Spencer Penrose, widely known philanthropist, financier, and mining man of Colorado, had designated that his estate of several millions of dollars be used for support of charitable, educational, and scientific purposes within the state of Colorado.

Many scientists throughout the country seemed to see this as my great opportunity. Twenty-three letters from eminent geologists and astronomers who shared my desire to see a vigorous and productive program instituted in meteoritics were gathered for submission to the Penrose *El Pomar* Foundation. Again, Dean Gillespie helped. All the old hopes and plans were dusted off, new ones were added, and the lot was bound together, complete with blueprints donated by a noted Denver architect, for a proposed Institute of Meteoritical Research. The package was transmitted by the Denver Chamber of Commerce to the Foundation.

The first two or three interviews with spokesmen of the *El Pomar* Foundation were most encouraging. Then, armed with my prospectus, I met with the Foundation's board of directors. They heard me favorably, but because of the absence of one of their members, they postponed definite action until a later meeting, to which they asked me to bring more detailed budget information. The next word from the board was that they had decided against any further consideration of the proposal for a meteoritical institute.

Plans for the Institute had included detailed proposals for field work, as well as for an educational program, research fellowships, cooperation with other interested agencies and individuals throughout the nation and overseas, exhibition of meteorites, publication of scientific papers and a monthly bulletin. The director of the Institute would have received a salary of $5,000 a year. The Nininger Collection was to have been donated to the Institute.

The extensive field program envisioned in 1939 was to provide for study of the relationships of meteoric phenomena with long-term weather forecasting, solar radiation and conditions of the upper atmosphere, study of the chemical composition of meteorites with respect to such problems as rust resistance of metals and formulae for steel alloys, the exhaustive study of the non-terrestrial minerals found in meteorites, the investigation of the explosive disintegration which marks the luminous flights of meteorites, and the detailed study of heat effects registered in the fragments recovered, to throw new light on the problems of air resistance to high velocities and of stratospheric transportation. The program was also slated to study the forms and markings of meteorites as an aid to high-velocity ballistics engineering, the possible relationships with cosmic rays, meteoritic dust and any effects it might have on the growth of vegetation and the healthy development of animal life, and the possible existence of "ethereal sound" in connection with the flights of great meteors.

The recovery program included extensive plans for the alerting of and cooperative efforts among a network of institutions and agencies so there might be prompt and efficient action on appearance of a fireball, not only to plot the probable locality of fall and alert both trained staff people and the general populace for the recovery of specimens, but to study the light phenomenon and its accompanying dust cloud and sound waves.

Such a comprehensive program was not to be in the 1940s, nor has such an extensive and coordinated plan ever been put into effect.

3

From the steps of the Denver Museum of Natural History, one can enjoy the finest landscape view of any museum in America, looking across to the west where the gigantic peaks of the Rockies rise in a succession of billowy folds of crustal rocks which have lain exposed to the forces of weathering through perhaps a hundred million years. In that amount of time, how often have meteorites fallen on those mountain peaks?

Perhaps one fall a year has landed within the area under observation. If these falls produced on average as large a number of fragments as have the falls recorded in man's experience, then a tremendous number of meteorites have landed in those mountains. If so, where are they now? Only five falls have been recovered out of the area, and of these, one was witnessed.

One must allow that most of the meteorites buried themselves so deeply at the time of the fall that if they were stones, they would have for the most part disintegrated before the process of erosion would reveal them. Or, if instead of striking soil, which would accommodate their internment, they struck solid rock, then they must have been shattered so badly that they would escape the notice of any passerby and within a few years would weather so as to be rendered indistinguishable from ordinary rocks. Then, too, perhaps less than one percent of the surface in the region is trodden by man in the course of a generation.

No meteorite whose fall occurred more than a few centuries ago would be recognizable now unless it were of the iron variety. On this point, there is interesting conjecture: Although man's experience during the past convinces us that about twenty-five stony meteorites fall for every iron, Western Colorado has yielded four iron meteorites representing as many falls and only one stony fall, the Johnstown shower of July 6, 1924.

It might well be assumed that mixed up with the sand, silt, and gravel swept out of the descending canyons of the Rocky Mountains and added to the soils of the Mississippi Valley over millions of years are the remains of millions of meteorites about which nothing more definite can ever be known.

If there indeed are meteorites in old formations, what are some of the reasons why such remnants go unrecognized, and in what ways might they be identified and recorded?

Various attempts have been made to explain this assumed "absence of meteorites from all but the most recent formations," one hypothesis advanced by several writers being predicated on assigning a comparatively late origin for these small additions to our planet. To account for such a brief and recent accretion, some radical changes in the history of the solar system have been suggested.

It is my belief that in the present state of our knowledge (or lack of it), we are not justified in considering the "absence of meteorites from the older sediments" a problem at all, for the very simple reason that we have no credible evidence that meteorites in a terrestrialized form are not present in all of the sediments. The oxidized meteorites I took from the Haviland, Kansas, crater were so altered as to be mistaken for ordinary iron concretions.

If the failure of geologists to find meteorites in old formations is taken as evidence that they are absent, then by the same token, we could have known in 1930 that there were no meteorites in or upon the soil of a 25,000-square-mile area of Texas, an area certainly far better known to geologists and to laymen than any pre-Pleistocene deposit of similar area anywhere in the world. Yet, beginning in 1933, my own efforts recovered from that Texas area within ten years more than a thousand meteorites, all of which evidently had been on the earth a long time.

A few months in the field observing mining operations, miners, road-building operations, and other excavation activities and talking to superintendents, foremen, and both skilled and unskilled operators, would firmly convince the observer that even though thousands of meteorites had been moved by such operations, the chances of a single one being found are very slim indeed.

No adequate search has ever been made to determine whether meteorites are or are not present in any of the older formations. Indeed, a search sufficient to warrant an answer to the question would be rather difficult to make unless it should turn out that the older formations are considerably richer in meteorites than are the recent alluvia and other top sediments. In the first place, only a very small portion of any of those formations is accessible, and in the second place, the geological profession has never made such a search possible by training its personnel in the art of identifying even fresh meteorites, to say nothing of those that have undergone long weathering.

The cretaceous chalk beds of western Kansas afford a very extensive exposure of an ancient formation, one that is easily accessible, and that has long been a favorite hunting ground for paleontologists. The chalk is of a color which contrasts strongly with that of meteorites as we know them. During the past eighty years, these beds have been visited by probably several thousand geologists, including students. I interviewed many of these geologists during my fourteen years of residence in Kansas and after. By their own admission, less than two percent had ever thought of looking for meteorites during field trips, and an equally small percentage felt that they would have recognized a meteorite had they seen one.

The likelihood of a meteorite being recognized in such a situation is lessened greatly by the fact that no one knows exactly what changes such material would undergo during 60 million or 100 million years. From observations on those that are known to have lain in the soil a mere twenty to fifty years, we may surmise that a 60-million-year-old specimen might appear as a rather ordinary concretion, if indeed it were distinguishable at all from its surroundings.

Why should meteorites not be preserved as well as organic remains which often are recognizable in the very oldest sediments?

Organic fossils are identifiable due to their form rather than their substance, and since meteorites have no distinctive form, any alteration in their composition and structure renders them less recognizable. Most petrified woods would not be easily recognizable except for the traces of cell structure or the forms of logs or twigs, and in many cases, all trace of the cell structure has disappeared.

The moon is covered with thousands of craters which astronomers now generally accept as produced by meteorite impacts. These appear to be of all ages, some being fresh-looking and others so old that only traces of their battered forms can be detected. One wonders if our weather-troubled planet may have been equally battered and if the evidence has been erased by erosion.

Naturally, if crater-forming giant meteorites have been assaulting our Earth, then others of small size also must have been pelting the terrestrial skin. We wonder again not *why* they have *not* been found but whether they *may have been* and have gone unrecognized.

Meteorites of even moderate age seldom are recognized by geologists and paleontologists who are the most logical individuals to be expected to report them in old sediments. Furthermore, the area of outcrops wherein fossil meteorites might be exposed is so infinitesimally small as compared to the soil-covered areas in which cultivation is likely to expose them that they could not with certainty be regarded as absent even had geologists been instructed in the art of identification and encouraged to seek meteorites as they hunted for rare minerals and fossils.

The record of billions of tons of coal having been mined without bringing to light a single authentic meteorite has been cited as evidence that meteorites are absent from the carboniferous formations. Even well-preserved meteorites often are difficult to recognize when plowed up in ordinary soil until they have been well washed, and because of the way coal is mined and sorted, it would be unlikely that even a fairly fresh meteorite could be singled out from the lumps of sulfide, concretions, slate, and other waste that plagues the industry. Fairly fresh meteorites, of course, are not going to be present. Even after the most thorough scrubbing, even an experienced meteoriticist would be hard-pressed to recognize a meteorite after it had been subjected to the same metamorphic agencies that change vegetation into coal.

If farmers working their fields by daylight fail to find meteorites until especially alerted, how could miners working underground or hurriedly sorting out waste be expected to distinguish between concretions and old, altered meteorites?

Geologists, paleontologists, even entomologists, could add to our knowledge of meteorites and to our representative collections of meteorites during the regular course of their own specialized field work.

Dr. George Sternberg of Fort Hays State College in Kansas had been collecting fossils for thirty years when I asked if he had ever found a meteorite. He said that he had not. I coached him on the recognition of meteorites, and within two years, he had found two meteorites while fossil hunting.

Amateur rockhounds, perhaps, are as good a source as any for problematical stones that may, when we know more about what forms to expect, turn out to be fossilized or altered meteorites. Very few of the old formations are accessible to view, but these are the canyon walls or the cutbanks of streams and roads which rockhounds frequent.

The great majority of the meteorites that are recognizable on our planet may logically be assumed to be in a state of nearly complete terrestrialization. Two of the very few completely oxidized siderites recorded on the entire planet have been found in the state of Kansas, an area of less than one-sixth of one percent of all land on Earth—a strong argument for geologists to be on the alert in other areas. An important problem that faces the geologist is the development of means to trace the final stages of terrestrialization so as to make possible the positive identification of the remains of meteorites.

***

Throughout geological literature, the recurrence of geological revolutions—periods of world-wide readjustments in the earth's crust—has been recognized. The fact of such great changes seems to be well-established, but there has been no adequate explanation of them.

The little planetoid Hermes missed our planet by a few hundred thousand miles in 1937. In May 1942, in a brief article, "Cataclysm and Evolution," published in *Popular Astronomy*,[12] I proposed that perhaps in the past such encounters had not always been near misses. I suggested that collisions with planetoids might explain geological revolutions:

> *It is not at all improbable that the Earth bears many scars of far greater dimensions than the largest known meteorite craters. Ten chances to one, these scars are concealed by water, jungle, desert, or arctic wastes, so that civilized man has never come upon them... There are of course a thousand chances to one that all of the encounters occurred too long ago for the scars to be preserved down to date.*
>
> *It seems that we have here an adequate explanation of those successive revolutionary movements in the Earth's crust that are so generally recognized as having taken place, and also of the*

---

[12] "Cataclysm and Evolution," *Popular Astronomy*, Vol. L, No. 5, May, 1942, pp. 270-272.

sudden blotting out of the fauna and flora of certain great areas which the fossil records suggest... If the dimensions of the lunar craters are to be taken as any indication of the sizes of the bodies that the Earth has encountered, then there must have occurred great changes in the shorelines, the elevation and depression of extensive areas, the disappearance of low-lying land masses, the creation of islands, the extension and withdrawal of seas, as well as widespread and protracted volcanism. Violent climatic changes would have resulted, locally at least, from the heat of the impacts and from changes in the content of the atmosphere. More general changes might have resulted from a possible shifting of the poles, in the cases of the largest impacts. These changes would have necessitated faunal and floral readjustments. Species would have disappeared and new ones would have developed to take their places. Changes in geographical range would have brought about new adaptations, and we should expect, in general, just those breaks in the series that are actually found in the rocks.

My brief paper stirred little attention. Over the course of the years, I made quite an effort, both in print and in conversation with scientists, to convince geologists and astronomers of the importance of meteorites to both sciences. Whether because I was ahead of my time, or because I didn't speak the language of mathematical formulae, general agreement with my view was a long time coming.

During the 1930s, I had done some experimenting with high-power rifle bullets in the formation of craters. To represent the stratified earth, I used alternate layers of sand and plaster of Paris. The contrasting colors and different textures enabled me to detect the nature of the results of a shot more readily. I gained some pretty clear notions of how Earth's crust must respond to a large impact.

PROSPECTS AND PROSPECTORS   361

I became convinced that all of the lunar features were the result, directly or indirectly, of meteoritic bombardment. This idea was mulled over in my mind for a decade before in March 1943, in a paper for *The Scientific Monthly* entitled "Meteorites and the Moon," I argued strongly for the meteoritic origin of all lunar features.[13] I also proposed that cometary encounters "may well be considered as having been responsible on the earth for the puzzling succession of geological revolutions."

I argued the proposition that large meteorites are a factor in dynamic geology more forcefully in "Geological Significance of Meteorites," published in the *American Journal of Science* in February 1948.[14] As originally drafted soon after the appearance of the "Cataclysm and Evolution" piece, it contained a much more elaborate treatment of the part cosmic collisions have played in the earth's development than eventually appeared in print. I had sent it in to the *Journal*, but it was returned on the grounds that I belabored geologists too much for conservatism. I filed it away, then after the epochal fall at Sikhote-Alin in Siberia in 1947, I resubmitted a modified version, and it was published by the very same journal that first rejected it. Before its appearance, Dr. Harlow Shapley, who I assumed must have read my manuscript, wrote to me to suggest that I read a recent paper by Dr. Reginald Daly on the Vredefort Structure of South Africa, published in *Journal of Geology*.[15] I sent for a copy. Dr. Daly described just such a feature as I had hypothesized as being responsible for small

---

[13] "Meteorites and the Moon," *The Scientific Monthly*, Vol. LVI, March, 1943, pp. 259-266.

[14] "Geological Significance of Meteorites," *American Journal of Science*, Vol. 246, No. 2, Feb., 1948, pp. 101-108.

[15] Daly, R. A., "The Vredefort Ring-Structure of South Africa," *Journal of Geology*, Vol. 55, No. 3, May, 1947, pp. 125-145.

362 IN SEARCH OF FALLING STARS

crustal movements in the process of mountain building. This was a major step in paving the way for an acceptance of the collision explanation of geological revolution. The notable field program of the Canadian Department of Mines and Special Surveys has provided more recent proof of the correctness of my contentions.

In my *American Journal of Science* article, I admitted to having often been regarded "as somewhat of a radical in respect to meteoritical questions," and added that in my "most reckless moments," I had not been prepared to recognize such a rate of encounter with large meteorites as was indicated by the two Siberian falls of 1908 and 1947. I considered the significance of large, stony masses that are disintegrated by the atmosphere and so do not form craters in most instances, as well as large-scale collisions:

> *For example, the Pasamonte, New Mexico, fall ... left a cloud of dust aggregating about 1,000 cubic miles which was formed at elevations ranging from 17 miles to about 48 miles. 1,000 cubic miles of air at 30 miles elevation represents a weight of about 3,000,000 tons. Only a few small stones from this fall are known to have reached the earth, but their texture is such that they may be rubbed away with the fingers.*
>
> *The mass which invaded the atmosphere on that occasion must have comprised many thousands of tons. And had it been one of the firm varieties of meteorites it would have doubtless formed a considerable crater. But 17 miles of atmosphere lay between the level of its disintegration and the soil of the earth. This tremendous shock-absorbing cushion robbed it of nearly all of its violence. ... There are very good reasons for believing that in space Pasamonte was a greater meteorite than that which the Russian scientists now report landing in southeastern Siberia and possibly also larger than the fall of 1908 which was estimated at*

40,000 tons.

Had the Pasamonte meteorite been of a metallic variety or even one of the more resistant stony varieties, it might have formed a sizable crater. As it was, all we can prove is that the massive cloud was formed in the upper atmosphere and that a few small stones were strewn over a strip of soil 28 miles in length. But almost certainly an enormous tonnage of minute particles were deposited over a much larger area.

The important point that we are trying to make is that here is a process, cosmo-geologic in nature, that must have played an important role in the history of our planet.

...it seems inevitable that large scale collisions have occurred throughout geologic time. The moon shows effects which can best be explained in terms of such collisions. On our planet where continents float on a more or less fluid sub-stratum, large scale collisions would logically initiate great continental movements such as are believed to have taken place during the various geologic revolutions.[16]

In 1949, in his book, *The Face of the Moon*, Dr. Ralph B. Baldwin suggested the magnitude of the effects on earth of impacts of bodies from space: "Written in the book of geology in still obscure characters are the records of hundreds of thousands of collisions of the earth and extra-terrestrial bodies."[17]

Although definition of the Vredefort Structure had a measurable effect on geologic thought, so far as I was able to see, my own two articles made no visible impression. I reemphasized the same points

---

[16] "Geological Significance of Meteorites," *American Journal of Science*, Vol. 246, No. 2, Feb., 1948, pp. 101-108.

[17] Baldwin, Ralph B., *The Face of the Moon*, University of Chicago Press, 1949, p. 113.

in a chapter on "Planetoidal Encounters" in my book, *Out of the Sky*, in 1952.[18]

The book *Target: Earth*,[19] by Allan O. Kelly and Frank Dachille, published in 1953, went all out for "collision geology," but it was not well received; indeed, it was simply ignored. I believed and still believe *Target: Earth* to be a great book that will mark the beginning of a new epoch in the study of geology. Today, the principal theme of the book is being argued in the best journals.

Our planet shows only a few recognizable scars to mark the points of our collision with masses too large for the atmosphere to effectively brake their speed, but we are beginning to learn more of such great pockmarks as space probes begin to photograph similar damage done to the moon and the planets.

---

[18] *Out of the Sky: An Introduction to Meteoritics*, University of Denver Press, 1952, pp. 274-279.

[19] Kelly, Allen C. and Frank Dachille, *Target Earth*, Carlsbad, CA, 1953.

13

Plateau

Wisdom is knowing what to do next.

—David Starr Jordan

The arrival of World War II meant an end to the kind of work I was doing, apart from my curatorship at the museum. Like everyone else, we lived on news, anxiety, and hope for an ending. Gasoline and tires were unavailable for anything but essential contribution to the war effort. Rationing ended the freedom of movement, the spur-of-the-moment investigations, and the extended travels on which my harvest of meteorites mostly had depended. For a time, I worked as a salvage investigator for the War Production Board, combing the backcountry for salvageable metals in the old mining towns, along the abandoned rail beds, and where machinery lay disused in farm lots.

In the fall of 1942, our youngest daughter left for college. We smiled and waved and made cheerful signs through the train window, but when the train pulled briskly away, very briskly indeed, we turned sober faces to one another. We walked along the platform, down the stairs, through the tunnel, and out to the street. We drove home with very little conversation to the emptiest house that we had known in nearly twenty-four years.

We had been waving one or two youngsters off to college each of the past six years. But before, there was always someone left to turn our attention to, provide for, and watch. Now even she had advanced to that same mature-ish going-away stage which brings such pride, such feeling that a parent is engaged in a vital program of producing. I recalled the going-away of our first; we felt so proud to have a boy of college age. For years, we had planned with him and looked forward to this milestone. On reaching home, we had realized with shock that we would not see much more of him, that this was the beginning of the parting of our ways. When our second left, that, too, was a new experience. When a daughter embarks on the sea of life, it is quite different from the going of a son. A father takes pride in feeling that his son has the strength and rugged courage to meet and overcome hardship. But a daughter, that's different. We feel the need to hold out a sheltering hand. We're a little less ready to turn her loose. With the last one's train speeding over the rails, we again faced life alone. Memory then reminded us that we had passed this way before, and it hadn't been bad at all. Behind, there were years of experience to guide us, and ahead lay the shadows of three new cottages along the trail.

I managed to enjoy scouring the mountains for old, disused railroad tracks, or swinging across some once-rich canyon in an old ore bucket to see what piles of rusting iron might remain at a broken shaft or mill. Later, I found challenge in oil exploration, working in and out of Denver and Albuquerque, going on into the developing southern New Mexico fields to assess possible locations and negotiate for leases. With our youngest away in college, Addie joined me, and we set up a temporary home in Artesia, New Mexico.

Although the work was full-time, this meant usually an eight-hour day or forty-hour week. This left a good deal of time for off-hours meteorite work, because I had always been accustomed to working sixty to seventy hours a week when self-employed. I used every small opportunity that presented itself to continue what field work I could in connection with meteorites.

It was my habit when traveling in the arid southwest to carry fruit and sandwiches, or cheese and crackers, for lunch along the way. I would take food in hand and stroll about, scanning the ground for meteorites. For some twenty years, I had done this off and on without results, but I felt that every effort decreased the mathematical odds against my finding something.

On May 17, 1944, I stopped for a supper break at about 6 p.m., approximately forty-nine miles south of Albuquerque, New Mexico. The site was near Bernardo and was recognizable by the overhead structure of an old-style bridge over the Rio Puerco, where it was then crossed by a former roadway of US Highway No. 85, now modernized and re-routed. The area is wind-swept, sandy, flat, almost bare of vegetation, but sprinkled rather generously with water-worn stones and small, dark pebbles.

I had eaten, washed in the river, and was about to return to my car when my attention was attracted by the end of a small, dark pebble protruding through the sand—a pebble that was similar to, but perhaps not quite like the others scattered around that had been the cause of my luckless and almost constant stooping for the past thirty minutes.

I nearly passed this one up, but then I bent my protesting back once more. When I loosened the pebble from the soil, I saw at once that it was a small, stony meteorite—a new one for the record books, since there was no previous report of a find anywhere in that region.

It had a reasonably fresh appearance, the crust bearing some rust spots and stained by the soil to a dark brown rather than black appearance.

Finding of this little aerolite, named *Puente-Ladron*, the size of a small pecan and weighing less than eight grams, was one of the great thrills of my life. I had found other meteorites, but this was the first of only two I came across in the course of my habitual scouting in areas where no meteorites were previously known to have fallen. About ten years later, in a rock-strewn field near Cottonwood, Arizona, where I had walked out to examine some unusual piles of boulders, I picked up a two-pound stony meteorite, the shape of a blunt cucumber.

No further specimens have been found at either of these locations. Because I believe that no stony meteorites penetrate our atmosphere unbroken, that they all must shatter, I am sure other stones from the Puente-Ladron and Cottonwood falls remain to be found.

On a warm day in 1943, I became thirsty while driving near Wilmot, in Cowley County, Kansas, and as I passed a farmstead, I noticed between the house and the barn an old-fashioned hand pump with a tin can hanging from it.

Without bothering to go to the house, I went to the well and pumped a canful of water. I was enjoying the refreshment when I noticed a few steps away, beneath some plum bushes, a rusty-looking stone. I stepped over and picked it up. It was an old, weathered stony meteorite.

I approached the house and knocked. The man of the house was not at home, but I showed the stone to the lady who came to the door, asking if she knew anything about it. She appeared puzzled that I should be interested in such a homely object. I told her it was a meteorite and asked if perhaps her husband had spoken of it. She said that he had not, and she was sure he was not responsible for its being where I found it, for they had rented the place only recently.

I asked to buy the stone, but she insisted that I take it along if it was of any use to me. I gave her $5 for it.

In Bethune, Colorado, in 1941, I recognized three meteorites mounted on a cement wall in a rock garden, and I found the Abernathy, Texas, stone in a rock garden while searching for meteorites from the Plainview fall.

Other meteorite finds that I made personally were at such sites of prior discoveries as Beardsley and Haviland, Kansas; Plainview and Odessa, Texas; Holbrook and Canyon Diablo, Arizona; and Xiquipilco in Mexico.

2

Just prior to the outbreak of the war, a boyhood friend sent to me some meteorites from a fall in 1938 at Pantar, Lanao, the Philippines, which he had witnessed. He and his wife were then interned by the Japanese until early 1945.

Herbert J. Detrick and I grew up on farms not far apart in Oklahoma. Herbert graduated somewhat earlier than I from McPherson College, and had gone to teach in the Philippines in 1908. I followed his career over the years as it was detailed in the college alumni bulletin. He and his wife, Lula, both were successful teachers. Herbert was named supervising teacher and assistant to the governor of Nueva Vizcaya Province and became governor of the province of Palawan in 1915. Later, he managed a large plantation and engaged in mining and lumbering, and in 1939, he opened a hotel in Dansalan, Lanao, Mindanao Island.

On the sunny morning of June 16, 1938, the rice stood lush and green on the hills of Lanao, and the people of the province went about their usual business. Storekeepers opened their doors, housewives instructed their cooks for the marketing, students scurried to their classrooms and the native Moro farmers bent over their rice fields.

370 IN SEARCH OF FALLING STARS

Suddenly, there were strange explosive sounds from the sky, and then an object was observed for at least two minutes as it came from the east, emitting ringlets of smoke, and then ended its passage with a violence of explosions and vibrations and a spreading dark cloud that persisted for more than a half-hour.

Excellent photographs of the cloud were obtained. At the village of Pantar, the Moros in their rice fields saw more than the smoking object and the great cloud; they saw fiery objects with tails of smoke shooting out from the cloud and many small objects actually falling into the fields, from which sixteen meteorites were recovered later from depths of up to twenty inches. At the same time, a pattering like hail sounded on the galvanized-iron roofs of several houses in the neighborhood.

I read accounts in the news that Detrick was tracing down and collecting the meteorites of the Pantar fall. I wrote to him and was surprised to receive the reply, "Believe it or not, I've been collecting these stones for you." He, too, had been reading the alumni bulletin. I advised him on what data to gather along with the stones, and for a year or two, he worked in his spare time, surveying, drawing maps, collecting specimens. When finally he sent me his collection of thirteen stones not long before Pearl Harbor, he instructed that I do as I pleased with them, keeping what I wanted, selling others if I wished.

After Pearl Harbor, when it was evident there would be invasion of the Philippines, we decided simply to hold the Pantar meteorites in safe storage. We had no word of the Detricks' fate. When the Philippines were liberated by MacArthur, still we learned nothing except that people of the Pantar area had been imprisoned. Finally, in the summer of 1945, we learned the Detricks had been interned in Los Banos prison and had been freed in February 1945. They returned to the United States nearly broken, emaciated and destitute. We proceeded to sell their specimens to help tide them over until they received partial settlement for their holdings in the Islands. In appreciation, the Detricks added to our collection a representative from the Pantar fall.

3

The war over, Addie and I were truly on our own again. We had been married thirty years. From Las Cruces, New Mexico, I sent to Addie, home on a visit to Denver, an oil painting of a desert landscape. Its portrayal of an austere but glowing natural scene seemed custom-ordered for our anniversary. Pictured were three stout, blooming yuccas, victorious in the desert over drought and storm. Our thirty years contained a lot of stake-driving and rope-tightening to hold the tent down while we scratched in poorly watered soil, but despite the skimpy going, life had been rather colorful, brightened by beautiful sunsets against rugged, forbidding barriers. Our own three sprouts were grown, and they were on their own.

The close of World War II meant a time for decision. My job with the oil company was at an end. After an interruption of four-and-a-half years, we were ready to go back to meteorites full-time. Addie and I had made up our minds to leave the museum in Denver before I entered into wartime occupations, but our plans, like the individual hopes of everyone else, had been set aside until war's end.

J. D. Figgins, who as director of the museum had arranged the terms of my curatorship with the board of trustees, had a high interest in meteorites. Figgins had served as Admiral Peary's right-hand man on the three expeditions which recovered and transported the three great Greenland meteorites put on display at the Hayden Planetarium in New York. During Figgins' leadership, the museum participated in a number of my field projects. This arrangement, during a period when my field program brought forth more meteorites than were being found by all other institutions throughout the world, rewarded the museum with a rapid growth in its own collection. Dr. Figgins retired in 1936.

His successor was Dr. Alfred M. Bailey, whose scientific and field interests lay along other lines. Under the stewardship of Dr. Bailey, the Denver museum, already recognized as one of the finest and most beautiful in the country, expanded and improved. Its emphasis also shifted to reflect his interests as one of the leading bird men in the country and a respected collector of avian and mammalian specimens from many parts of the world. I was invited to stay on as curator, but otherwise, the museum was not interested in further participation in my field program.

Other than the support and assistance from the museum, our work had depended on a host of friends.

"If you ever get into a tight spot, come to me. I can always raise a little cash."

Many a man had said that to me and meant it. I never was a good beggar, however. In fact, I never at any time felt I could accept outright gifts, much less ask for them, unless they came from established foundations which existed for the purpose of supporting work like mine. And when it came to foundations, I was not very successful. Time and again, I had applied for grants with rare success, and never received more than very small sums.

I was fifty-nine years old. It seemed important that in our return to an independent course of action, we should be assured some continuity and security for our declining years without the constant harrying necessity of funding and re-funding field projects as they presented themselves. In the final years before the war, the cost of our program had grown to between $4,000 and $5,000 a year.

We were the owners of one of the world's greatest meteorite collections. It had been amassed with a single purpose in mind: the advancement of knowledge through research and education. We had never discussed the possibility of turning the collection, in its entirety, into cash, except that we viewed it as our "life insurance" against emergency and old age. But we were staunchly opposed to accepting the kind of poverty-ridden, handicapped existence we knew some scientists were forced to accept from institutions as recompense for various kinds of natural history collections and research.

We had observed that great institutions have a way of persuading, virtually coercing, men of achievement in the natural sciences to donate priceless collections for the "honor" of having bestowed them. Then the names of the donors were modestly displayed on labels of the exhibition cases to pass unnoticed by any but specialists of the same field, while the building which may house the life work of a dozen devoted scholars bears, emblazoned across its facade, the name of one who might have donated a small fraction of his income.

Finally, Addie and I believed, we hit upon a means by which our collection and our knowledge of meteorites, for which we had been unable to obtain any institutional support, would provide our living with enough surplus to continue a research program. We would create our own institution.

For more than a decade, we had thought longingly of living near the great Arizona meteorite crater. So we decided the time had come for such a move.

This greatest of known meteorite craters had been virtually untouched by research during its entire known history of more than half a century. At that time, it was an open range for desultory collecting. Scientists who visited it rarely did much toward a scientific interpretation of the crater beyond bringing away a small meteorite, generally purchased from a custodian on the rim who was permitted by the owners to charge a small fee to sightseers for a view of the pit.

Barringer and his associates, with commercial mining as their objective, had made extensive explorations during the years from 1909 to 1931. After the failure of the commercial venture, none of our nation's great research or educational institutions had made so much as a casual survey of the crater. The University of Arizona had never published more than a cursory description of the feature nor sponsored any kind of survey.

Attempts I had made in 1939 to measure and map the distribution of small metallic fragments on the surrounding plain had suggested that further research could produce significant results. One delegation headed by W. T. Whitney of Pomona College searched for meteorites shortly after our work ended, but nothing of importance was reported.

This state of affairs seemed an invitation to move our collection into the area, where it could serve the dual purpose of educating the traveling public and providing a base for further research.

It was not the garish signs and gaudy tourist traps of Route 66 that attracted us. Our interest lay in the proximity of Meteor Crater and its emphatic testimonial to the effects on this earth of stones from space. Our institution would be a Meteorite Museum. Nothing else. No snakes, no skeletons, no stuffed birds, no wild animals, and no curios—just meteorites.

It would be the first and only museum of its kind, and it would be run on a high, educational plane. Not only would we put our collection on public display as one of the greatest in the world and unique in some aspects, we would also present it as effectively as we could and accompany the exhibit with educational lectures that could not be duplicated by most geology or astronomy departments at any of our finest universities.

Addie and I decided to take our trailer to Flagstaff and spend the summer of 1946 there while we looked around for a location for our museum before actually moving the collection.

The crater, owned by the Barringers, rises from the midst of a cattle ranch owned by the Tremaine family. Interspersed among sections of land owned outright by the ranch are sections of state land on which the Tremaines have grazing rights. For years, the Barringer family had been friendly. The crater owners offered no financial assistance but assured us verbally of close cooperation. Freedom to search for and collect meteorites on the Barringer property had always been extended. When we presented our plan to locate near the crater, I again was invited to search and collect freely, though I was assured that almost all of the meteorites were gone by then.

"Keep whatever you find, unless something of ten pounds or more turns up, in which case, we should like half. We like to keep a few pieces around to show or give to friends, and we are getting pretty low."

In practice, we divided all specimens we recovered on a basis of weight, a fourth going to the Barringers.

By April of 1946, we had been in touch with the Tremaine group also, and had been given permission to undertake an extensive survey for which funds again had come from the American Philosophical Society, which had contributed $250 toward the $1,000 cost of the 1939 research.

We settled in our trailer under tall Ponderosa pines at the foot of Mt. Elden, east of Flagstaff. Our younger daughter, Margaret, graduated from college in June and decided to spend the summer with us. We had a grand time, frequently visiting the crater and searching for small meteorites under the terms of our written permission from the Barringers' Standard Iron Company and our verbal understanding with the Bar T Bar ranch. The search was tedious and not very rewarding at first, but gradually, we became more efficient. Margaret found the largest specimen we recovered that summer—a two-pound iron that she pulled excitedly from the soil by one protruding corner.

Any recall of that summer brings to my mind a picture of Margaret, on her knees, digging like a dog after a ground squirrel. She had called to us but couldn't take time to look up from her digging. She had found an iron of several ounces in a little ditch, and by scratching uncovered a larger one and then a third, as I remember. She had the best luck of the day, and of the season, for that matter. We became so enthralled with the search that we made many trips to the crater, a distance of forty miles. Each of us carried a magnet mounted on the end of a cane. By this means, we could test an object without stooping to pick it up. If it clung to the magnet, it was either oxide or a metallic meteorite. The oxide was meteoritic also, but had been altered through weathering. The oxide chips were many times as numerous as the metallic fragments or irons.

The Niningers' temporary quarters in a trailer below Mt. Elden, near Flagstaff, Arizona, during the summer of 1946 while scouting for a location for a meteorite museum. Addie is in the doorway.

All in all, we collected about ten pounds during the summer, mostly of the order of a tenth of an ounce in size. When I reported to Brandon Barringer, he merely shrugged.

"Keep it!" he told me.

We were maintaining records of the sizes and distribution of these small iron meteorites around the rim and extending plain. Later in the summer, Ab Whelan, who had performed magnetometer work for me in New Mexico, came from Artesia to conduct a magnetometer survey in the crater environs, a project that unfortunately was never completed and that gave no definite indications of remaining large masses under the soil.

Prior to the war, we had discussed with members of the Barringer family our hope of moving into the area and had talked of establishing a museum on the very rim of the crater on some kind of partnership arrangement. We could not seem to convince them that the plan was workable, but in the summer of 1946, when further discussions were undertaken with the Barringers, and with spokesmen for the Tremaines' Bar T Bar ranch, which surrounded the Barringer property, we were surprised to learn that their own planning had progressed to the point where architectural sketches had been prepared for a building of striking design, hanging over the rim and balancing inward over the crater bowl. Signs and circulars had been prepared.

Burton Tremaine and I laid our cards on the table. He warned that they had the capability of going into the business alone and driving us out since they had the power to outdo us in advertising and could set up a larger institution. I pointed out in turn that it would be impossible for them to obtain another collection like ours and that our museum would in no way compete with or obstruct their tourist trade.

From our point of view, the meeting ended on a note of encouragement. We felt that we were now in a position to go ahead independently or cooperatively, but we had full confidence that once having proved ourselves, we would be offered a permanent place in the plans for the crater rim. However, a divergence of views arose, and eventually we went our separate ways. The Barringers and Tremaines operated for a time much as in the past, but ultimately built a handsome structure at the crater, where their lure to tourists was a closeup view of the meteor crater. They hoped to derive income which would support further exploration for the great mass they still believed to lie beneath the earth.

By the end of the summer of 1946, we had leased a building on Route 66 west of Winslow, Arizona, and north of the Arizona meteorite crater. Here we set about establishing our Meteorite Museum, where our appeal was to people interested in learning about meteorites by examination of one of the finest collections in the world and by exposure to educational lectures and materials.

14

A New Lease

The best-laid schemes o' mice an' men
Gang aft agley,
An' lea'e us nought but grief an' pain,
For promis'd joy.

—Robert Burns

The building we leased on Route 66 was situated on a small bluff just off the highway, looking across to the crater from a distance of five-and-a-half miles. It constituted a sort of landmark in itself. It never was beautiful, at least not to my thinking, but it was an interesting structure, faced with natural flagstone, with a heavy square tower jutting against the sky. Remains of the building still stand, lonely and deserted, a watchtower of rock in a rocky desert.

Prior to our taking it over, the building was known commonly as "the Observatory." Tourists climbed its tower for a "view" of the crater—which could be viewed just as well from its front porch.

From neither point could one see *into* the pit. Curios were sold, and a dingy, crude, plaster and chicken-wire mock-up of the crater purported to bare the secrets of its creation and the cache of iron treasure supposedly in its depths.

Although I have heard others describe the building as beautiful, to me it was an architectural monstrosity, but there can be no question as to the beauty of its location on a low hill above a nearly level plain dotted with grotesque shapes formed by nature from the same red sandstone that provided its walls.

We could look west through large picture windows forty miles to the San Francisco Peaks, the highest point in Arizona, and a host of smaller purple hills and mountains. To the east, if we stepped outside or climbed the tower, we could gaze toward the volcanic plugs and colorful mesas of the Painted Desert thirty miles away. Just a few miles to the south rose the rugged rim of the mile-wide meteorite crater, described by one of the wisest men of Europe, the great Swedish chemist and physicist, Svante Arrhenius, on a visit in the early 1900's, as the most interesting spot on earth.

The fact of the crater's presence was our reason for choosing the location. Most probably there never would have been a meteorite museum had there not first existed the meteorite crater. We felt that we could do without police protection. We could do without neighbors. We could get along without society and entertainment. Our location pleased us but also left us no choice. We had to get along without many accepted features of civilization.

In about 1930 or 1932, when its situation was even more isolated, our building had been erected by the bare hands of one white man and the help of some local Indians. The floor was laid with flagstone and although it was a little rough, I always thought it beautiful and practical. The walls were thick, but the native stone had been laid originally with the red mud of the hill on which the building stood. When the first heavy rain came, much of the mud washed out and the whole structure threatened to fall apart, whereupon its builder recognized his error and attempted to fill the resultant voids with good cement. This rendered the walls substantial enough, but additional labors never succeeded in making the structure waterproof, as we were to see later.

We signed our lease, parked our trailer in the back yard and arranged for expansion of the plumbing facilities to accommodate visitors. The landlord dug a well and installed a water tank. We cleared everything out of the building, polished our picture windows and hung new venetian blinds across them. Then we climbed into our 1942 Chevrolet and headed for Denver to bring back our meteorites.

Our idea was big, our plans were bigger, and our risks were huge. As had been the case so often before, we were striking out into uncharted territory. We had not heard of a museum of meteorites anywhere, much less one located on a lonely highway hundreds of miles from any heavily populated center and twenty and forty miles from the two closest towns.

We received no great encouragement from any of our friends, though we were flooded with good wishes which might have been received more enthusiastically had they communicated conviction that our venture would succeed.

One friend of ours in particular, an attorney, baffled me with his parting comment. Respected, successful, and philanthropic in his interests, this man had helped me with free legal advice on occasion. Now he listened sympathetically to my plan and asked a number of pertinent questions. He offered assistance if it should be needed and said his final words in fatherly tones:

"Now, don't you let those people down there hurt you."

As I walked to the elevator, I pondered his words, repeated them to myself, trying to accept his warning as a joke.

382 IN SEARCH OF FALLING STARS

With or without bright forecasts of success, we had made the only choice we could see open to us. Once more, we broke all formal ties and put our reliance on ourselves.

Dr. Bailey expressed appreciation for my years of association with the Colorado Museum of Natural History and directed his staff to assist in the arduous task of moving and loading our several tons of meteorites. Since the Denver museum was installing a series of new exhibit cases in the mineral department, Dr. Bailey kindly offered us some of the old ones that were being discarded, and these massive but usable pieces went with us to Arizona.

The moving of a 16,000-pound collection of meteorites a distance of 750 miles is no small chore. The packing required weeks of heavy labor. When eventually we loaded our collection into the moving van, there were nearly two hundred boxes and crates, including two very heavy ones containing our two large stony meteorites, both of which were somewhat fragile. Hugoton weighed nearly 800 pounds before crating; Morland only a hundred pounds less. Then there were eighteen heavy iron specimens, weighing from 150 to more than 1,400 pounds. These were not crated and were in no danger of being damaged, but they constituted a real danger to everything else in the van in case of an accident, since any one of them, careening loose from a heavy jolt, could smash through any box or crate in the cargo.

Most important were the thousands of small specimens, some of which were worth several times their weight in gold and many of which were so fragile that we had wrapped each one and its label in several thicknesses of tissue or toweling or both, then again in a generous amount of newspaper before stowing them carefully in wooden boxes.

This cargo was the entire Nininger collection as of the year 1946. It had been gathered over a period of twenty-three years and had absorbed every bit of our life's earnings. Except for some elderly furniture, a middle-aged car, and our clothing, it was literally all we had. So now all our eggs were in one basket, a trucking company's moving van, headed for a crude stone building 750 miles away on Route 66, miles from city protections and services and without close neighbors.

To us, moving the collection appeared as risky as transporting diamonds or delivering bags of money. We had selected the transfer agent cautiously. The contract contained a clause requiring the cargo to be locked into the van in Denver, not to be opened until it reached the front door of our building on Route 66. Insurance posed a problem because our cargo failed to fit any of the usual classifications. Coverage to the extent of actual monetary value was prohibitive in cost, and the monetary value was the least consideration. More important was the irreplaceable nature of the material. Since we couldn't raise money to cover the premiums anyway, we paid our thousand-dollar shipping costs, settled for nominal insurance and simply accepted the risk. Company spokesmen extended impressive assurances of competence and safety, we signed the agreement and started our life estate on its way. The company's estimate was that the truck would reach its destination twenty-four hours after starting time, traveling day and night with two drivers.

We agreed to be on-hand to greet the van. We gave ourselves a head start and drove steadily in order to have everything ready for the unloading, intending to supervise the operation personally. The van did not arrive as expected at the end of the second day, nor did it arrive the third day. Anxiously, we waited through another morning, scrutinizing each truck that approached from the east, watching it travel the considerable distance that lay within our view and sighing with disappointment as each passed on without pause. In the afternoon, we decided to drive east to Winslow to seek word, not yet having a telephone.

384 IN SEARCH OF FALLING STARS

We knew that van loads were seldom lost in transit, and we hardly dared think of an accident, yet fear was building that something had happened. About halfway to town, a sight loomed up a mile ahead at which our hearts plummeted. At the side of the road was a wrecked van, headed in our direction, from which small, heavy boxes were being transferred to a second truck. The wrecked vehicle carried the symbol of our transfer firm. We bore down upon the prospect of disaster, our thoughts racing. Had we only invested in insurance! How much of our collection would be salvageable? What of the small, fragile, rare specimens—why had we not sent them by registered mail or by express? How much loss of labels and identification might there be?

Usually when some unexpected calamity strikes, my mind works at its best, but as we approached that mess in the road ahead of us, I felt a little sick. As we pulled up and stopped, a man on the ground handed up a broken box to the man in the relief van. I could see then that the box contained heavy nuts and bolts.

The drivers showed concern for our problem, but they had no information about our truck. In town, we could learn nothing. A call to the Denver office of the firm brought no satisfaction, either. There was nothing to do but return to our empty building and wait. All the next day we waited, almost in a panic. In mid-afternoon, a van mounted our hill and backed up to our door. It was not the same van we had seen loaded in Denver, nor was the driver the same. When the door was opened, my heart sank again. Crates were broken, boxes we had stacked so carefully in such precise order were helter-skelter. The great, fragile Hugoton stone was half out of its broken crate. Some boxes were open. It was evident that in the shift to another truck, things had not been handled carefully at all.

The driver had no explanation other than that all cargoes must be transferred in Albuquerque. He might as well have been hauling stove wood so far as he was concerned. He shrugged. He only "acted on orders." And his orders were to deliver that truck to Los Angeles by morning. "So let's get unloaded."

After I unloaded my feelings, we fell to unloading meteorites. Two men were to have accompanied the truck to handle the unloading, but there was only the driver and a small, 67-year-old Indian he had picked up as a hitchhiker. There was no equipment to handle heavy pieces.

The Hokanson family, friends from Flagstaff, had driven out to visit us, and for three hours, Elmer Hokanson, the driver, the Indian and I, with assorted help from the ladies and children, worked until we had unloaded and carried inside some five tons of meteorites. In addition to all 189 crates, bags, and boxes, we unloaded the several large show cases from the museum.

There remained about three tons of heavy irons, including one weighing 1,406 pounds, that had been shipped without crating. These we simply rolled out of the van into the yard.

The unloading went fairly well, and as we approached the finish, things looked brighter. The sky was brightening also. This was the evening of October 9, 1946, a notable date in the history of meteorites and astronomy for a reason more profound than the moving of our collection. All over the planet, men were alerted for the earth's probable encounter with a swarm or stream of meteorites that very evening. The Giacobinnid-Zinner comet had passed eight days earlier, leaving a trail of debris through which our earth would pass. The Hokansons had planned to enjoy an evening of meteor-watching with us in the desert. The show had already begun during the final forty-five minutes of our unloading chore.

We were a wilted, tired lot when finally we spread blankets on the ground, turned our eyes to the sky, and munched sandwiches the women had prepared. We appointed a timekeeper and began counting.

The number of meteors per minute climbed gradually from twenty-six in a minute to thirty, to thirty-six, to forty. As the tempo of meteor flashes across the sky increased, we recorded counts of sixty, seventy, eighty. Once, the count passed the hundred mark.

Although the majority were rather small, there were numerous meteors of first and second magnitude, and some even brighter. We noted at least a dozen that were more than twice as bright as Venus at her best seasons.

We interpreted the celestial display of October 9 as a fitting prelude to the opening of our museum. And it could be said truthfully that some three tons of meteorites fell in the front yard of our meteorite museum during the Giacobinnid-Zinner shower. It was quite an opening event.

Billy, the Indian who had come along with the truck, weighed about 100 pounds and was about the size of the average twelve-year-old, but what he lacked in strength, he made up for with willingness. When it came time for the truck to leave, he didn't want to go with it and insisted on staying with us.

There we were, in the middle of a desert, and Billy's only reference was the driver of the truck that had brought him off the highway. The driver could not tell us much, but we could not very well get rid of Billy, or at least we had not the heart to, for he had worked well and asked for nothing but shelter. We came to the conclusion that he could be of great help to us in getting our material unpacked and in order, so we let him stay. He proved to be a good worker. Things went along all right for Billy until a sign construction crew brought liquor onto the premises. This, it turned out, along with an inclination to dependency, was Billy's weakness.

The first days on Route 66 were crowded with painfully laborious tasks of shifting, uncrating, lifting, unwrapping, placing, labeling, rearranging. My arms ached, for it seemed as though all 16,000 pounds of meteorites had to be moved two or three times before they found a permanent resting place.

Billy washed down the interior walls and painted them. The front veranda was stacked high with boxes for storage, the museum room was crowded with display cases and specimens helter-skelter, the living quarters were full of furniture waiting placement, dishes overflowed the cupboards, and kitchen utensils cluttered the drainboard. The plumbers, who were to have finished the day after we left for Denver, still hadn't finished when we got back. Washbowls and toilets decorated the back yard. Our trailer seemed a haven of refuge.

Billy unpacked so many odds and ends of dishes and knickknacks that he shook his head.

"Why don't you live like the Indians? Then you don't have so much work."

When he got a few dollars ahead, Billy caught a ride into town. For a time we knew nothing of him, but he returned later, only to disappear and reappear from time to time. Usually he came only to pay friendly visits. We would let him work if we needed him but never hired him again except on a day-to-day basis. Once he wrote from Oklahoma, explaining that he had been in the hospital, and asked for bus fare to Arizona. We sent him the money. He came, visited, and went on his way. Again we had word from Denver, where he had given our name as his only friendly reference. We did as much as we felt was advisable for him or for us, but eventually lost track of him, holding only a hope that he prosper wherever he might be.

As soon as a considerable part of the collection was housed under glass, we decided to open our doors. All the major specimens were in place except Hugoton, its 770 pounds still secure in a plaster jacket. The derrick and hoist were set up, ready to finish the job. All the big irons from Canyon Diablo had been placed on a central platform where they made a fine focal point.

Visitors had been coming up the hill ever since we arrived, approaching our door or peering through the windows, and on October 19, we put up our little sign, "Open."

This was our first taste of serving the public, of being on the inside of the counter rather than standing outside to be served. We were to find that such service sometimes can grow tiresome, that some people can be rude, and that we could burn with indignation at actions we considered to be indignities. We never grew tired, however, of our basic purpose—to share the knowledge we had about meteorites in the hope that doing so would spur the growth of the science of meteoritics.

Our hours were governed by the presence or absence of customers, and by daylight, for as yet we had no electricity. As soon as there was good light, we were ready to open, and as soon as good light was gone, we closed, though occasionally we guided persistent guests by lantern light.

Our display room took up most of the space of our building, twenty by forty feet. The tower, with some storage space to the side, opened off of one end of the room. Our living quarters were one room, about sixteen by eighteen feet, with a tiny bathroom adjacent to it. Public restrooms had been installed in former storage space and opened to the exterior of the building.

We set up book shelves to partially separate our sleeping quarters from the kitchen. There was a small window on the west, at the foot of our bed, and another on the north, over the gas cooking range, which was fed by butane gas from a tank outside the window.

Evenings found us in our back room, cooking and eating and reading by kerosene or gas lantern. We stocked our disconnected electric refrigerator with cakes of ice from Winslow, nineteen miles away. Addie managed with old-fashioned sadirons, and washed by hand or used the laundry service in town. The only way we could receive radio news was on the car radio, which at times picked up a pretty good signal, but was not dependable.

The first rain proved that the builder had never succeeded in replacing all of the red mud with cement. After every storm, we had to go around with a shovel or dust pan and bucket and mop. There were always a dozen places—and not always the same ones—where mud piles would accumulate. The most unsightly part was the walls. They had been plastered, and we had painted them white so they might reflect as much light as possible for the exhibits. But the muddy water that ran down the walls left glaring red streaks. With all of our efforts to stop the leaks, we were never entirely successful. We would climb a ladder to plug a hole, but at the next rain, one or two new leaks would appear. Most came from where the ceiling joined the walls.

The rains did not bring our only problems. The winds often whipped our exposed hill unmercifully. The same red dirt that was running down the insides of our walls in wet streaks when it rained was also blown miserably around our refuge, pitting the paint on our car and trailers, sifting through windows. Snows, when they came, were often also borne on cruel blasts of wind.

Winter could be very cold on Route 66, and the summer, hot. Nevertheless, we nurtured a small oasis of Chinese elm trees and some cacti, and we thought our surroundings were beautiful.

Coyotes sang regularly from nearby stations and our dog would chase rabbits any time she chose. We had other animal neighbors, as well. One morning, I came out after breakfast to unlock the front door. The mechanical cash register rested on a counter that ended against the wall about two feet to the side of the door. I undid the padlock and turned to go back when an odd, indefinite image of the cash register seemed caught in my memory. The roof above the door was supported by a log pillar that branched at head-height. Had I

The American Meteorite Museum on Route 66 near the Arizona crater. The museum operated in this location from 1946-1953.

Postcard showing the interior of the American Meteorite Museum on Route 66, c. 1947. The Niningers' daughter, Doris, is on the left.

seen something strange there? I turned to look again.

Draped across the register keys and then spanning the two-foot gap to the fork in the log pillar, from where his beady eyes stared at me, was a huge bull snake. His tail dangled from the "Total" key about eight inches. I could not resist calling Addie. "Come see! I think there's something wrong with the cash register."

She did not appreciate the joke.

Another time, Addie met a large king snake on the tower stairs, and several times, we dealt with rattlesnakes. During one of Margaret's visits with us, I had stepped outside the kitchen door as she and Addie returned from Winslow. They stopped the car just beside our dog Blondie's doghouse. When Margaret opened the car door, Blondie rushed in between her and the little house, barking fiercely. Margaret was a favorite of Blondie's, so we couldn't understand the dog's behavior, but Blondie wouldn't be persuaded to withdraw. She kept ducking her head under her house as if to sniff under it. Finally, we were convinced that she sensed danger. We lifted the dog house, and there lay two rattlesnakes. Blondie dispatched them furiously.

We also occasionally found some of the small but dangerous scorpions. When Shorty, a ranch hand, was replacing railroad ties that roofed our "garage-workshop," he came across numerous six-inch centipedes. When we left the desert after seven years to move our museum to a new building in Sedona, in Oak Creek Canyon, the last farewell was issued by a rattlesnake flashing its head in and out of a chink in the flagstone wall just opposite the knob on the kitchen door as we were preparing to lock up for the last time.

Half the tourists who drove up our hill would read our little sign listing admission fees of twenty-five cents for adults and fifteen for children, then turn and leave. Some would only look from the car and then drive on. Some would drive in, around the building, and out without stopping at all. A few would come in without reading the sign and then stalk out when fees were mentioned.

In spite of this, the number of customers our first day totaled sixty, and most seemed very pleased. Admissions increased steadily overall, though there were occasional days when the number would drop to a dozen or less. Once in a while, the number surpassed a hundred.

As spring opened up, the flow of traffic increased, and the hours of daylight lengthened. We were encouraged enough by our volume of admissions and by the additional income brought in through sales of literature and specimens, both directly to museum customers and by mail order, to decide that we could afford to hire help. The seven-day weeks, with lengthening hours, were beginning to tell on us.

We offered jobs to George (Don) Thompson of Denver, and his wife, Ruth, whom we had known for years as the son and daughter-in-law of our late friend, Art Thompson, and his wife, Miz. Don and Ruth joined us in April of 1947 and set up housekeeping in our twenty-two-foot trailer.

Ruth helped Addie with the bookkeeping chores. We all shared duty as museum guides and lecturers, though I carried the burden of the latter. Don, in addition to museum duties, performed shop work.

Harvey Nininger at the American Meteorite Museum on Route 66, c. 1947.

Like his father, Don was handy with gadgets, and he brought a mechanical aptitude which was sorely needed in our situation, located as we were, miles from any city services or parts suppliers.

We had a gasoline-powered water pump to worry about; plumbing, which suffered sometimes at the whims or carelessness of tourists; a cesspool to manage; a parking area to keep clear and provide with barriers; butane tanks for cooking stoves to maintain; oil tanks for heating stoves to fill; and the cash register to keep in running order. My method is to attack things mechanical with dogged determination and any tools I can lay my hands on, but without much natural talent. So Don's aptitude was welcome, and so far as was feasible, mechanical chores fell to his lot.

Don and Ruth studied hard to master a practical knowledge of meteorites and the most effective means of presenting that knowledge to the public. The facts about meteorites are themselves so interesting and unusual that an accurate portrayal is all that is necessary to attain the objective of interest.

The visiting public included many and varied personalities. Some brought weird and sundry stories concerning meteorites that would at once be recognized as honest mistakes. A sense of humor was required to deal with some of these in a way that would correct the error without offending the customer.

In the museum, a visitor could heft in his two hands a piece of matter from outer space. He could himself touch and wonder at these particles of matter that a scientist can weigh, can analyze, and can study, and from which he can gain information just as significant as that which the astronomer gains when he catches in his telescope light that has been on its way to Earth for hundreds of millions of years.

We frequently received groups of touring school children or older students. We quickened the youngsters' interest by inventing "treasure" hunts with clues by which they could locate particular displays or specimens. Our visitors often included scientists of note, many of whom I had corresponded with or dealt with in the past, but some of whom I met first through our museum. Often, former students from the various colleges where I had taught would stop to view the collection and to visit.

Once in a while, it was difficult to draw the line between "paying visitor" and "guest," but most of our friends recognized that we were in the somewhat perilous business of making a living by exhibiting our collection, and they supported our endeavor while they brightened our days with their visits.

The closing of the books on the first twelve months of operation showed more than 33,000 paid admissions. The visitors represented every state in the Union and forty-three foreign countries. There had been classes from fifteen colleges and high schools, a few groups of scientists, and several other miscellaneous travel groups. We had distributed more than 5,000 books and pamphlets.

Three- to five-minute talks were presented throughout the day. Additional attentions—often half-hour lectures—were delivered to groups or particularly interested visitors. At invitation, I had spoken before numerous schools and clubs in neighboring Arizona towns.

One day, the casual mention of a fireball that had been seen the

Harvey Nininger lecturing to a school group at the American Meteorite Museum on Route 66, c. 1947.

evening before brought two reports of the same fireball from museum visitors who had each witnessed it from 300 miles apart, one headed west and the other, east. Thus, quite by accident, we discovered that our location was well-suited to receiving reports on fireballs. We would post a bulletin and could expect additional reports from visitors within a few hours.

Our museum was small, the collection large. It was organized to include illustrative exhibits to demonstrate seventeen aspects of meteorites and related phenomena: 1) the classification of meteorites (we were able to display most of the eighty known varieties); 2) surface features and problems of air resistance at high velocities; 3) the structure of stony meteorites and their bearing on the origin of meteorites; 4) the structure of metallic meteorites in relation to chemical composition and possible bearing on origin; 5) the shapes of meteorites (with 3,000 specimens available for study by interested researchers); 6) the disruption of aerolites during flight and effect on total weight of annual accretion; 7) the approximate frequency of "shooting stars;" 8) the distribution of meteorites of a shower and significance relative to magnitude of parent mass; 9) the nature and cause of disruptions; 10) the relation of depths of penetration by various members of a shower; 11) the average probable composition of meteorites in space relative to fireball phenomena, arrival on the soil, and in comparison with composition of the earth; 12) meteoritic oxides and their importance to astronomy and geology; 13) lunar craters and meteorites; 14) lunar craters and tektites; 15) ballistical implications in the forms of certain meteorites; 16) influence of meteorites on the lives of primitive peoples; and 17) meteorite craters— their present known distribution, the probable abundance of concealed craters, their geological significance, and their relation to regional mineralization.

One day, a man stepped to the cashier's window and with a

skeptical expression paid his admission, then came inside. He glanced about, read several of the monitor cards posted on the walls, examined the case nearest him, then turned to the desk.

"I'd like to call my wife who is out in the car. It is evident you have something worthwhile here."

The two of them examined the displays. Finally, the man turned again to me.

"Is this a state-supported or government-supported institution?"

I told him that it was privately owned, privately operated.

"Well, how was such an enormous collection ever assembled?"

This question, for which an answer involved some brief resumé of the story of my life, came to us repeatedly during the fourteen years we operated the museum.

In only three cities of the United States were comparable exhibits to be seen—Chicago, New York, and Washington, D.C. But even in these great institutions—The Field Museum of Natural History, the American Museum of Natural History, and the United States National Museum—there were no similarly extensive presentations demonstrating significance, distribution, varieties, surface features, effects of weathering, and so on, nor was there a competent guide to explain and answer questions.

I fully understood the visitor's puzzlement. From only a slightly different angle, I had puzzled over the meteorite problem thirty-five years before. Only what had puzzled me was not the presence of a great collection before me, but the lack of such a collection in any of the institutions wherein I had studied or taught.

Aside from light, meteorites furnished the astronomer, during those years, his most intimate contact with the universe surrounding our tiny planet. Yet a man might have earned a doctor's degree in astronomy in almost any American university without hearing a single lecture or reading even a chapter on the subject of meteorites.

Meteorites constituted the sole material evidence of the *modus operandi* of earth growth, yet one might earn a doctor's degree in geology without ever learning even to recognize a meteorite in the field or laboratory.

This situation has only to some extent been ameliorated since Sputnik jump-started the Space Age.

We were convinced that in putting our collection on public view and charging a fee, we were serving a purpose beyond our personal needs for self-support; that we were fulfilling an educational function. During each year of the museum's operation, more hours of instruction were given to more people on the subject of meteorites than were given on this subject in any university in the land. There was no other institution where an individual instructor was devoting his entire time to the subject of meteorites.

But on Route 66, stretching across several states with one "gyp" joint after another, where the gullible paid a quarter, a half-dollar, or a dollar to view faked cave dwellings, stuffed animals, or snakes or Gila monsters that he could view as well in public zoos, it was hard to convince many visitors that our museum was legitimate.

One man who drove up with his family turned away when he saw there was an admission charge. When I asked him why, he replied in effect that I had no right to charge for an exhibit advertised as educational.

"Education in our country is free," he stated.

Of course, he was mistaken. Education is not free. It must be paid for through taxation, endowment, or in some other way. The fact that it is provided on a theoretically equal basis to all children does not make it "free" in actuality. The burden is spread in some way among taxpayers. The great museums of our cities, the art galleries, the botanical gardens, all are paid for in some manner. The fact that one enters without an admission charge does not mean that the institution is operated without cost. It is supported either by tax money or by gift. Some of the great museums charge nominal admission fees which represent only a small fraction of the cost of operation.

Schools were not always tax-supported. There was a time when pupils paid fees for the privilege of attending classes, when the teacher collected his salary or wages directly from the pupils he taught or from their parents. The establishment of a tax-supported school system followed realization by the public of the importance of education to the general welfare.

In grandfather's day, no one thought of studying music, painting, or subjects other than the three R's in the public school. For them, he had to obtain and pay a private teacher. Gradually, as the importance of these other studies was realized, they, too, were added to the curriculum and paid for out of the People's taxes. But before this could happen, someone had to pioneer in the teaching of these various subjects in order to demonstrate their importance.

When I first became interested in meteorites in 1923, the principal sources of information were a few great collections in tax-supported museums and technical papers not readily available. In the process of studying such collections and as much literature as I could find, I acquired a desire to see something done to bring this aspect of our universe to the public. I learned that people who had opportunity to handle meteorites, or who had witnessed their fall, had a great yearning for greater knowledge about these lumps of rock and iron that occasionally plunk themselves down on our earth. When I could not find an institution to support such an aim, I did what I conceived as the next best thing. I set up my own museum, and since I couldn't levy taxes, nor establish an academic course and charge tuition, I imposed admission fees, feeling not much different than I had during the sixteen years when I taught in colleges or universities and went monthly to the business office to pick up my check.

2

When I was combining study with teaching at McPherson College, a lad used to hang around my laboratory often out of a burning interest in the work we were doing with bugs, birds, and small mammals. He hung around so much that sometimes he got a little in the way, but he was a likeable kid, and one day, when he was underfoot, I tried a diversionary tactic.

"Take this reading glass and go out and study that ant hill there in the yard," I told him.

The task so interested him that he spent most of the rest of the day on his knees, glass in hand, watching the comings and goings of those tiny communal creatures.

That boy was John Hilton, the now noted artist, naturalist, and author. The Hiltons moved from McPherson not long after my first acquaintance with John, and I didn't see him again until many years later. By then, he had made his name in the artists' and writers' world, and when I read one of his articles in the *Saturday Evening Post*, I decided to look him up, wondering if he was the youngster I'd known in McPherson. In 1938, we visited him in his desert studio in California.

"You know," he told me, "you are responsible for all of my interest in science." And he reminded me of the ant hill.

John Hilton described for me a brilliant fireball he had seen streak across the horizon. He told me excitedly how, while driving on a lonely road, he had seen the nocturnal landscape rendered as bright as day, and how the shadows of the sagebrush rotated as the great light passed. Then he had cursed himself for not thinking to use the camera that was hanging from a neck strap.

"Well," I said, "you are an artist. Paint me a picture of that, and I'll trade you meteorites for it." That picture hangs in our home.

400   IN SEARCH OF FALLING STARS

John did a lot of scouting around the desert and, recognizing a possible meteorite finder when I saw one, I coached John on what to look for and how to look. In the years afterward, he sent me a report from time to time, but until 1948, none of his leads proved good. Then John wrote to me from Mexico that he had found a meteorite. He added that he had checked the maps and lists of meteorite finds in my book, *Our Stone-Pelted Planet*, and he was sure this was a new meteorite, since it was 200 miles from the nearest listed location.

When I wrote John to say how delighted I was at his news, I cautioned him not to be too sure it was indeed a new fall.

I had a hunch about his meteorite, and when I saw him, I told him the story. Back in the 1890s, when the Arispe meteorite first was found about forty miles south of Cananea, several sizable masses were brought out, and there was a tale among the old-timers that there was another big piece, one that had been kept for a time in a local shop, then loaded by its owner on a burro and taken away. They thought the meteorite had been taken to Magdalena. On my first hunt for Arispe irons in 1927, I had followed the burro and its burden as far as Magdalena but had lost track there. This was halfway to where John Hilton had found his meteorite, and I had an idea his might be the donkey-borne Arispe specimen.

John and I cut off a little corner and polished and etched it—and there was the story. The structural markings of a meteorite are like a signature, and this was Arispe, sure enough. John got more kick out of playing detective than he had out of finding the meteorite in the first place, and he promptly wrote another magazine story, this one about structural patterns of meteorites.

Hilton's meteorite had been used as an anvil for thirty-five years on a Mexican hacienda, mounted among the roots of an overturned stump. It weighs 269 pounds and bears a natural flat surface, eleven by five inches. The Arispe anvil piece is testimonial to the toughness of natural meteoritic steel; though pounded on for a generation, it shows no hammer marks. Several meteorites have served as anvils on farms. One of the Tucson irons was used by a professional blacksmith for this purpose for a number of years. Another homely need

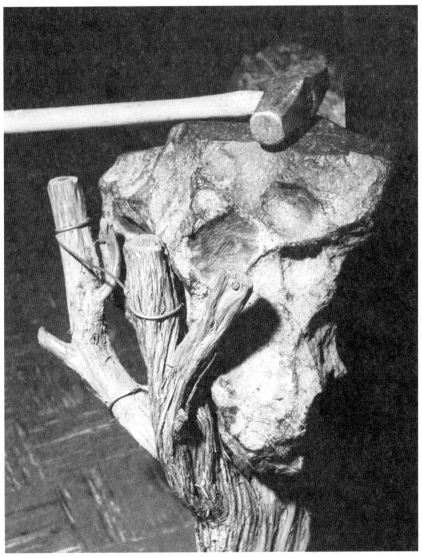

A 269-pound Arispe, Mexico, meteorite that was used as an anvil on a Mexican hacienda for thirty-five years. It was recovered in 1948.

was filled by the Estacado stone, which was employed as a wash stand outside a kitchen door before it was sold for $1000.

***

No meteorite collector ever had better luck working near the Arizona crater than the late O. J. "Monte" Walters. Monte had a sharp eye and a persistent mind. Sometimes I thought he must also have a special instinct, a nose for meteoritic material, for I never knew him to return from a collecting trip empty-handed.

In the early post-war days, surplus mine detectors were easily available, and I obtained a couple of these for use in locating buried irons. For me, their performance was erratic. We used an M625 standard war surplus detector. This detector consisted of an electric coil carried in a disk at the foot of a rod and connected to batteries borne in a pack. The electric current creates a magnetic field around the coil, and this field is affected by ordinary soil in such a way as to produce a faint steady hum in an earphone worn by the searcher. When the detector passes over a metallic object, the steady hum suddenly intensifies to a whistle or buzz, sometimes almost a bark. Unfortunately, the detector cannot distinguish the difference between a meteorite and a tin can, a piece of wire, a bottle cap, a nail, or any other object made of ferrous metal. The detector is sensitive to metallic meteorites of average size—about a half-pound—to a depth of ten inches, and can indicate a two-pound iron buried as deeply as two feet.

Carrying and manipulating the machine requires some physical effort, but more than that, its effectiveness depends in large part on the aptitude of the operator. Monte Walters had a knack with these instruments. When he went out with a detector, the thing would hum busily, then go into a gleeful buzz that had Monte wielding his shovel at frequent intervals.

In 1946, the persons most familiar with the Arizona crater informed me that the land had been pretty well stripped of meteorites. One veteran hunter told me he had personally covered every foot of a zone three miles in width adjoining and encircling the crater pit with a metal detector. I did not doubt this Arizonan's integrity, but I judged him fallible. My experience with detectors, and my calculations years before of the man-hours necessary to cover such and such an area foot-by-foot, told me it would take about thirty-two years of steady searching, eight hours daily, to cover the area he described.

Monte Walters gave me no such picture of the "worked-out" nature of the crater vicinity.

"I can still find meteorites," he told me.

So I agreed to take all that he could find off his hands at the then-going price per pound. I had taken on a tough obligation because Monte proceeded to deliver. Sometimes it was difficult to keep our side of the agreement, and Monte would have to wait for his money, but he never complained, and never doubted. In fourteen years, he delivered to our museum approximately 4,000 pounds of meteorites, comprising some 12,000 specimens, for which we paid him between a dollar and ten dollars a pound.

Walters had spent his early years as a cowhand in the neighborhood of the crater and had become familiar with the distribution of meteorites around the rim. Our agreement covered hunting on state land, and he searched on several sections on which we had filed mining claims, and on two or three others on which he had filed at my suggestion.

Walters was a fine friend, and in my scale of values, he was a great man. His devotion to his family, particularly his relationship to his sons, was admirable. Deer hunting, fishing, and meteorite hunting were the weekend recreational activities for the boys and their dad. During all of the many hours that I spent in their presence, I never heard a cross word spoken. Monte was always firm and knew his own mind, but he also respected the minds of his sons, and their admiration and respect for him was almost worshipful. Monte died in a veterans' hospital July 8, 1961.

With a staff of two to four besides ourselves, I had time in 1947 to pursue work at the crater under my permit. Despite the estimates that the great amount of collecting done over the previous four or five decades had nearly stripped the environs of meteorites, we were anxious to make a thorough search.

A retired navy captain guided us to our most efficient metal detector. Captain Harold Draeger, M.D., had taken up meteorites as a hobby in his retirement. He came to our museum with a mine detector he wished to try out at the crater. The first day he visited our museum, I had a firm commitment that prevented my accompanying him, so Addie joined the captain and his wife on an excursion to the crater. I designated a location where I thought he might find something, and off they went.

When they returned that evening, there was great excitement. The captain had located meteorites faster than the two women could dig them out from depths of a few inches, and when the catch was spread out on the floor of our living quarters, we were indeed an excited group.

We at once purchased a detector like Captain Draeger's. Don Thompson was a good operator, and we kept it busy. He and I took turns and set out to give the ground a good going over with the expectation of being able to furnish a useful map of distribution of fragments.

3

For our first year on the desert, we were busy and buoyed up by seeming success.

In the fall of 1947, I chased once more after the Danforth puzzle in the Maine woods, but returned feeling that perhaps it should simply be marked off my book for good.

During a long Sunday afternoon off, Addie and I drove to Long Valley and came back over the Mogollon Rim through a hundred miles of dense forest where occasional patches of aspen gleamed like candles in the dark.

We knew there were plans to change the course of Route 66. Stories about the impending move of the old and dangerous narrow two-lane road, with its dips, decaying shoulders, and high incidence of traffic accidents, carried changing rumors as to the path the new highway would take. Finally it was settled that the new road indeed would bypass us, running a half-mile away and to our rear, with only a dirt access road to our hill.

Still, the time for this happening seemed far ahead. Besides, we were scouting possible sites for a new museum building, and we had not yet given up hope for an eventual institute and home on the crater rim.

Power company poles had begun marching over the desert in our direction, though they were not coming close very fast, and we still depended on gas and kerosene light.

We hired a new, temporary employee. One of our visitors, Fred Boyer, showed so much interest in the place that when we found he was footloose, we simply asked him, on a hunch, if he would stay for a while to help. Fred and Don put in some much-needed improvements. They hauled away rocks and dirt to make a driveway, set up parking barriers, and built a garage of unique construction, a simple arrangement of old square-stemmed telegraph poles, purchased from Western Union, unloaded along the railroad and trucked in to us.

The visitors' count for November 1947 passed 1,100. Our mail orders included a welcome $500 order for meteorites to be used in nuclear studies. Average daily attendance of sixty to ninety people, with sales of literature and specimens, brought us, after sales and admission taxes, something over thirty dollars on an average day. Out of this, we had to support ourselves and a staff, pay operating expenses, and retire heavy indebtedness.

Don and Ruth had begun preliminary work on an illustrated catalog of our collection, which was intended to be an historical account of its accumulation as well as a listing of its contents.

We began to dream that perhaps our museum would grow enough that it could support a research department manned mostly by college students and new graduates under one or two trained supervisors. Much of the research, those projects involving chemistry for example, could be farmed out to other institutions.

Perhaps our glasses were rose-colored. We learned that the custodian at the crater rim had complained about our collecting, but we continued working in areas for which we had written permission. We began to notice that other parties were searching the same sector of the rim without filing the reports that previously had been made regularly to us for our records of any hunting done by visiting collectors and scientists. My map could have no meaning unless all finds by all parties were reported. Finally we abandoned the map but continued to work the rim under our permit. The permit was renewed for six months, ending in October of 1948. Addie and I filed for mineral rights on some of the intervening sections of state land within the cattle ranch. We staked claims and laboriously built stone monuments to mark them.

By summer of 1948, the power line still had not reached us. The cost was going to be high. We installed a Wincharger—another mechanical device to watch after—and stored its rows of batteries in our telegraph-pole garage to power our own electricity. The Wincharger provided direct, not alternating, current. Addie still couldn't use her electric iron or washing machine. The new highway snaked closer and closer to us and to the by-pass that would maroon our museum.

All of our efforts to have our program taken over by some respected institution had failed. When time seemed to be running out for us, we had come with great reluctance to the conclusion that we must either abandon our program and our dream or find a way to make our great collection support the program. When we had decided that the environs of the great meteorite crater would be the best place to try, we had faced the fact that this probably would be our last move, that we would spend the rest of our days in the desert. What final disposition we would make of the collection was yet to be decided. When we had moved to Arizona, I had thought it probable that ultimately it would become the property of the Society for Research on Meteorites. It was my hope that such also would be the fate of the great crater, although I had no intimation of this possibility from its owners.

My interests had been bound closely with those of the Society, and my association with its leadership had been close. I did not foresee that time would change this situation, nor that a space race eventually would transform the collection into a negotiable asset.

***

At Norton, Kansas, in 1948, Addie and I found and lost, almost simultaneously, one of the great meteorites of modern times.

On February 18, 1948, at about 5 p.m., a great fireball, variously reported as a rocket, a bomb, a space ship, or a burning plane, was sighted by people in several states. *The Denver Post* telephoned our museum. From the teletype machines, the numerous reports from all over Kansas, Oklahoma, New Mexico, Nebraska and Colorado,

408   IN SEARCH OF FALLING STARS

including Denver, were read to me while I took notes. I promised to call back after making an analysis.

In less than an hour, I returned the call, saying that my wife and I would leave within the hour to begin a survey, and giving my judgment that the object unquestionably was a meteor and that a meteorite must have landed somewhere in the region of Norton County, Kansas. Similar estimates of the meteor's destination were given to the press the same day, February 19, by other scientists, including Dr. Lincoln LaPaz, of the University of New Mexico, and Oscar Monnig.

Addie and I drove late into the night of February 19, and before we reached Tucumcari, New Mexico, we were beginning to meet reports of the meteor. We stopped to question a bridge-building crew, and they were able to give us a good description. We drove on across the Oklahoma panhandle and into Kansas. In Oklahoma, we talked with some men who were constructing a grain elevator. Yes, they had seen the big fireball, or whatever it was. They had supposed it to be a burning plane and had raced up over a nearby hill only to find that the "smoke" was farther away than before. They were surprised to learn that the same phenomenon had been seen in Denver.

When we reached the area around Norton, Kansas, the second day, we found people still excited by the spectacle. For fifty miles around, the blast-like explosion had been heard, followed by protracted rumbling, and a dust cloud had been left hanging in the sky on a cloudless day.

Before undertaking a survey, I inquired whether any other scientist had come to Norton or had been working on the fall. The editor of the Norton paper said no, but Dr. LaPaz had wired that he would come if anything were found.

Our friend H. O. Stockwell was in the area when we arrived, but he had found no use for his detecting machine and announced he would return home. We agreed on a plan of cooperation. For two days, I interviewed people within about a fifty-mile radius of the town, and finally put together data as to direction, elevation, angle of descent, appearance of the cloud, and so on, and determined the point of fall as well as I could.

The next day, I lectured in the high school of Norton, telling several hundred people my estimation of what had happened and instructing them as to what sort of gray or black-crusted stone to look for, and I spoke again that night to a hundred or so men and boys attending a father-and-son banquet in the neighboring town of Norcatur. Specimens were exhibited at both of these free lectures. Also, I left a sketch map for exhibition in the high school at Norton to guide local residents in their search for small fragments, it being our custom first to concentrate on the search for small pieces, lying on the surface, and to let those finds guide the later search for large fragments, which might be buried deeply. Our plan from this point on was to keep people in the area aware of the need for search and alerted to be on the lookout for strange stones.

The fourth day, a local pilot took me up to make an aerial search for any sizable scar, but a storm blew in and we were forced down. Addie and I had to abandon our hunt, getting away just as highways were threatening to close. The storm continued for days and deposited a snow blanket of twenty-five inches—the deepest in years.

During our four days in the Norton area, we gave newspapers and the public all the information we had gathered as to what to look for and where, and then waited for some response, having asked that any suspect stone be sent to us for examination.

As spring approached and we thought snow surely must be disappearing from the fields, we waited anxiously in Arizona for word to come from Norton. None came, which seemed strange, for we had learned always to expect numerous samples to be sent in after such a campaign as we had carried out. Sometimes they would be meteorites

410 IN SEARCH OF FALLING STARS

and many times not, but we had not received a single sample by the end of March. We debated whether we should return and give some more lectures, but the expense was a barrier. It would be necessary to hire help at the museum during our absence and this investigation had already cost us heavily.

Then one evening in late April, we heard the voice of Dr. LaPaz in a radio report from a town near Norton, announcing that he and his search crew had recovered several hundred fragments of a very rare type of meteorite that had fallen from the great meteor of February 18.

We hurried to Norton, confident the populace we had alerted and instructed would rally to our support. When we arrived, we went first to the local editor, who had been most cooperative during our previous visit, for information and advice. He had been a student at McPherson College, and his father was a long-time friend. But on this second visit, he was very discouraged. He told us he did not believe we would be able to work in the community at all. He said that he was afraid I had not a friend left among the farmers. The new investigator, the "discoverer" of the fragments, had represented himself as working with the military and had said that this meteoritic material was of vital importance for the safeguarding of the lives of servicemen. I was described as purely "commercial," with no interest in science, and it was predicted that if I were to obtain any of the meteorite, I would probably cut it all up into little pieces and make jewelry out of it.

I couldn't believe my ears. I said this couldn't be done to me in the state of Kansas, where I had lived and worked so many years!

The editor shook his head. "I hope you are right."

But I was wrong. Men with whom I had been friendly and whom had been appreciative of the instructions I had given freely, turned away, refusing even to talk to me. One mother was vehement. "That man said he represented the American armed forces and that you were merely a peddler, and that if I were to let you have even a tiny scrap of that meteorite, it might cost some poor American soldier his life."

I found one small fragment of the stone in a field, and when I went to the landowner to ask if I might buy it, he said he had been "told about me," and when I handed the piece over to him, he took it and simply turned away.

The other group had gathered about 150 pounds of the meteorite. I went a few miles farther on into territory where I suspected a larger mass might have fallen. After several interviews with people who had not been contacted otherwise, I became convinced that a large fragment had been heard to descend nearby. I proceeded to alert the people in the area.

There was a wheat field just across the state line in Furnas County, Nebraska, which my survey indicated was within the target area, but the wheat was quite tall at that time (the ninth day of May), and a search was impractical. I talked with the owner of the wheat— not of the land, for he had leased the field—Mr. Harold Hahn, who was running a tractor in an adjoining field. I told him my conclusion that the meteorite might lie amid his wheat, and I asked him to be on the lookout when he harvested that field.

I returned home again to Arizona. I prepared a summary of my survey of the Norton fireball that was due for publication in the October issue of *Sky and Telescope*.[20] That article carried as an addendum, submitted just at press time, a brief account of Harold Hahn's notification to me that he had found a great stone in his wheat field, and of our plans to make a third trip to the area.

It was a fruitless journey. On August 16, Hahn telephoned me at the museum in Arizona. He reminded me of our conversation in the

---

[20] "Tracing the Norton, Kansas, Meteorite Fall," *Sky and Telescope*, Vol. VII, October, 1948, p. 293.

corner of his wheat field and of my suggestion that he keep on the lookout during harvest. Of course, I remembered.

"Well, we found a hole about eight feet across, and last Sunday, some of us went out and dug down and found a meteorite that measures thirty-nine inches across."

Hahn added that he felt I had gotten a bad deal on the earlier finds and was calling me first. I told him that we would start at once for Norton, and I instructed him to cover the hole to protect the meteorite from vandalism.

When we arrived, we learned that one of Hahn's neighbors had taken it upon himself to notify Dr. LaPaz, and that he would be there too. Early the next morning, we met Mr. Hahn and inspected the hole and the meteorite. He and a few neighbors had some grain sacks half-filled with fragments they had gathered from around the large stone, which had been pretty badly mutilated during the digging. Addie and I spent some time examining those fragments and protecting some of them by careful wrapping. My ten-power lens revealed some very unusual minerals which I had never observed in any other meteorite.

I suggested that Mr. Hahn and I climb down into the hole and wrap the small fragments, of which there seemed to be hundreds, lying between the big stone and the wall of the hole so that they might be carefully preserved, no matter who might ultimately turn out to be the purchaser of the meteorite. There was barely room to crouch, but we squeezed ourselves in beside the big stone and for an hour or two wrapped specimens in tissue and newspaper and paper towels. We filled various containers with these carefully wrapped specimens and passed them up through a grating that had been placed over the hole. Addie and others carried them to Mr. Hahn's car.

Suddenly I heard familiar voices and looked up to see the faces of Dr. LaPaz and Frederick Leonard.

We had been aware that Mr. Hahn, who owned only the wheat crop

Harvey Nininger inspecting fragments of the Norton, Kansas, meteorite, 1948. Addie, left, looks on.

on the land, had no claim on the meteorite. It belonged to the owner of the land on which it had fallen, but we had hoped that preference would be given to us when and if the owner was ready to sell the meteorite. The landowner and her attorney, according to press accounts, were persuaded that I was "interested from a commercial standpoint as distinguished from a scientific point of view," and determined that the meteorite should be sold at auction to the highest bidder.

The University of New Mexico, represented by LaPaz, and the University of Nebraska combined offers. Stockwell joined Addie and me in bidding up to a certain point, but our opponents kept raising their bid until we were forced to give up.

The Denver Post reported afterward that there was "very little difference" between our final effort and the successful bid of "slightly more than $3,500," and quoted the attorney for the owner as saying he did not know what scientific use the universities planned for the meteorite. Thus did two great institutions join their resources against an individual on grounds of "commercialization." In the auctioning of the Norton meteorite, the "pure science" advocates had themselves set a new price level high enough to lead the way toward the further commercialization of meteorites.

The only specimens of Norton that I ever obtained came from certain sympathetic farmers who had secretly withheld their finds until they had opportunity to transfer them to me.

I wrote to the University of New Mexico about the possibility of securing specimens of the Norton meteorite. In response, I received a brief letter explaining that "For reasons made clear in the enclosed *Preliminary Application Form*, such a request should be accompanied by copies of this form in triplicate, duly executed and witnessed." Enclosed was one copy of the two-page form.

The form dealt in detail on the need for taking "every step possible to protect against diversion of its meteoritical resources to those interested in meteorites from a personal or commercial rather than a scientific point of view." It went on to state that a "first *necessary* condition for obtaining favorable consideration of an application... for loan and/or donation of meteoritical materials shall be satisfactory answers" to three questions and proper certification of the form.

The detailed questions had to do, briefly, with whether an admission fee to exhibits of meteoritical material was charged, had ever been charged or might be charged, by the applicant; whether the applicant or the institution represented was engaged or ever had been engaged or contemplated becoming engaged in the sale of jewelry and other items made from meteorites; and whether the applicant would agree that all information obtained would, if published at all, be released through recognized scientific channels and would not be incorporated into leaflets and similar publications intended for advertisement or sale.

The application form certainly counted me out. I could not help but wonder, though, about other institutions. Many major museums charge admission to exhibits. And many engage in sale of various souvenir items and develop and sell informative publications.

The Norton meteorite never was widely distributed for study, and rather than becoming one of the most easily obtainable stony meteorites as its great weight and number of fragments would suggest, it has remained for the researcher one of the rarest. The total weight of the fall is recorded at more than 2,375 pounds—nearly a third heavier than the next largest known stony meteorite fall.

The most important lesson to be learned from Norton is the emphatic indication that we are not yet qualified to develop a definite concept of the word *meteorite*, so far as structure and composition are concerned. The fall of such a great mass of a type of meteorite that up until then had been considered extremely rare is proof that generalizations cannot be safely made as to the "average" or "general" composition of meteorites, nor as to the relative frequency of certain structural characteristics.

All of the meteorites with which man has been able to acquaint himself constitute the equivalent of not even five-hundredths of one percent of the number which are encountered by the earth during a single revolution around the sun. A fall like Norton, had it gone unseen by man and undiscovered, soon would have been terrestrialized into an unidentifiable state, leaving the records unbalanced as to the relative rarity of its characteristics.

**

There were days when there was little traffic on the road below and even less coming up our hill. There were days when heavy traffic passed on the highway—and passed right on.

Many cars continued to come up our hill only to swing away and go down again without a stop. Others brought the kind of troubles that people who deal with tourists complain about everywhere: visitors interested only in restrooms, which often were badly abused, and careless drivers who dislocated or smashed the parking barriers or who swept through our back yard with no care for our privacy.

And there were those occasions we so regretted when a visitor would pay, then obviously take no interest in our exhibits or our story. We had to follow some sort of established policy—if we didn't collect at the door, we wouldn't collect even $10 in a day—but we hated having anyone leave feeling their money had been misspent. Fortunately, those who took the time to look and listen seemed to go away pleased.

Ab Whelan, the geophysicist who had made magnetometer surveys in New Mexico oil fields for me, came to conduct instrumental searches for sizable meteorite fragments in the environs of the crater. I had suggested to Ab long ago that with all his field work, he undoubtedly had encountered a meteorite at one time or another without being aware of it.

Ab had never reported anything, but when he came to do the crater work, he told me he had found a "big old rock" that he had thrown into his pickup truck and brought home to Artesia, but had just never gotten around to writing to me about it. He said he didn't suppose it was anything, but I told him to send me a sample when he got back to New Mexico. He did so. His "big old rock" was a meteorite that weighed 167 pounds. It had been resting in his yard for two years since he had found it lying out on a deserted ranch, near what had been the foundation of a house, not far from Acme, New Mexico. It was impossible to determine whether the meteorite had been brought in, or lay where it fell, but searches failed to produce any further specimens.

We measured our days by changes in the weather and in the flow of traffic, and by the ritual totaling of the cash register tape each evening. Winds blew so fiercely that our signs were toppled more than once, the garage roof had to be battened down and the trailer tied to keep it from blowing over, and at times, we had to shove heavy meteorites against the doors to hold them shut.

Electrical storms occasionally put our batteries out of order, leaving us just enough power for one small light at night. In winter, our water pipes would freeze for a week at a time, thawing out some evenings just enough for us to draw and store the next day's water supply. Snows would swirl about the building and drift on the road up the hill so as to discourage visiting by all but four or six people, who perhaps stopped chiefly for brief refuge from the storm-driven highway.

On some days when slow traffic or bad weather brought a mere handful of visitors through our doors, better-than-usual sales of specimens and books would bring the total up to an average or even record level, while at other times only our mail order business made the difference by which we were able to survive.

Compared to 1947, business in 1948 all along Route 66 seemed to be generally down, but for us, on the whole, it was better, perhaps because we had established ourselves as a going concern, or perhaps because 1948 brought us unusually good press, with our museum featured in several major magazines and on newspaper pages throughout the United States and even abroad.

Because there was such a dip in business along the highway between the heavy traffic of summer and the slack of winter, Addie and I decided we would take a lecture-field trip over the winter season of 1948 and 1949 while Don and Ruth cared for the museum.

We headed for California to fill lecture engagements and enjoy a little warm weather. We parked our trailer at Palmdale, California, one afternoon, and the next morning, we found ourselves surrounded by twelve inches of snow with more coming down. It fell until there was an eighteen-inch blanket over the Mojave Desert. I rearranged lecture dates on foot and by telephone to accommodate the weather, and it was three weeks before we could move our trailer.

However, we had a pretty good winter among high schools, colleges, and rock clubs, the income from which, combined with loans on two of my insurance policies, enabled us to carry the museum deficit and pay off a $3,000 note. We spent five months in the Coachella Valley, the Mojave Desert, and the San Joaquin and Inyokern Valleys. Our points of contact were chiefly the mineral societies. In addition, we lectured to thousands of school children, many of whom lived on farms and ranches or on the desert and therefore had ample opportunity to observe and handle rocks.

We financed our travel by charging a small fee for lectures and through the sales of meteorite specimens and books. The cooperating mineral societies received fifteen percent of the proceeds for their own treasuries. Our overhead of car expense—pulling the trailer and going off on chases into rough country—was a burden on our margin of profit.

We returned to the museum in the middle of March. Don and Ruth had struggled through a pretty tough winter, but we all looked forward to the upswing in business and weather that spring should bring.

4

Early in 1949, a press announcement came saying that the University of New Mexico henceforth was granted "exclusive rights" to conduct meteorite surveys and to recover meteoritic materials at the Arizona crater. The announcement was made jointly by the University, the Tremaines, and the Barringers.

There were three separate contracts. The Barringers in theirs specified a prohibition against activities that might deface the scenic beauty of the crater or damage permanently its scientific value.

The press announcement also mentioned that Dr. Otto Struve, president of the American Astronomical Society, had compared the Arizona crater with the site of the February, 1947, fall in Siberia, which he reported had been turned into a national monument by the Russians. From this time on, I confined my work to such of the checkerboard squares of state land sections in the environs of the crater on which Addie and I already had constructed our stone monuments or would continue to select for filing of mineral claims. At the time of the announcement, I knew I was being criticized for "commercializing" the subject of meteorites by insisting that somehow I must collect a living wage from the thousands of people who were visiting our educational exhibit, hearing our lectures, and purchasing our literature or a few specimens. For years, our collection of more than 5,000 specimens representing 526 falls was made available every day of the year for inspection by any and all persons willing to pay twenty-five or fifty cents to make use of it. Reduced rates were available to groups.

My signal offense seemed to be that our museum sold specimens to collectors, colleges, universities, and museums. Also, on a much smaller scale, we sold jewelry made from the small nickel-iron meteorites we had retained from among those collected around the Arizona meteorite crater. In no case did we allocate for jewelry any specimen for which there could be any conceivable scientific need, nor did the jewelry constitute more than a very small item in our volume of sales.

A collection of meteorites properly labeled constitutes a more reliable source of information than any printed treatise on the subject. The shadings are slight between the relative grades of scientific respectability represented in receiving a price for a cabinet specimen for a scientific collection and receiving royalties on a textbook.

Obviously, meteorites are too important scientifically and much too rare to justify their general use for ornaments or any other commercial use. So far, in man's experience, meteoritical material is several times scarcer than gold, and it is possible that a greater tonnage of diamonds has been recovered than the total of meteorites in all the museums of the world.

Because of their potential yield of information concerning the universe outside the earth, all meteorites should be strictly preserved for educational and scientific purposes except in those rare instances where the representative samples of a single fall exceed the demand by institutions of learning.

In the vicinity of the Arizona crater, so many nickel-iron meteorites were recovered during the first three decades of the twentieth century that they had become practically a dead item in the museum supply houses of the country. On the other hand, they were commonly offered to tourists in souvenir houses along the highways of northern Arizona as mere curios. No one ever will know how many thousands of these were sold between the time of the discovery of the crater in 1891 and 1946, when we began an effort in cooperation with the crater owners to keep some kind of record of the number of such specimens recovered and disposed of.

In 1946, we developed ways to mount some of the small meteorites in the form of costume jewelry, sectioning and etching suitable specimens and polishing others along contour lines. The process was tedious and expensive, but it removed them from the "common rock" classification which they carried when sold in the rough as souvenirs, at perhaps a quarter each, most often to be tossed aside shortly and lost. A small meteorite turned from a nondescript "rock" into a $6.00 jewelry item surely would receive more care and attract more attention. Families of scientists in particular seemed attracted to our unique jewelry. We provided a proper label and descriptive certificate so that each piece carried with it a bit of knowledge that most certainly would be passed on to others.

The limited marketing of these small meteorites as jewelry and two other types of jewelry created from meteorite products—metallic spheroids encased in plastic and polished, and bits of jet-black polished oxide—may have made the general public more aware of meteorites than the combined efforts of institutional departments of astronomy and geology.

How "pure" is institutional science? How "commercial" must a private enterprise be? Every university employee, whether teacher or researcher, is paid a salary which is made possible for the most part either by taxpayers' monies or by the contributions and endowments received by private institutions. In spite of this support, these institutions charge enrollees some sort of tuition or fees.

Sometimes I would wish that the critic of my "commercialism" could be forced to spend a few weeks in my shop preparing paper weights, a pair of bookends, or a series of specimens for some enthusiastic collector. One simply cannot work with meteorites day after day without absorbing a lot of information. Preparatory work upon an extensive series of specimens constitutes a better textbook than has yet been written on meteorite structures.

If the order being filled is intended to serve as a study collection for students, or as a general reference collection, its chief function will be realized at its destination, where any number of individuals may be instructed by it. Nearly all of the contributions to meteoritical knowledge in America during the nineteenth century emanated from the pens of men who were associated with collections of meteorites. Could such collections have been developed without a great amount of preparation and direction? Can such preparational activities be undertaken without there being remuneration?

During the 1920s and 1930s, I visited practically all of our leading museums and universities to inspect a sufficiently wide variety of specimens to meet my needs in respect to structure, classification, and identification, and especially in relation to my favorite problem,

422 IN SEARCH OF FALLING STARS

the surface features of meteorites. One of the most impressive facts of these visits, emphasized repeatedly during my explorations, was this: Practically all of the specimens in those collections, both great and small, had at some time passed through the hands of Dr. Harry Ward of Rochester, New York, or Dr. A. E. Foote, of Philadelphia, both of whom were out and out "dealers" in minerals and meteorites.

Our "commercialism" differed from that of these men in one important particular: Whereas they assigned research a minor and secondary role in their program, it was given first place in ours.

Since our small operation was limited strictly to meteorites, we seldom employed more than two people. Several thousand specimens of several hundred falls were cut and polished in our laboratory. Each of those was inspected carefully under a hand lens, and many were examined under high-powered microscopes. We etched more than a thousand slices of nickel-iron meteorites. Many were re-polished and re-etched for purposes of research. All specimens were checked and re-checked by me personally. On several occasions, important discoveries were made during the course of preparing a pair of bookends or a paper weight from one of the Canyon Diablo irons. I had developed a number of my scientific papers directly from observations made during just such preparations of commercial objects.

Exchange of specimens between museums and collectors is by dictionary definition a form of commerce, yet without this kind of activity, knowledge about meteorites would have come about at an even slower pace. When one collection contains several stones of the same fall and another possesses a similar duplication from another fall, an exchange is to the obvious advantage of the scientists associated with both collections, as well as to the public visitors who view them. The practice of division and exchange has been followed by many who have contributed most to our present knowledge of meteorites. Those collectors who have refused steadfastly to allow their meteorites to be cut and studied, or who refuse to make exchanges, seldom have added much to the world's fund of information.

My research plan would have foundered at its very beginning had I not been able to turn a part of its products into bread and butter, shelter, clothing, and education for my family.

5

By all appearances, the new road had been finished for some time by the end of June 1949, but it was not yet opened for traffic. We had become accustomed to the situation, and business was going along as usual. Then one day, Don and I were both on duty, and the day had been going pretty well when there came a sudden, noticeable lull. We were standing among the display cases, discussing something or other, when one of us remarked about the lag.

"This has turned out to be a dull day. I wonder why."

I walked over to the northwest window. The new highway bore a stream of traffic. We had anticipated the change, but when it came, it was hard to believe. Our records for that summer showed that from that day on, our income was just about cut in half.

There we were, landlocked in our desert, with the new traffic pattern eating away at our success. We were still badly in debt. Now it was evident that we could expect no more than the money needed to pay the rent and the living expenses of two people.

The work was enough to be arduous for two, but attendance was insufficient to carry the salaries and expenses of four. Reluctantly, with Don and Ruth, we faced the fact that we simply could not justify or carry the added expense for regular help. Don and Ruth returned to Denver, and our remaining time on Route 66 was faced largely alone. We grew accustomed to caution about leaving the building alone even briefly. Addie and I regularly traded the chores of going into town or staying on duty alone, and we always arranged for someone to be at the museum when we were planning to be gone together, even for an afternoon or an evening. We found reliable and interested people who would care for the museum on weekends or during more extended periods when we wished to be away for field trips or lecture tours.

We refused to give up for two reasons. One of these was fact; the other was an old dream. The fact was that we knew our collection was of great value, and if placed on the market gradually, it would be sufficient for us to live on for the rest of our lives. We had never put any liens on the collection. Our experience had convinced us that it could not be sold in its entirety as a collection, but nevertheless, I knew its value, and if worse came to worst, we could give up our dream. And that dream was of a meteorite institute where a collection like ours would supply information both to researchers and to the general public until the subject that had most interested us should find a place in high school and college curricula.

Those were lonely and sober days. Our entire lives had been committed to this program. The venture had proven itself effective and had been self-supporting. Nothing was being done anywhere in the world that even in a small way duplicated our effort, so far as we knew, and we had expected to spend the remainder of our active years in this program. But we were under such worries and pressure that we feared our health could not stand up for long. I was having cold sweats at night. Bitterness and fear had been growing in me— emotions I had hardly known before.

In spite of its isolation, I loved our location on the desert. The atmosphere was wonderfully invigorating, especially in early morning and evening. I was usually up before daylight and by early dawn was ready for a run with our dog, Blondie. She was hardy and active. She loved to dash about among the red sandstone boulders chasing rabbits, coyotes, or whatever else she could stir up. Those morning romps with Blondie eased the worry and stress. I would leave the house on the run, jump the parking barrier, dash down the hill to the old, abandoned highway and then walk for a mile or two, watching the sunrise. Nothing ever helped my nerves as much as a morning hike.

Along with the exercise, I did a lot of thinking. Certain bits of literature memorized during college days bought comfort. I am sure that I recited Kipling's poem *If* to myself a hundred times on those walks. Henry Van Dyke's *God of the Open Air* was another favorite. I would remind myself that truth is the greatest good, and I would recall that old definition of faith that had followed me since 1910: "Faith is the momentum of a righteous life, the distance into the future a soul can run that has been running right in the past."

***

In the summer of 1950, there was great activity on the crater rim as a new building went up and huge signs advertising the crater blossomed all along Route 66.

We were still thinking of constructing a new building that would front on the new highway. We sought more and more to bolster our backsliding admissions income by sales of meteorites. Ward's Natural Science Establishment, our first large customer so many years ago, again undertook sales on consignment to an extent that was accounting for between a third and a half of our income. We were selling off duplicate materials and were even beginning to offer specimens that we never would have dreamed of letting go. At the same time, we began to make efforts to find a purchaser for all of our collection, or for substantial divisions of the whole.

Margaret returned to Arizona in the fall of 1949 and spent nearly a year with us, assisting in final preparation of *The Nininger Collection*, the catalog of our meteorites, published in 1950. By the time the catalog reached print, we were despairing our survival as an institution. We made up our minds that if a buyer for the collection did not appear shortly following publication of the catalog, we would have to split the collection into several sections to be sold separately, or close shop and dispose of it piecemeal. Two things saved the museum from being closed. The first was our inability to find a buyer for the collection. The second, the unexpected but controlling factor, was that we found we could not set ourselves free of our lease.

When we had not found a buyer by the summer of 1950, we decided we simply must quit. At the end of the season, we would store our collection and close the museum, sell enough meteorites to build ourselves a home, and somehow work out a new life.

Our lease had been expensive to purchase, and we had been careful to include a clause allowing either party to cancel on sixty days' notice, but somehow, the clause was left out of the final draft, and Addie and I failed to note its absence.

We sent our sixty days' notice only to be countered with a statement that we would be bound by an automatic renewal provision, paying our monthly rental whether we occupied the building or not, until July 1, 1953.

As in previous crises, we tightened our belts. We did some close figuring, changed our program here and there, cashed in one insurance policy, borrowed on the others, and made do.

Although looking ahead, those next three years seemed endless, in reality, they were in some ways less difficult than the preceding years, except for our feeling that we were not free to come and go at will without a constant presence at the museum.

Because business was lighter, the work was less, despite the fact that we were carrying it alone for the most part. By bringing in temporary help, we were able to attend a number of important scientific meetings that were welcome breaks in our routine. Arrangements could be made easily when the events were close by, and when they occurred at great distance, I would manage to line up lectures nearby to offset expenses and would carry specimens to sell or, where possible, follow up on a report in the area that might bring in a new specimen.

Addie and Harvey Nininger in front of the American Meteorite Museum on Route 66, c. 1951.

Our museum was frequently visited by scientists of national and international repute who wished to examine our collection. Usually, I would accompany such guests on a visit to the crater to explain my most recent research.

I was beginning to have a whale of a time with these crater studies. Day and night, I worked with soil samples gathered at the crater, sifting and sorting for hours at a time, with each operation adding a little more to the solution I was formulating for the puzzle of what exactly had happened out there on the desert eons ago. It seemed that I would never get through my work at the crater. Never had I been so busy in all my life.

The excitement and stimulation counteracted the tensions and problems in our daily life. Growing scientific notice of what I was doing, and the resulting attention in the press, were reflected in the attitudes of townspeople and seemed even to somewhat influence the flow of visitors to the museum.

Addie Nininger at the sales counter of the American Meteorite Museum on Route 66, c. 1951.

15

Discovery

No one regards what is before his feet;
we all gaze at the stars.

—Iphigenia, Quoted by Cicero in *De Divinatione*

Years before, when Dr. F. R. Moulton visited Denver in 1936, he and I lunched together at one of the local hotels. Our talk turned to my book, *Our Stone-Pelted Planet*, published three years earlier.

"I see that you hold to the view that there is a big meteorite in the bottom of the Arizona crater," Dr. Moulton remarked.

I replied that D. M. Barringer had decided so, and that according to his records, such a mass had been encountered by drill.

Placing his hands on the edge of the table, Moulton leaned toward me with a smile. "Nininger, there cannot be a meteorite in that crater."

"Why not?"

He explained that he had investigated this whole matter mathematically and had concluded that it was impossible for a mass of any

430 IN SEARCH OF FALLING STARS

such magnitude as that which had produced the crater to stop suddenly and remain intact. On impact, it would have to be transformed into gas; it would explode.

I admitted that Dr. Moulton had a much better right to an opinion than I, but said I still believed the mass must be in the crater.

Moulton had been employed in 1929 by Mr. Barringer's organization to give them a valid estimate of the tonnage of metal that they might expect to recover in a proposed mining operation. The professional scientists in the Barringer group had considered Moulton to be the man best fitted to make such an estimate, which was to serve as a basis for seeking the necessary financial backing for the project. These men had been thinking on the order of ten to twenty million tons of metal. When Dr. Moulton theorized that only three million tons, or perhaps as little as fifty thousand tons, could have survived the impact, the financial sponsors panicked and withdrew their support.

Dr. Moulton did not tell me all of this background, which I learned later. He only told me his conclusion. "There cannot be a meteorite in that crater. It is mathematically impossible."

As the years passed, I tended to agree more and more that Moulton might be right. Being a field and laboratory man, however, I felt I simply must have more than figures to satisfy me on a problem which involved matter colliding with matter. The Arizona event had taken place recently enough that a very raw scar still marked the site. The answer as to the fate of the impacting meteorite must still be available if a sufficient search were made.

I said to Dr. Moulton, "If you will secure some funds for me, I'll go to Arizona and prove your theory, if it is correct, or if you are wrong, I think I can prove that too." He was noncommittal.

Later, I wrote to Dr. Harlow Shapley of Harvard to the effect that I believed the vast amount of exploration at the Arizona crater had been centered mistakenly within the crater itself, and that if the correct answer to that puzzling feature ever were to be found, it would depend on exploration around the crater rather than within it. If Moulton's theory was right, then certainly the proof of it must be found, if at all, on the outside.

In 1939, through the good offices of Harlow Shapley, I received $250 from the American Philosophical Society, conditional on my securing a matching amount, to conduct a search for small fragments at the Arizona crater. Dean Gillespie provided the matching amount, and I designed and had constructed a magnetic rake to be drawn as a trailer behind our Studebaker.

With my son Bob and Irving Hoglin, I drove down to the crater and we combed some twenty-three acres of the more barren terrain within the two-and-one-half-mile zone surrounding the pit. Our harvest was some 12,000 small nickel-iron fragments—total weight, forty-two and a half pounds. A pattern of a radial distribution was indicated, furnishing support for Dr. Moulton's explosion theory.

Bob and Harvey Nininger stand beside a magnetic rake constructed on a trailer and towed behind their Studebaker, 1939. This was used to search for small meteorite fragments surrounding the Arizona crater.

During the next decade, I carried out various other research projects in the vicinity. Cattlemen who held grazing leases on state land as well as private holdings objected to further use of the magnetic rake. Magnetometer search, too, was stopped short of completion by objection of property owners.

During the frequent meteorite-hunting excursions Addie, Margaret, and I made in the summer of 1946, my interest turned more to the dust and small particles of sand-grain size that continually collected on our magnets. Back at the trailer below Mt. Elden, I spent many hours going over these small particles which en masse always gave a strong chemical reaction for nickel. Much of it I could recognize as small chips of oxide and an even larger portion seemed to be volcanic ash.

I suspected during that summer that I was finding the condensation products which Merrill had declared were lacking and which the Moulton explosion theory seemed to require for its verification, but our trailer facilities were not adequate for the complete examination which the several types of material we were collecting required. Such study would have to await our move and the opening of our museum, which had been set for September.

When we finally got our collection moved into the old rock building on Route 66, there were months of installation problems even after we opened, and to our gratification, we found our time very much occupied with caring for visitors. The research on crater materials still had to wait. When museum attendance reached the stage where we could employ help, then all my spare time was devoted to research problems.

During the life of our permit to search, we collected nearly 6,000 specimens from the outer slopes of the northeast sector of the crater rim. Only a few of these weighed as much as ten pounds. The two largest weighed fifteen and seventeen pounds. The average weight was about four ounces.

In the course of collecting these small specimens, we took numbers of soil samples by means of a hand magnet mounted at the tip of a cane. These magnetic soil samples were sealed in envelopes, marked by location, and set aside for study when time would permit.

After October 1948, when the withdrawal of my permissions for research made it impossible to continue exhaustive and definitive collecting, I concentrated my efforts on the alternate sections of public lands on which we obtained rights, and also began to take the bags and boxes of soil samples collected as far back as 1939 and laid away for analysis out of storage.

In the sorting of fragments gathered during the magnetic rake survey of 1939, we had discarded all material that would pass an eight-mesh screen. As we sifted out and cast away this great bulk of fine particles, I had begun to wonder whether perhaps we were culling the most important part of our harvest, and so I had bagged up a goodly sample of the discard for laboratory study and carried it back to Denver, but I had not got around to examining it.

The greatest keys to my evolving theory about the crater were found in my boxes of soil samples. I was beginning to be certain that neither I nor anyone else had given to the soil in the environs of the crater the careful attention it deserved, and when I turned my focus to the small particles of various descriptions that crowded the soil, the most exciting avenues of exploration and speculation opened up to me.

When a sample of magnetics from the crater area was examined under a lens, the view was of a confusing variety of particles of many shapes, colors, and descriptions. Some shine, some are dull. Spherical particles are not uncommon, nor are ovals, biscuit shapes, pear shapes, perfect droplets.

We had found that the hand magnet would pick up from ant hills richer loads of the little pellets of rounded or oval shape. Perhaps the ants tended to favor the rounded or oval heavy grains in building their hills, but more probably the wind had blown away the lighter volcanic ash and had left behind a disproportionate amount of the heavier metallic grains.

Barringer and B. C. Tilghman in 1905 had reported finding an abundance of small, black magnetic particles in their work at the crater. Tilghman described these as "blackish-gray" in color and of "torn, irregular" shape. He stated this material was "absolutely universal over the whole locality inside the hole and out for as far as observed, somewhat over two miles from the hole."[21]

The material constantly adhering to our magnets agreed with this description for the most part, but as I pored over the stuff, picking out particles one at a time, I found some that did not conform. These were rounded, gray or brown in color, and they responded positively to nickel tests. When I attempted to grind them in a mortar, I found they had malleable metallic cores.

By early 1948, I had concluded that here was one of the condensation products necessary to Moulton's explosion theory. In 1949, I reported to the American Philosophical Society the results of oxidation studies I had made under a society grant. I described these metal-center pellets and also the finding of oxide droplets, rich in nickel and black in color, but without the metal centers. Continuing studies revealed other condensation particles—reticulated pellets consisting of soil and sand particles bound together by a reticulum of nickel-iron oxides, and globules of silica glass coated with oxides.

When we first began studying these magnetics in the soil samples from the crater area, the little metal-center pellets with their coverings of sand and soil were the most apparent. It seemed that these

---

[21] Tilghman, B. C., "Coon Butte, Arizona," *Proceedings of the Academy of Natural Sciences of Philadelphia*, Vol. LVII, 1905, p. 898.

must have been formed by condensation from metallic vapors in the absence of oxygen and arrived on earth as raw metal, and had then bound to sand and soil grains by the cementing action of oxides. But if they indeed had been formed in the absence of oxygen, the process must have taken place in the heart of a large cloud of metallic vapors or other gases from the core of which oxygen was excluded. What, then, had happened to the large enveloping portion of the cloud? There must be other particles present in the soil.

The metal-center pellets made up but a small part of the magnetics I gathered. They were easily separated out, being one to two millimeters in size, but even without them, our samples consistently showed the presence of nickel. Isolating another kind of nickel-bearing particle was a real puzzle. Attempts to separate and test groups of similar-appearing particles failed, and finally it seemed the only answer was to test individual particles for nickel.

At least a hundred particles must have been tested, individually, before finally, when one more particle was dissolved and subjected to the dimethylglyoxime test, a beautiful strawberry red appeared in the test tube.

Surprisingly, the nickel-bearing particle was neither round nor black, as I had expected; it was brown and had a somewhat lumpy surface, though it tended toward the spherical in shape.

Once we had learned how to recognize these and separate them from the several other kinds of magnetic particles, we took hundreds of samples on all sides of the crater and found that these metallic droplets were present in the topsoil to an extent that projected to thousands of tons! Here, then, was the proof of Dr. Moulton's explosion theory. I could think of no other way to account for these little droplets, two and a half times as rich in nickel as similarly sized irregular fragments, than the explanation that they had condensed from a meteoritic cloud produced by the explosion which Moulton had hypothesized. I named these droplets *metallic spheroids*.

Harvey Nininger collecting magnetics at the Arizona crater, c. 1951.

Further testing and chemical analysis by F. G. Hawley showed the metallic cores were extraordinarily rich in the percentage of nickel—17.30—and these particles, even as small as .1 mm. to .2 mm. in diameter, bore only a thin coating of oxide. It was evident that the idea that all small particles resulting from the disintegration of large meteorites would undergo immediate oxidation was seriously in error.

The survival of these tiny bits of pure metal was exciting. The spheroids, so tiny that 280,000 of them would weigh an ounce, were found as far from the crater as five miles. At that distance, they were distributed at a rate of a few hundred pounds to the square mile. Close to the crater rim, the concentration was on the order of a thousand tons per square mile. We measured the quantity of spheroids in sixty different locations on state lands and estimated that between 4,000 and 8,000 tons must be present in the upper four inches of soil within an average radius of two and one-half miles from the crater rim. By projection, taking into account the processes of weathering, it is not unreasonable to estimate that the original deposit totaled as much as 100,000 to 200,000 tons.

I worked a very long time at the problem of separating the little metallic spheroids from among magnetic particles of volcanic ash, magnetic sand, oxide scale, and several other substances. Since we had no shining laboratories with complicated technological apparatus at our outpost on Route 66, I was working with the most rudimentary of equipment. Some of our visiting scientists took amusement in my primitive ways, and some of them carried back to their universities and industrial plants raw materials that I gave them, wishing to test or demonstrate ideas or techniques of their own, but none of their methods proved to be practical.

Meanwhile, with no complicated gadgets at hand and with no mechanical skill worth mentioning, I continued experimenting until I could make a perfect separation on a small scale using nothing more complicated than a few small aluminum baking pans (actually, toy cake pans), a magnet, a series of sieves, and some cardboard boxes anyone could pick up at a dry goods store.

My simple set of procedures was tedious, but it worked, and that is something that none of the highly technical laboratory and university scientists could claim for their schemes. One noted scientist laughed heartily as he watched me slide my pans and boxes back and forth over a magnet, making capital of the fact that the particles were both magnetic and rounded and thus could be sorted out by their responses to the forces of magnetism and gravity and the principle of the inclined plane, but when my visitor tried his own method, the result was failure.

Spurred on by discovery of the little metallic spheroids, I searched

438   IN SEARCH OF FALLING STARS

and searched everywhere on all sides of the crater, on the rim and on the plain beyond, examining everything more critically than ever before, and one day made an even more exciting find.

I had stopped to examine a gravel pit dug into the crater rim by the state highway commission. I found a few crushed bits of brownish, yellow-green slag. Some showed a gray outer crust. I looked for more, and soon picked up a small tear-shaped piece. It appeared the same color as the light gray dust and gravel among which it lay, but the rockhound's licking test revealed a dark greenish-gray color under the dust. A canteen of water dashed onto the gravel made it easier to identify a number of such small "bombs" of various shapes and sizes. All of these, when broken, were seen to be of a spongy structure, but composed of brittle, glassy material. When I ground the bits of slag on a sheet of carborundum cloth from my supplies in the trunk of my car, and then held them under a pocket lens, they showed small, imbedded metallic particles, bright as chrome steel.

As I drove hurriedly back to the museum on Route 66 to make a nickel test, I puzzled over various questions. Could these be mere volcanic cinders? Could lava fragments carry such imbedded metallic particles? If these indeed were bomblets created by the impact of the meteorite, why had they never before been discovered?

Then I remembered that on the far side of the crater rim, the side nearest to the closest volcanic mountain of the area, lay scattered a number of volcanic "clinkers" that were similar to, though larger than, the bits of volcanic ash familiar on all sides of the crater. Every scientific and other visitor to the crater had walked over these heedlessly. Some scientists had picked them up, casually examined them, and tossed them away, pronouncing them volcanic lapilli. It seemed not illogical that the cinders nearest the source would be larger than those on areas farther from the volcanic mountain.

I, like everyone else, had trod over these "lava bombs" with complete disinterest, but now I was ready to look at them with somewhat more respect. The nickel tests I ran that evening on my little bomblets were positive. The next morning, I drove to the crater to study the clinkers on the southeastern rim.

Mingled with the rubble of the crater rim were tiny droplets of melted country rock, some as round as bird shot, others pear-shaped, oval, cylindrical with rounded ends, and still others of almost any imaginable shape and ranging from microscopic to the size of walnuts! One could kneel in a single spot and pick up 200 pieces without even moving. Some looked just like those from the gravel pit, but others seemed to be mere volcanic cinders. All, when ground on my carborundum cloth, showed metallic grains. Here again was the question of the possibility of metallic grains within the volcanic ash.

I began walking toward the volcanic mountain seven miles away, stopping frequently to search, keeping a keen lookout all the time. The bomblets thinned rapidly beyond the first 1,500 feet from the crater, and hardly any were found farther than a mile from the rim crest. I walked more than halfway to the mountain, and as I went, the fine-grained volcanic ash became more and more abundant and slightly coarser, but nothing was found to match my glassy-metallic slag of the gravel pit and crater rim.

At the Henbury meteorite craters in Australia, and at Wabar in Arabia, explosion products had been found, little glass bombs shot through with tiny, nickel-iron spherules, to which had been given the name *impactite* in reference to their origin: glass bomblets formed by a melting of rock by the meteorite explosion and scattered through a mist of nickel-iron.

The riddle had been asked often about the Arizona crater: Why were there no glass bombs like those at the Australian and Arabian craters? It was always assumed in these discussions that the Barringer group would have found impactite had it been present. The mathematical theorists always had an answer based on proper mathematical formulae, and those of us who were not mathematicians listened. I don't recall ever hearing anyone suggest that it might be advisable to make a search. The situation reminds me of the classical illustration of medieval logic—the argument over the number of teeth of a horse, a dispute that waxed so hot as to end almost in violence and that was left unsettled because no one thought to look in a horse's mouth.

Scientists had speculated and estimated in their attempts to explain the crater, but had never had done much looking. Now we had an answer: There were such glass bombs all the time. Arizona had its impactite, but we all had been walking over it so carelessly, we had not noticed.

Perhaps there is a lesson to be learned from this story of Dr. Moulton, the Arizona crater owners, and me: that team work between theorists and scientific fact-gatherers is a necessary part of good research.

Moulton should have insisted on a more thorough job of fact-gathering before accepting the assignment to furnish the Barringer organization with an estimate of mass within the crater. And he should have visited the crater.

On the part of the Barringer group, the failings were the very inadequate fact-finding survey of the crater and its environs, and the later and less-excusable failure to view the work of others with an open mind, which left for the public a confused picture of what happened some 20,000 to 50,000 years ago on the southwestern American desert in what is now Arizona.

Looking back on our luncheon of 1936 from the perspective of many years, I wondered if my failure to grasp Moulton's point of view may have been the reason that his proposed efforts on my behalf had come to naught, if perhaps he had decided quietly, then and there, that I was not capable of usefully employing an endowment for research. If this were true, he gave no indication of it and, in fact, continued visible efforts for several years beyond the occasion.

My work more and more convinced me that Moulton's explosion theory was correct and that the great meteorite that dug the Arizona crater had vaporized on impact. My report to the American Philosophical Society in 1949 pointed in this direction, and in October, 1950, I made public announcement to the press of this belief.

Many scientists have wondered if large meteorite impacts may be accompanied by atomic fission. Calculations by British scientists led to the conclusion that such impacts should produce temperatures comparable to those developed in the explosion of atomic bombs. It is well known that the surface of the soil was left glazed by the heat of the first atomic blast in New Mexico in 1945. In December of 1953, I was permitted to visit the atomic testing ground at Yucca Flats, Nevada. Not only was the surface of the terrain under those blasts glazed, but several thousand tons of surface rock had been melted and reduced to vesicular blobs of slag and glass, scattered by the fury of the blast to great distances. All of the various shapes and sizes of impactite bombs found at the Arizona meteorite crater were duplicated at Yucca Flats to the most minute detail of structure and form, except that those of Yucca Flats contained no nickel-iron particles.

No longer is the Arizona crater the largest known meteorite crater: A recently recognized Canadian crater gapes seven miles wide, and there are others also that are larger, but none boasts the fortunate combination of size, accessibility, and freshness of the Arizona pit.

It is recognized now that there are undoubtedly many major and unexplained topographical and geological features whose origin was the impact of giant meteorites, comets, or asteroids, and that other such features have been obliterated by ages of geologic change and erosion past recognition. The great "canyons" of the ocean floors may well be of impact origin; fossil lava flows may be the result of ooze from the plastic interior of our earth which spread from wounds gouged out by invading bodies from space.

When we moved near the crater in 1946, our friends voiced skepticism. Hadn't all angles of this fifty-year-old discovery been investigated? Certainly nothing remained to be discovered where a million dollars had been poured into exploration.

I believed that all of the most important facts had been observed and recorded, though I suspected there might be some need to reclassify and reexamine certain accepted theories. I was amazed to discover that there were basic inadequacies in the manner of attack on the crater problem, or problems.

First, each of the writings on composition and structure of the meteorites had been based upon the study of one or a very few specimens. It seems obvious that no safe generalization could be made relative to the nature of 100,000 individual specimens simply by the study of a few.

Second, no thought seemed to have been given to the possibility that specimens taken in different locations relative to the crater might show different and significant characteristics.

Third, nearly all of the monies expended—in today's dollars probably an amount equivalent to $1.5 to $2 million—had been devoted to exploring the crater pit on the assumption that the principal bulk of the colliding mass resided in its depths, despite the conclusion by many scientists that such a huge mass would of necessity explode on impact. If the colliding mass had exploded, the logical place to look for the evidence would be not the bottom of the pit, where it was necessary to work under several hundred feet of rubble, quicksand, and water, but rather on the terrain surrounding the crater.

There was no indication that any careful study had been made of the soil of the surrounding plain nor of the outer rim slope. Our subsequent investigations demonstrated that an expenditure of a few thousand dollars on that terrain would have supplied facts more in harmony with the best scientific theory than had any of the extensive and expensive efforts inside the pit.

The prime purpose to which our collecting efforts was directed, and which would have been served by the comprehensive mapping originally intended, was to prove or disprove a pattern for the distribution of fragments about the crater.

It is unfortunate that it was never possible to complete some of the projects I started. However, the specimens collected, their characteristics, and the information I had as to their sources made possible some studies and conclusions.

Despite the interruption of the mapping activity, I knew the quarter section from which any one of half of all the fragments we collected had come. I knew still more specifically the locations of 5,000 specimens, and the depth of the layer of soil from which any of them had been dug. From the cowboys and other collectors of earlier days, I had learned the approximate locations where hundreds more of the larger specimens had been collected.

Our 1939 survey had suggested strongly the radial distribution of small fragments around the crater. I made it a point to ask the cowboys and other meteorite hunters if there seemed to be any pattern of distribution. Their answers revealed that these men regarded as guides to good hunting areas lines drawn from the center of the crater through any locations where finds had been made. This evidence of radial distribution was offset by distortions of the pattern, notably a concentration of small irons on the northeastern rim of the crater.

Altogether, I spent more than 25,000 daylight hours within sight of the great crater in Arizona and put in more than 2,000 hours of work in and around it.

I had scouted the crater's interior and exterior and had studied it from the air. An excellent set of aerial photographs was made available by the photography school of Lowry Air Force Base at Denver. A study of prevailing wind patterns and the examination of the crater out-throw as shown on these photographs supported an explanation that the heavy concentration of fragments on the northeastern rim was due to wind.

Approximately 500 of the meteorites we collected had been cut into sections, either by me personally or under my personal supervision. I had etched and studied some 1,500 sections under magnification.

In 1936, I had discovered among several etched specimens from around the crater one that evidenced by its structure and composition an origin different from the rest. This I named Canyon Diablo No. 2. Then, as we sectioned and studied more and more of the crater irons, we soon learned that not all of these specimens had come from the same meteorite, for their "signatures" did not conform to a single pattern. As more and more were cut, we discovered not only additional representatives of Canyon Diablo No. 2, but at least four more types—evidence that either a swarm or a small system of discrete bodies had traveled together, rather than just one large mass.

When the question was raised as to whether the crater was formed by one mass or a great swarm of meteorites, all of the attempts to answer were based on mathematical investigations as to the mechanical possibilities and probabilities. No one seemed to think of hunting for material evidence to support either hypothesis.

All of the small meteorites taken on the crater rim showed evidence of having been altered by heat, while fragments of similar size taken from the plains beyond the crater rim showed no such evidence.

My examinations of etched specimens under microscope revealed thousands of inclusions, including forty groups of the famous *carbonados*, or black diamonds, found not infrequently in Canyon Diablo specimens and occasionally in other meteorites. I made an attentive study of the carbonados and of the incidence and relative abundance of inclusions of cohenite and schreibersite, troilite and kamacite.

Fitting together various findings, I developed a theory as to the formation of the crater. The evidence of a composite fall, the distortions of a radial pattern of distribution, the evidence of heat alteration in small fragments found on the rim and not elsewhere, the distribution and nature of the so-called shale balls, found mostly on the crater rim itself, and the finding of explosion products—metallic spheroids and impactite—seemed to me to indicate the impact of a planet-satellite group of meteorites, which on first striking the earth produced a shattering of the outer zone of the mass wherever it contained brittle inclusions, while the body of the meteorite bored inward into the earth to a final greater explosion of the violence of a hydrogen bomb. Remnants not vaporized shot downward and deep into the fractured rock or upward as red-hot and white-hot slugs—the rim specimens that showed heat alteration.

The most significant conclusion was this: The Arizona crater is a great part of our national heritage and should have been treated as such from the start. Since it was not, certainly now it should be acquired and given to the public as part of the national parks system.

The Antiquities Act of 1906 has met with the approval of practically every citizen in a near unanimity of opinion not often given to congressional legislation. Only the Antiquities Act prevented the complete destruction of the now world-famous Petrified Forest when, in 1906, plans were being perfected by commercial organizations for the crushing of the beautiful agatized logs for use in the making of sand paper, grinding wheels, and other abrasives. Such scientific marvels as the great Natural Bridges, Zion Canyon, Muir Woods, Rainbow Bridge, Craters of the Moon, and early vestiges of civilization like Montezuma Castle and Mesa Verde would have suffered mutilation or destruction but for the protective hand of the National Park Service.

It has become a part of the American way of life to set aside those outstanding bits of creation so endowed by nature as to contribute significantly to man's understanding of his environment. Thus our national government protects the heritage of all her citizens of present and unborn generations. But because man gains knowledge of his environment mostly by short steps, each new advance paving way for the next, certain features unrelated to their immediate surroundings and without counterpart in man's previous experience have gone unrecognized for generations.

Such was the situation with the Arizona crater. Regarded by scientists for a half century as of questionable import, this feature of the Arizona landscape now looms preeminent as a potential storehouse of vital information. The world's finest example of meteorite craters, regarded by some as easily the most significant scientific marvel of the American continent, passed into private ownership by virtue of mistaken identity and compounded error.

The crater was interpreted in 1891 as of volcanic origin, rather than formed by meteorite impact, by G. K. Gilbert of the United States Geological Survey. Dr. Gilbert was a very learned man of his day. He had considered the idea of impact, but he'd taken the more conservative stand and pronounced the pit a volcanic blowout, which it resembled in many ways. After A. E. Foote, Gilbert probably was the first man to suspect the existence of meteorite craters, but no one could know very well what the appearance of such craters might be. In any event, his mistaken determination of volcanic origin set the stage for the next blunder.

When a mining engineer learned twelve years later of the crater and of the persisting story that it had been formed by a meteorite, he conducted some explorations and then promptly filed a mining claim on the land. The United States Department of the Interior committed the second error when it accepted the Barringer estimate that the crater contained mineable ore and granted title as mining claims. Although law required proof of the existence of ore in mineable quantities before application for title, the crater ownership through half a century has failed to yield a single ton of such ore.

But now the presence of a large body of meteorite has been shown to be an impossibility. Instead of being buried in the pit, the meteorite material is disseminated in the soil around the crater and in the fill within. This conclusion is reinforced by all other investigations of meteorite craters in other parts of the world: Meteorites large enough to produce craters larger than a few hundred feet in diameter simply do not reside in the craters they produce but are reduced instead to gases and dust, blowing themselves to bits out of the craters they have dug.

The now-accepted explosion theory that debunks the claims for mining wealth within the crater is in itself an overriding reason for preservation of the crater as a national treasure. The study of explosion products and the effects of the explosion on surrounding rocks looms as an extremely important area of investigation. We do not know, for instance, the extent of alterations in the materials of the earth's crust produced by the brief application of tremendous pressures beyond compare in any other phase of the earth's behavior.

Within the last few years, two completely new minerals have been produced by application of high pressure to common minerals of the earth's crust, and a subsequent search proved that both of these minerals are present in the Arizona crater. It seems regrettable to allow this tremendously challenging laboratory to lie unused any longer.[22] Extensive explorations should be made in the surrounding rocks to ascertain what changes occurred as a result of the impact, and how far-reaching they were.

---

[22] The increasing interest in meteorites in recent years has changed this situation. The crater is now actively used for research and as a teaching tool for students interested in meteorites. *Ed.*

448 IN SEARCH OF FALLING STARS

Accessible as it is and of such large size, this crater should be studied thoroughly in every aspect, and serve as a standard by which to evaluate other discovered craters less well-preserved and either larger or smaller than the Arizona "type" specimen.

The expensive explorations carried out by mining interests since the Arizona crater passed into private hands in 1903 had the avowed purpose of exploitation and resulted in some degree of mutilation of some of the most meaningful parts of this magnificent product of cosmic collision. To serve its greatest usefulness, it should have been kept under the strictest supervision by a skilled staff of scientists, every square foot of the great pit and its surrounding uplifted rim, as well as the out-thrown rubble and the surrounding plains, being made accessible to all interested citizens without endangering its preservation.

When the 1947 meteorite shower in the Sikhote-Alin mountains of southeastern Siberia resulted in formation of some third-rate craters (in point of size), Russian scientists attacked the problem with a corps of specialists, including astronomers, geologists, meteoriticists, metallurgists, geographers, photographers, artists, and surveyors. Buildings were constructed over some of the Russian craters to preserve them in undisturbed condition for future generations.

The United States trailed far behind the Soviet Union in this sort of study. Russia has had a National Committee on Meteorites for nearly forty years. This committee has attacked the problem of meteorite craters in a manner of thoroughness which far surpasses anything done in the United States.

The craters produced by the fall of 1947 in Siberia were not large and therefore did not yield all the different types of information which could be gleaned by study of larger craters, but they did produce some rather striking results and without question lent impetus to the ballistics program of the USSR.

Canadian government surveys indicate that old meteorite scars of large size on the earth may be innumerable. Canadian scientists have found at least twenty almost certainly proven craters larger than the one in Arizona. One of these, about seven miles in diameter, is a conspicuous feature, and others up to thirty miles in diameter have been proven to be of impact origin by core drilling.

Our own great nation had the first introduction to this very important question of impact scars and still possesses the finest example, but our country is far behind in the study of this aspect of the earth's relation to the solar system. As a National Monument—a recognized part of our heritage, receiving the scientific attention it deserves and eliciting the respect and awe of traveling Americans—the crater would bring honor to the owning family and constitute the greatest possible tribute to the late Daniel Moreau Barringer, who in the face of frustrating opposition continued to work in support of an unwelcome and controversial interpretation of the crater as of cosmic origin.

16

Home

He hath no leisure who uses it not.

—George Herbert, *Jacula Prudentum*

There were tense moments, hours, and days as we looked from our hill and saw the traffic speeding on the new highway with seldom a vehicle approaching by our narrow dirt road. But when we had recovered from the shock, we turned to the tasks at hand. One of our chief objectives in coming to this place had been to give us an opportunity to carry on crater investigations. We now set about re-ordering priorities. We sold some oil leases which I had picked up during my war job with the Solar Oil Company. We upped our offerings of sales material (meteorites), and I devoted much of my time to crater investigations. During those forced three years, I made some of my most important discoveries.

As our time on Route 66 began to reach an end, we were still uncertain as to the future. We were counting the months and weeks

Addie and Harvey Nininger with their dog, Blondie, at the American Meteorite Museum on Route 66, c. 1951.

until our long lease should expire, and at the same time were studying and discarding one idea after another. Finally we decided to build a new museum. We chose as a site the town of Sedona, Arizona, just south of Flagstaff and about thirty miles to the southwest. New friends lent us money for the construction of a suitable building. While they supervised its construction, we took care of the summer tourist trade on Route 66.

In September 1953, we transported our tons of meteorites down the steep, twisting Oak Creek Canyon road to our new building. It lacked the picturesqueness of the old "observatory," but was fresh and clean and bright, and the downstairs apartment seemed to us the ultimate in civilized comfort after our seven years of mud walls and one-room housekeeping. We had nurtured the few green growing things—cactus and Chinese elm—on our wind-torn hillside; now we had a real garden area, with space for grass, for flowers and vegetables, with plenty of precious water to support them. We had ordinary, dependable alternating current electricity; Addie was reintroduced to the ease of housekeeping equipment with plugged-in power. After seven years, we were again part of a community, with neighbors, shops, and conveniences.

We entered our new venture with a more relaxed attitude than we had known in any of our years on Route 66. By then, we were accustomed to the strains and surprises of dealing with changing crowds of people, and we had grown bold enough to occasionally put a "closed" sign in our window. Until such time as we could hire help, we were prepared to care for the museum's needs ourselves. Our business started out well enough to encourage us that the Sedona move would prove to have been wise, but we were not so busy, nor were the burdens so heavy, as on Route 66 even after the move of the road.

In 1954, we received a copy of *Minerals for Atomic Energy*, written by our son Bob, then deputy assistant director for exploration of

The American Meteorite Museum in Sedona, Arizona. The museum operated at this location from 1953-1960.

Interior of the American Meteorite Museum in Sedona, Arizona. Harvey Nininger is on the right, behind the counter, c. 1954.

the division of raw materials of the Atomic Energy Commission, and it seemed more of a thrill than when my own first book was published.

In 1955, Margaret and her husband, Glenn Huss, came to Arizona to assist us. As well as sharing the load of conducting visitors through the museum, they helped us with a new book, *Arizona's Meteorite Crater*, and a revision of *A Comet Strikes the Earth*.

When he was a boy, Glenn Huss lived in Horace, near Tribune, Kansas, while I was lecturing in that part of western Kansas and at times had Alex Richards lecturing in schools there. On April 17, 1937, Glenn wrote me concerning what he thought was a meteorite that he and another boy had found.

At the time, Glenn Huss was no more important to us than a hundred other boys who'd gotten our message, but in about 1950, he went to work for the University of Denver Press, and he met Margaret, who was also working there. Glenn had a master's degree in

English, and the two were married two years later. Glenn had been very interested in meteorites ever since hearing the school lecture fifteen years earlier.

When we needed help in the museum in Sedona, we offered him a job, and he took to the work as a duck takes to water. He was an avid reader, had a marvelous memory, and had a natural liking for chemistry and all kinds of science.

In *Arizona's Meteorite Crater*, I retraced the feature's past history, outlined my conclusions as to its formation, and reiterated my hope that the crater one day would belong to the nation. I suggested twenty-eight specific research projects to be undertaken to evaluate the great event of ages past in relation to scientific undertakings of the future.

In 1953, Dr. Loring Coes, Jr., while subjecting quartz to intense heat and exceedingly high pressures, discovered that he had produced a new mineral harder and more dense than quartz by 13.5 percent. This was named coesite. I was putting the finishing touches on my manuscript for *Arizona's Meteorite Crater* when I read the announcement of Dr. Coes' discovery. I read also comments by eminent geologists to the effect that coesite could never be found in nature, that it could not be produced short of sixty to 100 miles below the surface of the earth because only at such depths could sufficient pressures exist.

These men were overlooking a point. Many physicists had calculated that a meteorite striking the earth, even at a minimum speed of five miles per second, would exert a pressure far greater than that produced in Coes' laboratory. Consequently, I appended a suggestion which appears as a footnote on p. 50 of *Arizona's Meteorite Crater* and recommended again on p. 154 that a search for this mineral should be made in and under the rim of that crater. It was with a feeling of great satisfaction that I read in *Science*, July 22, 1960, of the discovery of coesite in shocked sandstone from the Arizona crater. The discovery of coesite triggered further investigations into the changes wrought on the rocks of the earth by the exceedingly high pressures exerted by crater-forming meteorites, and some of these investigations turned up another new mineral, stishovite.

In *Arizona's Meteorite Crater*, I suggested that the diamonds found frequently in Canyon Diablo meteorites may have been formed at the instant of impact instead of arriving from space in the meteorite. During some twenty years of handling and cutting irons from the crater, I had noticed that all of the diamonds encountered were in "rim" specimens, those found on and near the crater rim, all of which showed effects of high temperatures, so I raised the question of whether the diamonds might be the result of high pressures and temperatures. I theorized that these rim specimens came out of the very final explosive fragmentation that had taken place in the depths of the pit, where most of the meteorite was reduced to gases and dust and the remnant in the form of small pieces had been shot high into the air and then had fallen in close.

Dr. Michael Lipschutz and Dr. Edward Anders of the Fermi Institute, whose meticulous research proved the hypothesis, credited me for the origin of the idea and the loan of diamond-bearing specimens to explore it.

I was just completing *Arizona's Meteorite Crater* when late in 1952, I discovered the presence of impactite on and in the crater rim. My manuscript preparation was interrupted by some two years of investigations into the nature, distribution, and quantity of this fascinating material.

2

On October 10, 1957, when Sputnik was about one week old and still dominating our nation's news pages, a huge, brilliant fireball flashed across the morning sky of three western states. A highly trained pilot, a lieutenant commander of the navy, reported that he had dodged the fireball and then watched it disintegrate beneath him when he was flying at 17,000 feet.

My survey of this meteor showed that the point of disintegration was at a height of several miles over the central Colorado mountains, more than 150 miles from the plane. In a Sputnik-oriented world, this pilot's error was a serious blunder. For if the observations of highly trained pilots and military men were no more accurate, it would be as easy to confuse a meteoritic fireball with an intercontinental ballistic missile as it would be to repeat this obvious mistake in location, distance, and nature of the object observed. But the navy pilot had made his report: "I was flying at 17,000 feet near Myton, Utah, when I saw this flaming object coming from my left in what I deemed to be a collision course. Quickly, I banked and turned. As I did so, the great fireball disintegrated under me in a shower of sparks. I resumed my course, and when I looked back, I saw that several small fires had been started."

Far to the north in Wyoming, men on night duty in an oil field reported that a blazing object as large as one of their huge field tanks had passed over them, narrowly missing some of the storage tanks. At Grand Junction, Colorado, far to the southwest, observers trembled as they saw this dazzling light glide to a landing on the slope of Grand Mesa in western Colorado. From Alamosa, 200 miles south of Denver and slightly west, it was seen to disappear behind the railroad watering tank. At the same moment, men in the control tower at Lowry Air Force Base in Denver saw the object disappear over the western part of that city.

A Denver resident, one of the few abroad that morning at 4:07 when the meteor swept in from the northeast and disintegrated over the central part of Colorado, described his reactions to me:

"I was standing right here. Had just come out to get in my truck.

You see, I haul trash for people, and I get out early. I had just opened the door of the truck to get in, and there she come! I'm telling you it was the scariest thing I ever seen, twice as big as the sun and ten times as bright! Sputterin', throwin' off sparks and big hunks of white-hot stuff. I was scared to death. Tried to call my wife. She was right there in the kitchen, but I couldn't talk. I stood here trembling and pointing. It was all over in a few seconds. Passed right over that telephone post and only about ten feet above it.

"Listen, they say it was a meter or something like that. I know better. I seen it. I know. That was Sputnik. And I knew that he could turn a death ray on me and just evaporate me. I want to tell you, I never been so scared in my whole life."

The office of the North American Air Defense Command at Colorado Springs reviewed all of these conflicting reports. Charged as it was with the nation's protection against intercontinental ballistic missiles as well as all other forms of sneak attack, this office could not allow itself to be misled by the conflicting, fear-inspired reports of a lot of frightened laymen. The military concentrated on the report of the navy pilot. A commission was dispatched to Vernal, Utah, the region where the officer had made his observation.

The commission began its investigation by flying over the area where the object was supposed to have disintegrated under the navy plane. No evidence of any fires was found. Nor was any debris seen that could be considered the remains of the fiery object that had frightened citizens of three states.

The commission then inquired of citizens in the little city of Vernal. Yes, one testified to seeing the object disappear about seven or eight miles southeast of the town. Out went the officers, but they seemed to approach no nearer to their goal, for residents here pointed still farther to the southeast.

"Right on the north slope of that mesa. I saw it strike, and there was a great shower of sparks when it hit."

The officers went to the mesa slope but found nothing. They queried residents of the vicinity. Again they were told that the object had struck farther to the southeast. Was this a hoax, or a hallucination? Whatever these people had experienced must be an entirely different phenomenon from what the navy officer had reported.

Back in Vernal, more witnesses were questioned.

"Go see Bill Higgins. Bill is a down-to-earth, level-headed rancher who is honest as the day is long."

Yes, Bill Higgins told the investigating officers, he had seen the strange fireball. It landed within a quarter mile of his house, but it was not southeast of town by any manner of means. It was southwest about fifteen miles.

"It struck right behind a little hill just back of my corrals. Looked like it would blow up the whole earth, but I never heard a sound."

Close further questioning convinced the commission they now had been pointed to the right spot. But there must have been more than one object, Bill added, for a friend of his out north of town said he saw one hit not far from his house at about the same time.

The more the commission investigated, the greater grew the confusion. After searching a few locations, they returned to headquarters and filed their report as an unsolved mystery.

Had these men gone with Bill Higgins to his ranch and asked him to point out for them the spot where he had seen the object strike, he would have pointed toward the southeast, and his friend north of town would have pointed in the same direction. The men in Alamosa would have pointed north and a little west. Those in Grand Junction saw the fireball a little north of east. Witnesses in Estes Park saw it pass on the west of them, going south, while the men in the oil field in Wyoming saw it in the direction of their oil tanks south of them, seemingly so near that they feared it would hit them and start a disastrous fire. Actually, the meteor was forty miles high, but to them it appeared no higher than it had to the trash hauler in Denver.

The great fireball was of such magnitude that if it had passed within a half mile of a man, he probably would not have lived to report it. But it never came lower than about twelve to fifteen miles above the earth. Most of the witnesses, including the navy pilot, had seen the object more than 150 miles from them. Only in the vicinity of Eagle, Colorado, had people seen it from less than forty miles. At about twelve to fifteen miles altitude, some ten to fifteen miles south of that village, the mass disintegrated. Those who were awake within a twenty-mile radius of Eagle heard a tremendous blast, followed by a thunderous roar and rumble, and a few in that sparsely settled region who had been awakened by the glare and blast reported they actually heard stones thudding down among the dense covering of scrub.

Investigation depends on such simple tools as protractor, ruler, pencil, map, and a listening ear. I never definitely concluded that any of the many fireballs I surveyed failed to deposit meteorites, even though meteorites were found in only a fifth of the surveys made. When a fireball vanishes at a height of five to fifteen miles, which is the rule with those that produce meteorites, it is impossible to designate a target for the surviving fragments smaller than eight or ten square miles before any recoveries have been made. To search such an area for objects the size of a walnut or a brickbat is like searching a haystack for the needle. Additional problems are rugged or highly vegetated country and the possibility that the fallen fragments are the size of peas or grape seeds or only sand-size particles.

3

One of the twentieth century's great scientists, truly the father of space flight, was Robert Hutchings Goddard. I have a very clear mental picture of Dr. Goddard on the occasion of a visit to his laboratory near Roswell, New Mexico, late in 1930, when he had just started his liquid propellant rocket research. The professor showed a very modest attitude toward his work, as there leaned in a corner a somewhat battered length of what appeared to be aluminum tubing, nearly double the height of a man, resembling somewhat an oversized stovepipe. Professor Goddard explained this cigar-shaped rocket had just been recovered from a flight that reached a height of two thousand feet—the first verified successful flight of his new experiments in propulsion, guidance, and the recovery of rockets. Goddard's patented gyro was the heart of the rudimentary rocket, guided by two sets of vanes and with a parachute contained in the nose cone to ease its return. His stovepipe was the forerunner of all the great space-probing giants of today.

Asked just what his objective in these tests was, which the press had been playing up as his "futile" efforts to reach the moon, he said very modestly that his hope was to explore the upper atmosphere. I was sure in my own mind that Goddard's ultimate aim was to help make space exploration possible. That modest scientist who blazed the trail for space exploration throughout the world was financed so stingily that one wonders how he accomplished anything at all.

Goddard began his experiments with solid propellant rockets during the last of World War I, when with funds from the Smithsonian he developed his first single-charge rocket in 1918 at the Aberdeen proving grounds. With the end of the war, the project was shelved, but it fathered the bazooka of the Second World War. Dr. Goddard then joined Clark University in Worcester, Massachusetts, and alternated service there for some years with his experiments in New Mexico, which were suspended on a couple of occasions for lack of funds. In 1939, by which time he had succeeded in sending a sixteen-foot rocket aloft 8,000 or 9,000 feet, he tried to interest the

Army Air Corps, but the only future the military saw for Goddard's rockets was the development of jet-assisted take-off. So Goddard aided his country in two wars, but he was too far ahead of his time for his real hopes to begin to be realized before his death.

Rocketry was not related to my line of study, but as I studied the flight patterns of meteorites, which encountered the same forces and impediments as rockets, and as I examined and studied the surface markings of thousands of meteorites, I came to the conclusion that when and if projectiles were developed with speeds approaching those of meteorites, blunt or rounded noses would have to be provided instead of the sharp, pointed front ends being promoted in the first years of rocket and missile research.

While on a lecture tour in the east in the mid-1930s, I went to see the Yale collection of meteorites. Records showed that a large number of the small pellets that had accompanied the fall of larger stony-iron fragments near Estherville, Iowa, in 1879 were preserved in the Yale museum, and I asked to see these. Dr. William E. Ford, in charge of the department of mineralogy and of the meteorite collection, was most helpful. For two or three days, I was allowed to study as I wished. In this time, I examined more than 600 of the little pellets.

Various scientists before my time had described the Estherville fall, one of the greatest falls in history, but the significance of the rain of small particles was an aspect that seemed to have been overlooked. Dr. Ford told me I was the first person to show any interest in Yale's Estherville pellets in the fifty years they had lain in the museum. These small irons had been set free as irregular lumps of metal in the disintegration of the large mass of stone and metal. A veritable hail storm of little nickel-iron nodules was described by a small boy who at the moment of the fall was driving the family herd of milk cows toward home past a small, shallow lake. He said the surface of the lake "was peppered like in a hail storm," and that "the cattle had stampeded in all directions." It was learned later that the little metallic pellets had rained down over an area some six or seven miles in length. Several thousand were picked up by curious people who took the time to hunt for them. The pellets varied in size but averaged about the size of hazelnuts.

I had studied the structure of the main mass of Estherville through specimens on display at various museums. This meteorite has a sort of fruitcake structure consisting of a matrix of very brittle stone in which are imbedded metallic nuggets of various sizes down to microscopic grains. The rain of pellets had resulted from the crushing of a large mass by atmospheric resistance, thus setting free the little nodules. A great "smoke cloud"—dust, of course—had been reported at a height of some fourteen miles.

Apart from this unique shower, very few metallic meteorites had ever been recovered promptly enough for a detailed study of surface features as they existed at the end of flight. Oxidation soon begins to modify this aspect of a body that has not been exposed previously to the active ingredients of the earth's atmosphere.

About thirty percent of the pellets I examined showed signs of "orientation." They had fallen without whirling motion and with sufficient force and speed to produce frictional shaping—a melting away and molding of the forward part while the protected rear portion remained relatively unaffected.

It was this group of small, oriented specimens that intrigued me most. Their flight markings showed they had traveled in a stabilized position without tumbling action, and their noses were almost perfect hemispheres. Similar rounded noses had been among the stones gathered from the Pasamonte fall. They did not conform to the ideas then set forth by writers on meteorites as to shapes to be expected in stones fallen from space.

This discrepancy led me to devote a good deal of time over

several years, beginning in 1934 or 1935, to search among the other groups of meteorite specimens for that small percentage which showed this characteristic orientation, and to carefully examine the remaining evidence of violent and fiery passage through the earth's atmosphere.

Among the 600 Estherville pellets at Yale's Peabody Museum, the shape of one tiny, blunt, but streamlined specimen caught my special attention. This significant little meteorite later was given to me by Dr. Ford.

The idea of the *blunt nose* for missiles was born that day in the museum at Yale but was obliged to wait for more than twenty years. Over two decades, the experts called for appropriations for the construction of greater and more powerful wind tunnels for use in discovering the most stable and least resistant designs for shaping the missile nose, yet all the while, tested models from the greatest wind tunnel of all were lying unused in the meteorite collections of the world. Every meteorite that tunnels its way through the atmosphere to land on or in the soil brings with it a record of its struggle against an air current more powerful than can be supplied by the most modern wind tunnel. These tested models stare up at visitors from museum cases and although they have penetrated the equivalent of several yards of rock, sometimes it seems their messages are unable to penetrate the gray matter of the engineers who look at them.

In 1935 and 1936, I published three papers describing the surface features of the Lafayette, Bruno, and Pasamonte meteorites,[23] in each case discussing the effects in terms of flight markings of their atmospheric travails.

The Lafayette specimen is one of the best examples of the results

---

[23] "The Lafayette Meteorite," *Popular Astronomy*, Vol. XLIII, No. 7, Aug.-Sept., 1935, pp. 404-408; "The Bruno Meteorite," *American Journal of Science*, Vol. s5-31, No. 183, March 1936, pp. 209-222; "The Pasamonte, New Mexico, Meteorite," *Popular Astronomy*, Vol. XLIV, No. 6, June-July, 1936, pp. 331-338.

464 IN SEARCH OF FALLING STARS

of an oriented flight through the atmosphere. The position of the stone in flight is quite evident. The spheroidal front is abundantly and evenly beset by a host of fine, crinkly ridges of blackish glass radiating from its central point. Describing it in 1935, I wrote:

Altogether the lined surface of the stone gives one the impression that it has been formed by the cooling down from a condition in which the surface of the entire front was in a liquid state to a temperature below the melting point of the stone which allowed the molten matter to congeal while in the process of being swept away ... the almost flat base ... appears to have been developed in a situation where almost no atmospheric disturbance existed. ... Here, in the wake of the moving mass, was apparently an almost perfect vacuum. ... We may think of this rear side of the meteorite as a furnace, for over its edges came the violently heated blast of air which closed in to fill the constantly forming vacuum which was always advancing rapidly enough to elude the inrushing air ... a very effective furnace ... which differed from the front exposure of the meteorite mainly in being in a state of calm, or rather, being in a vacuum.[24]

Examination of thousands of completely encrusted individual meteorites convinced me that there is a critical velocity beyond which aerial friction operates differently from its effect at lower velocities.

In meteorites traveling at several miles per second, we encounter a condition which differs notably from that which it obtains at lower velocities. The drop-shape with a rounded front and a tapering rear,

---

[24] "The Lafayette Meteorite," *Popular Astronomy*, Vol. XLIII, No. 7, Aug.-Sept., 1935, pp. 406-407.

which is considered the least resistant form in mechanical devices, is seldom attained in meteorites, probably for the reason that the velocity is too great. Consequently, the destructive erosion which would be expected on the lateral and leeward slopes is not in evidence. In fact, the drop-shape is notably absent in meteorites. In the few instances where found, such as the Boogaldi and Charlotte irons, we find the evidence of aerial conflict limited to the <u>rounded large end which was the forward end in flight probably because of its form rather than that the form resulted from this position.</u>[25]

The oriented Estherville and Pasamonte stones did not conform to the idea that a falling meteorite must be cone-shaped that was then prevalent among aeronautical and ballistics engineers and among writers on meteorites. In 1934, I heard a presentation by an aeronautical engineer who suggested, as applied to automobiles, a pointed tail and blunt front. The only consistent feature I found among oriented meteorites was a dome-shaped or hemispherical front end; the remainder of the meteorite might be almost any shape.

The most perfect example of oriented flight markings on a meteorite of some size is the twenty-eight-pound Bruno, Saskatchewan, Canada, meteorite. Its graphic flight markings indicate a rocking motion of the meteorite during a portion of its flight and, again, a rounded nose at the front with a low pyramidal base which at no time received the force of the air blast.

Surface features of individual specimens had been dealt with in the various descriptive papers of those meteorites, but so far as I knew, no writer had undertaken to survey a large number for the purpose of group study. I examined a total of nearly 7,000 meteorites,

---

[25] "The Lafayette Meteorite," *Popular Astronomy*, Vol. XLIII, No. 7, Aug.-Sept., 1935, pp. 407-408.

466 IN SEARCH OF FALLING STARS

including 800 metallic or partly metallic specimens, in eighteen institutions and two private collections in addition to my own, and including two institutions in Mexico and one in Canada. My studies of surface features were reported to the second and third annual meetings of the young Society for Research in Meteorites in 1934 and 1935 and published in the press.

The 1935 paper, "The Surface Features of Meteorites," dealt particularly with surface melt and erosion during flight through the atmosphere.[26] The 1936 paper, "Further Studies in the Surface Features of Meteorites" discussed orientation, form, and pitting of meteorites.[27] In the 1935 presentation to the Society, I stated:

Without question here is a field which should prove a fruitful source of information on questions of aerodynamics and ballistics. The engineers who are concerned with the problem of stratospheric transportation would doubtless find much to interest them in the study of meteorites which have survived a flight through the upper regions of the atmosphere.

In a progress report on our program published in *The Scientific Monthly* of August 1938, I predicted that "Aeronautical engineers will learn much from the study of meteorites before they master the stratosphere."[28]

In the prospectus for a meteoritical institute submitted to the Penrose Foundation in 1940 was a section predicting "practical benefits"

---

[26] "The Surface Features of Meteorites," *Popular Astronomy*, Vol. XLIII, No. 2, Feb., 1935, pp. 121–126.

[27] "Further Studies in the Surface Features of Meteorites," *American Journal of Science*, Vol. s5-32, July, 1936, pp.1–20.

[28] "A Meteorite Survey," *The Scientific Monthly*, Vol. XLVII, No. 2, Aug., 1938, p. 141.

of such an institution. In listing a dozen such advantages, I included the following paragraphs as numbers 8 and 9:

8) Thorough investigation of the explosive disintegration which marks the finish of the luminous flights of meteorites, coupled with a detailed study of the heat effects registered in the fragments collected from these falls and consideration of the forms which characterize surviving fragments, should throw new light on the problem of air resistance to high velocities. Here is an opportunity for aeronautical engineering to gain new light on problems connected with stratospheric transportation. Problems which face stratospheric flying are difficult to solve chiefly for the reason that they lie beyond the reach of experimentation. Meteorites are constantly meeting with those conditions as they land on the planet. By a careful study of them, together with the light phenomena which mark their arrival, a better understanding may be had of the difficulties which flying craft will sometime have to meet.

9) The science of ballistics will need more and more to study the forms and markings of meteorites as the velocities attempted by ballistics engineers approach those of the invaders from space.[29]

My findings in the areas of orientation and surface features were reported in detail and generously illustrated in my book, *Out of the Sky*, in 1952, and additional photographs of oriented specimens were included in the catalog, *The Nininger Collection of Meteorites*, in 1950.

Perhaps I have been as much interested in the space program as many of those actively engaged in it. All meteoritical studies are

---

[29] During the 1960s, experiments by Dr. Dean Chapman, using modern wind tunnels, clearly demonstrated the formation of spheroidal fronts on glass and other substances.

directly or indirectly related to space. Apparently my papers on oriented meteorites and the nature of their flight were not read, at least not at the time of publication, by persons prepared to give practical effect to the information contained in them.

As experimentation with missiles reached the point where engineers were attempting supersonic velocities, it seemed clear to me that they could well copy the contours of oriented meteorites—that is, in place of the needle nose which seemed to be their ultimate objective, they should adopt the dome-shaped, or blunt nose, which I had concluded must be developed by any meteorite that achieved stabilized flight through a considerable portion of its atmospheric transit. In other words, one that had maintained an oriented flight.

As we entered World War II and our government was calling for ideas from the public which might expedite our military preparations, I took up the matter with Dr. Forest Ray Moulton, whom I knew to be one of the chief advisers on ballistics, and who, of course, had worked on meteorite problems as well as in the fields of astronomy and mathematics. But I found that Dr. Moulton disagreed. We argued at length, but he remained unconvinced.

In July 1942, I sent to the National Defense Research Committee of the Office of Scientific Research and Development a memorandum outlining facts about meteorites which I believed to have an important bearing on ballistics. I described, briefly, the matter of orientation:

These meteorites have been irregular in form at the time they were set free from the parent mass. Yet strangely enough, each of those which maintains an oriented flight acquires on its forward end or side a form which is closely similar in at least ninety percent of the cases. This form closely approximates a hemisphere and is not cone-shaped as much of the literature on meteorites would lead one to believe. My statements are based upon my personal scrutiny of nearly eight thousand specimens of fresh meteorites which had been collected before terrestrial weathering had modified them.[30]

I further stated my belief that when a body moves through a gas at such a sufficiently high speed that its surface is constantly melting, it then behaves like a body moving through a liquid and acquires the form of least resistance for a solid passing through liquids, and I recommended that a projectile might be provided with an alloy jacket of a low melting point.

In response to a request from the Ballistic Research Laboratories at Aberdeen Proving Ground in 1948, I listed my papers on surface features and offered to make available for study at the museum my examples of oriented meteorites, adding that I was glad the writer shared my conclusion that ballistics engineers would find it profitable to study oriented meteorites.

The cover of *Science News Letter*, July 26, 1952, depicted a needle-nosed model missile streaking through the Ames Aeronautical Laboratory wind tunnel at Moffett Field, California. The accompanying article stated that such was to be the shape of new, faster-than-sound super planes, yet to be built. Man, the article added, probably never would fly at such speeds, except for momentary tests, because of speeds, heats, and altitudes the human body could not stand. Heating was listed as the newest formidable barrier.

Under the dateline of August 1952, I sent to the Ames Laboratory a letter suggesting comparison of the pictured needle nose with the "blunt nose of my little Estherville meteorite which was shaped by aerial friction," of which I enclosed a photograph. I recommended that a foil of the same shape be subjected in the wind tunnel to the

---

[30] I had by then examined about one thousand additional specimens since my observations in the mid-1930s.

same velocity mentioned in the *Science News Letter* story. The reply I received, written by an "information specialist," was that wind tunnel experiments would not be feasible and that he had discussed the matter with several staff scientists, who believed "that the rounded shape of the meteorite's nose is due as much to thermodynamic action as to aerodynamic forces."

I wrote again to the effect that "the thermodynamic factor was directly and wholly dependent upon the aerodynamic force" and that "I still feel that these little pellets record some facts that will prove vital to progress in your line of research."

I did not pretend to be able to decipher the message that I was so sure meteorites carried for the space effort and the drive to develop long-range missiles. What I sought to convey was that there was a message, and I suggested again and again and again that it was in the national interest to find out what that message was.

On June 1, 1957, the cover of *Science News Letter* carried a drawing of a new "nose for missiles"—the blunt nose. The shape conformed almost exactly to that of my little Estherville meteorite. The accompanying article announced that the blunt shape "helps to beat the problem of excessive heat that is generated when the hypersonic missile re-enters the atmosphere" and revealed that the Distinguished Service Medal of the National Advisory Committee for Aeronautics had been awarded to H. Julian Allen for the discovery.

Allen had reached his conclusion mathematically, reasoning that the stronger shock waves of a blunt nose would carry off almost all of the frictional heat. It was also stated that the blunt nose had been adopted for offering more resistance rather than less. In a personal encounter some years later, Allen cordially convinced me he had reached his conclusion independently.

The evidence I had seen written on the blunt faces of meteorites, ridged and flowed over with molten flow lines, led also to my submitting a suggestion that a low-melt jacket be placed on all high-speed missiles.

There was a period in the development of spacecraft when it seemed an apparent impasse had been reached because engineers saw little, if any, hope that a craft could be designed that would not burn up during its return passage through the atmosphere or at least develop a temperature lethal for any human occupant. Over a considerable period, our museum was visited by numerous representatives of the space program who were looking for information regarding some metal that could be relied upon to resist the heat of friction. These men reasoned that if the meteorite had survived its passage through the atmosphere, it therefore possessed a peculiar type or degree of resistance to heat.

Two engineers from one of the principal government research centers in the east launched into a discussion of the re-entry problem, explaining why they thought that I, having spent many years studying meteorites, might know what substance or substances in meteorites were responsible for their great heat-resistant qualities. As I told them that they had been entertaining a completely false conception, that meteorites possess no such quality, I could not be sure whether their faces were merely registering let-down or contempt for my ignorance. I explained that meteorites survive to reach the soil simply because there is sufficient material on entering from space that some is left over when deceleration is accomplished. Then I told them that the Bruno specimen beside which they were standing did have an important message, that meteorites in reacting to frictional heat develop a form which should be used as a guide to engineers in the shaping of nose cones. I proceeded to a discussion of the blunt nose versus the needle nose then in use on all missiles and rockets, and showed a number of examples, among them the little Estherville irons.

After several such encounters, I concluded there are certain mathematically minded individuals who can only read mathematical formulae—that natural objects make little impression upon them unless accompanied by a graph or a formula. I, on the other hand, am perfectly blind to their brand of mathematical interpretation. In short, I can read meteorites; they can read formulae. We do not speak a common language.

Perhaps my awkward, non-mathematical presentation did find its mark in some of the many discussions we had in our little museum. I could never trace such an influence, but it would be my hope that some of my lectures and conversations stimulated some young fellows to thinking along lines that led them to work in the space program.

Certainly meteorites carry the message of the blunt nose plainly enough, having written the formula on their own noses in flight, and certainly the examples in my museum, clearly labeled and described, were examined by many aeronautical engineers and missile and rocket men during the years between 1946 and 1957.

4

During one of the off times in Sedona when I was busy with research problems and not on duty in the exhibit room, I was interrupted to greet a visitor from Jordan. It was Dr. Dashani, director of antiquities of that country, who was visiting the United States under auspices of the state department. He seemed greatly intrigued by the meteorites and began asking questions as to why we were operating the museum, how and where we had acquired the collection. I told him that we were just about ready to bring the museum operation to a close.

"Why? Why do you have to do that? You say you are not able to make it pay. Why doesn't your government give money for it? It gives money to our country to finance various cultural activities. Surely your government would finance your museum?"

It was rather difficult to discuss this matter with a stranger from another country which received aid funds from our own, but actually he was speaking thoughts I held in my own mind from time to time.

I went through again in my mind all the reasons I had so often laid out as to why the museum should not be sold: Activities associated with its operation and the field program which preceded it had been responsible for recovering at least half of all the meteorites discovered in America since 1923. The collection had furnished a major share of the meteoritic materials used by many institutions in recent research concerned with the space age. More information about meteorites had been presented to the general public, and in an easily understood manner, than was offered by any other American institution, and here, too, had been a source of information for specialists— geologists, astronomers, and aeronautical and astronautical engineers. The exhibit materials constituted an ideal "textbook" to prepare teachers for the presentation of a basic course in meteorites which could correct the current unfortunate and total lack of elementary instruction in this important subject.

In the exhibit cases and in our storage boxes were ample materials for investigations currently being carried out and for research that should be undertaken. In the files lay data which for lack of funds remained uninvestigated and which might well yield returns in meteorites recovered at least as great as had been achieved through all our previous field work. There was, in these post-Sputnik years, advanced research into various aspects of meteorites going on in many laboratories. Some of these new studies required the destruction, or partial destruction, of meteorite specimens, and thus the present limited supply was being used up rapidly. Nothing, so far as I could tell, was being done to replenish that supply other than the program we had started so many years before. And finally, there was no adequate integration of various research projects, no coordination of them with the elemental field facts which make meteorites such an important link in the chain of our understanding of the earth and its parent solar system.

I sharpened a pencil and once again outlined the staff and budgetary needs for a properly functioning meteoritical institute. I figured that an endowment of $2.5 million would be needed to support a staff of a dozen modestly salaried employees, ranging from a janitor and secretaries to the director at a recommended $12,000 stipend, plus an additional $300,000 for building and equipment. The annual budget for the proposed institute of 1939 had been less than half the $97,000 figure I estimated now. Money had not been forthcoming for a plan that would have put the Nininger collection "at the disposal of the Institute for exhibition and research purposes."

It would have been a great boon to see my museum program continued. The first years at Sedona had held promise, but by the fall of 1957, it had become evident that the winter months were never going to become fruitful enough to support more than minimum expenses of the museum, nor could the summer provide carry-over for the off-season. Addie and I spent the winter of 1957–58 in California again with our trailer, supporting ourselves by sales, lectures, and field work without draining the small income from the museum, where Margaret and Glenn remained on duty.

Once more, we were faced with the old questions: whether and how to keep the museum, and whether and how to dispose of the collection. To depend substantially on field work and lectures was too strenuous for me beyond the age of seventy.

When we had found ourselves stranded for three more years at our old location away from the new highway, I had written to a number of leading scientists with a known interest in meteorites, disclosing that our situation would require us either to divide the collection or sell it outright.

They had answered, "No. Don't sell the collection unless you can sell it whole." The response uniformly signified that the welfare of science would best be served by keeping the collection intact, if possible, and by placing it, if possible, with a university where it could serve research as well as remain on display.

But we had not been able to sell the collection whole, or even bit by bit, so we ventured to borrow heavily again and built our new exhibition building in Sedona on the assumption that this beautiful little town would become a mecca for nature-minded tourists. The move was good, the town boomed, but the boom was in terms of retired couples and art-oriented people, mostly. It was apparent that we were at a stalemate. We had assumed that as the town grew, and awareness of our museum's location grew, business would increase, but it stayed about the same.

My friend Gerard P. Kuiper of the University of Arizona, formerly of Yerkes Observatory at the University of Chicago, brought an Italian astronomer to visit our museum and, although we had closed for dinner and I was alone, I opened up, and we had quite a lengthy session. Neither of the two visitors felt they could accept my theory of the lunar origin of tektites. The next day, Kuiper came again, and I read to him two pages of facts about the moon and meteorites that I had jotted down as an aftermath of our discussion. These seemed to weaken his disbelief.

He asked me what support we were receiving from the government or from universities. I told him, "None."

"I think it is entirely proper that you receive the help that will enable you to finish off some of your important research projects and to write them up. You have more information than any other man in this field, and it would be a shame for you to pass on without leaving this in a usable form for future generations."

476   IN SEARCH OF FALLING STARS

He encouraged me to approach the National Science Foundation for a grant for research with a part-time salary. Other scientist friends recommended that I seek support from the Office of Naval Research or the Army Ordnance Corps. Both of these departments indicated interest, but nothing ever developed. Another proposal was that I apply for funds to make a round-the-world flight in search of craters.

I was tired of applying for grants, but I decided to make these new efforts anyway. They all came to naught, but the suggestions were indicative of an awakening interest in all things meteoritic.

With a generosity of understanding, coupled with a regret that Harvard found itself unable to acquire the collection, Fred Whipple had written, in 1951:

I think it would be very unfortunate were it necessary for you to decimate your collection in order that you yourself should survive. Obviously it is more important still that you survive.

In late 1957 and early 1958, it seemed that we finally had arrived at the point when we would have to put our survival first.

In 1956, Max Hey, curator of meteorites of the British Museum of Natural History, had attended a meeting of the International Geological Congress in Mexico, and had stopped at Sedona to examine the collection. He suggested that the British Museum might desire to acquire a large part of it. The following year, that institution asked us to submit an extensive price list of specimens that would amount to a "vertical split" of our collection.

In January 1952, we had invoiced the specimens listed in our *Catalog*, exclusive of some duplicate and storage material, at $258,000. In applying the British Museum's request for a vertical split, taking into account the expenses of cutting and preparation and the fact that the division of material did not destroy by half the value of the resulting parts, we selected 276 out of our 680 falls, of which we offered divisions of one-fourth to one-half, or smaller, at a total price of $155,000.

The British Museum was interested in obtaining material both for display and for various research projects. Those responsible for meteorites at the British institution informed us that our list of offerings with prices was acceptable but asked for time to look into the problem of raising money for the purchase. We agreed to await their final decision, which would be determined by their success in searching for necessary funds.

Before final acceptance by the British Museum, there suddenly was a surge of interest on the part of both Arizona State University and the Smithsonian Institution. The two American institutions were also dependent upon finding funds. Arizona State particularly had no resources at its disposal and would have to locate outside assistance. The Smithsonian indicated an interest in buying the collection outright, and with no time to make any revised inventory, we simply guessed that $200,000 would be about right for a discounted price for the collection as a whole, but we made it clear that a prior offer was pending for a portion of the collection.

All three parties were informed of the various offers, without any of them being named, and were told that we would deal with the first to come forward with a firm proposition. There was no attempt to induce any competitive bidding, but our hope was that the collection would remain in the United States, and our preference was that it stay in Arizona because of the importance of the crater to meteoritical research.

E. P. Henderson, curator of meteorites for the United States National Museum (the museum of the Smithsonian Institution) prepared a prospectus seeking support of the National Science Foundation for the purchase of the Nininger Collection. The prospectus

478 IN SEARCH OF FALLING STARS

described in detail the contents of the collection: the total number of falls ("more than 727 meteorites, almost as many as are in the National Collection [850]");[31] the number of witnessed falls—about 150; the inclusion of many stony meteorites, two of which, at weights above 700 pounds, were larger than any in the National Collection; and other features.

The collection contains a number of described specimens and many that have been so widely illustrated that such specimens are well-known to students of meteorites.

It contains numerous individuals showing flight markings. These features are of considerable interest because they help us understand the flight of solid objects through space. This subject is of special importance in guided missile studies.

The collection contains a number of meteorites showing how heat penetrated them during their flight through our atmosphere.

The Nininger Collection contains a fine series of specimens showing the effects of weathering.

It was gratifying to see the points made in favor of my series in flight markings, weathering, and heat effects. Henderson's report also included a note on "The Building of the Nininger Collection":

The Nininger Collection represents a lifetime of effort on the part of its owner. Against the advice of many of the best students of meteorites, when he began his life work in the 1920s, Nininger demonstrated that he could support himself by prospecting for meteorites. This was a courageous undertaking because meteorites were considered to be very rare, widely scattered, and extremely hard to find.

---

[31] We claimed only 680 falls.

Nininger proved that they were widely scattered but could be found if an intelligent search was made.

Nininger, and his wife, went into the field and lectured to local groups on meteorites. ... The Nininger family lived by the sale of some of the material collected. Thus, they were able to continue their search for meteorites and to build up this great collection. Their living was never without financial difficulties and after twenty-five years,[32] Nininger is faced with the need of selling all or part of his collection to provide his economic security.

It should be remembered that, although Nininger supported himself on the sale of duplicate material and sections of some of his unique specimens, he faithfully saved the best material for his collection. The only exception to this policy was the sale of an occasional specimen to the United States National Museum.

The prospectus included two comments that were, to my view at least, eyebrow-raising:

Mr. Nininger has passed 70 and is in failing health, thus his active pursuit of meteorites is over.[33]

and

Personal bias or animosity has no place in the management of the National Museum. Thus, qualified persons cannot and will not

---

[32] Thirty-five years was the correct figure.

[33] This prognosis proved to be wildly inaccurate. In the following decade, Nininger made eleven strenuous trips outside the contiguous United States to the Far East, Australia a second time, Europe twice, the Mediterranean, Central America, Baja California twice, Alaska, and British Columbia twice. He maintained his involvement in meteoritics for the remainder of his life, eventually passing away shortly after his 99th birthday. *Ed.*

480 IN SEARCH OF FALLING STARS

be denied access to the National Collections. Denial of access is possible if the Nininger Collection remains in private hands.

Henderson found "fair and reasonable" the asking price of $200,000, payable in annual installments of $20,000. He gave as his own rough estimate of the value of the collection a figure just under $300,000, and added that "By individually pricing this collection of meteorites and adjusting values for the deflated dollar, a fantastically high figure can be obtained."

After correspondence and delays and cross-correspondence and more delays, it began to appear to us that the American institutions were going to be too late.

A letter under the date of June 13, 1958 came from G. F. Claringbull, keeper of the British Museum, with a firm offer to buy the approximately 1,200 specimens of the vertical split—about 21 percent of our entire collection—at our named price less a 7.5 percent discount—$140,000. We cabled confirmation as requested. Also, we promptly withdrew all other offers.

Addie and I were somewhat shaken. After so many years of financial strain, and after having been driven to seek an institutional home for our collection, offering it even at a fraction of its value with no takers, suddenly it appeared that we were to be able to have our cake and eat it, too. The price we received for a portion of the material would be sufficient to pay off our debt and still leave working capital that might make it possible to continue operating our museum.

We had never desired to see any considerable part of our meteorite collection go out of the country. We expected that we might face criticism. We were not prepared, however, for a telephone call from a man claiming to be a representative of the Smithsonian Institution in Washington warning against our removal of supposedly "strategic" materials from the country, for we knew that officials of

HOME 481

the British Museum had written to the United States National Museum to ask specifically if there was objection to British purchase of a part of our collection.

This telephone call, with the implied threat of an embargo against shipment of meteorites, came in the midst of our crating and packing of specimens. There was no further interference. The call was apparently intended as a bluff.

Meteorites were not then, nor are they now, classified as strategic materials; their export is therefore subject neither to license nor embargo. This knowledge, however, did not prevent feelings of shock and some despair on top of what already had been a somewhat rending experience of giving up a large portion of a collection that had taken half a lifetime to accumulate. It had not been many years since, in desperation, I had offered the entire collection to this same institution at a fraction of what now was to be received for a fifth of it.

Shipping preparations took the entire summer of 1958. We could not agree to a literal, complete vertical split of the entire collection. Many of the precious and scientifically most important specimens were too small to share. We divided many of the larger individuals, and we sent the main mass—more than 600 pounds—of Morland, retaining our other large stony meteorite, Hugoton.

The division of the several large stones and irons meant weeks of sawing. Since we had given up our large saw upon leaving Denver, we had to haul a dozen of the largest to a California laboratory capable of handling them. Cutting was not the greatest chore. Everything had to be carefully weighed and labeled. Our catalog records had to be amended. Our contract called for compensation for loss in cutting, grinding, and polishing. This meant measuring and re-weighing everything and calculating the difference. Since some of these small specimens were priced at more than their weight in gold, this had to be done with precision. Each specimen had to be individually wrapped in a manner suited to its particular requirement. Some meteorites are so fragile they require utmost protection. Boxes had to be constructed to conform to the requirements of different types and sizes of specimens.

We handled the big Morland stone like a baby. We sheathed it in a plaster of Paris jacket, reinforced with burlap, and then built a crate around it. The plaster had to be kept from making direct contact with the stone, or it would be practically impossible to clean. Too, plaster reacts with the metal in meteorites and sometimes causes difficulties. So before the plaster coat was applied, the big stone was carefully swaddled in plastic.

Our decision had been made; the summer-long activity necessary to implement it gave us not much leisure, but a constant parade of recollection of finds and purchases, incidents and research.

When the first partial shipment of meteorites was in transit to England, we turned our thoughts to the comfort that for the first time in our lives, we would be able to pay all bills due without using every available cent. Over the years, we had built up a considerable file of investigations we would like to make overseas, and of meteorite collections and craters we would like to view. I was in my seventy-second year, but it was not my thought that my days of collecting and studying meteorites were over. We were determined to take prompt advantage of our time and fortune, and forthwith, we laid out an itinerary of prime target areas for investigations, then talked with travel agencies to learn ways to reach our objectives and sightsee as we went. We made our plans, obtained passports and inoculations, made reservations, and purchased tickets at what seemed a whirlwind rate. When the first payment arrived on September 11, 1958, Addie wrote checks until she risked writer's cramp. We enjoyed a social hour to celebrate paying off the mortgage, and after writing just a few more checks, we were out of debt for the first time in thirty-five years.

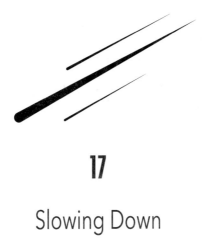

17

Slowing Down

Traveling is no fool's errand to him who carries his eyes and itinerary along with him.

—Amos Bronson Alcott, *Table-Talk*

After shipping the remainder of the British Museum purchase, on October 26, 1958, we sailed from San Francisco for a six-month Far East tour heavy with stopovers for field work and visits to institutions.

Shipboard is a nice way to travel, but we found ourselves growing a bit lazy. I seemed able to sleep almost any hour of the day and to sleep all night. I lectured twice aboard the ship and suggested to the captain that shipboard might be a good base from which to measure meteoritic dust. He provided a dozen unused white plastic waste baskets which the ship's carpenter helped anchor forward of the bridge. These were left undisturbed for six days. When I carefully examined them, I found an amazing amount of dust—too much, I thought. I told the captain the results were almost too good to be true,

484 IN SEARCH OF FALLING STARS

and the material would of course have to be investigated later in a laboratory, but we agreed it was only fair that I should report the results to his staff. I emphasized the importance of any positive results and their possible bearing on our space exploration, and I praised the men for their cooperation. After dismissal of the group, one of the men asked to see me privately and, very apologetically, spilled the fact that he had salted my catch with some dust from the machine shop!

***

Throughout the Orient, we were met, entertained, and escorted by local scientists or university or museum men. In a Japanese home in Kasamatsu, Hashima, Gifu Prefecture, we were served tea by a lady who displayed a meteorite that had fallen into her house in 1938 and showed us where it had gone through the roof. We toured museums and universities everywhere.

Secondary to our interest in displays of meteorites and purchase of some specimens was our interest in tektites, those strange blobs of glass that are found only in certain locations on earth. I first acquired a small collection of tektites in the early 1940s. Many explanations have been advanced for the origin of tektites, including theories that they may be of human manufacture, glass meteorites, nodules from "light metal" meteorites, products of lunar volcanoes, or remnants from the deeper crustal layer of earth set free by the hypothesized separation of the moon from our planet.

In 1938, Dr. Virgil Barnes proposed that tektites were the product of lightning striking in certain types of soil on the earth, and this theory enjoyed a considerable following for several years. Dr. L. J. Spencer of London theorized that tektites were a form of impactite produced by crater-forming meteorites on the earth. A number of eminent physicists now think that very ancient and very large meteorite impacts on the earth produced tektite glass which was thrown into temporary orbit and returned as tektites.[34]

Tektites have been known to occur on only a few areas of the earth's surface, all of these locations, with one exception, lying within an equatorial belt eighty degrees in total width. Though they have differences one from another, tektites are strongly similar, and those found in a particular region have common characteristics which differ somewhat from those found in other localities. It seemed to me that tektites fit into the picture of intense and violent bombardment of the moon by meteorites, with resultant scattering of the lunite rock to great distances and at great velocities. In February and March, 1943, *Sky and Telescope* carried a two-part article, "The Moon as a Source of Tektites," in which I outlined my theory that tektites were fragments of lunite (the shattered lunar rock comprising the surface of the moon) which had been set free from the gravitational force of the moon and then had fallen into the stronger gravitational pull of the earth.[35] In 1947, I set forth this theory in expanded form in a paperback booklet, *Chips from the Moon*, published by our museum.

These odd disks and buttons and dumbbells were high on our list of things to explore, and our Far East itinerary included three of the most heavily strewn tektite areas.

It was an exciting privilege to meet, in Manila, Dr. H. Otley Beyer of the University of the Philippines. Trained as a geologist, Dr. Beyer was sidetracked into archeology via ethnology and then, while pursuing a career as a professional archeologist, stumbled into

---

[34] The consensus among scientists today is that tektites are pieces of impact glass formed during the impacts of large meteorites on Earth. *Ed.*

[35] "The Moon as a Source of Tektites, I," *Sky and Telescope*, Vol. II, No. 16, February, 1943, p. 12; "The Moon as a Source of Tektites, II," *Sky and Telescope*, Vol. II, No. 17, March, 1943, p. 8.

tektites, first as a hobby and then so deeply that when we met him in his seventy-sixth year in 1958, he had in his possession 500,000 tektites. He contributed some of the best papers to have appeared on tektites and all the while retained his reputation as the top man in Philippine archaeology.

Dr. Beyer above all else was a field man. I had become somewhat impatient with the rapid growth of tektite literature, mainly produced by men who had studied a dozen or so tektites in the laboratory. It was very refreshing to meet a man who had collected a half million tektites from thirty or more locations on a single island and several hundreds more from locations scattered through a half dozen other countries, and who was not yet ready to offer a theory as to their origin.

The reason for this seeming tardiness in theorizing became obvious as one conversed with him and observed his work in field and laboratory: He knew too many facts. Always, just as he had about finished aligning his facts, he found it necessary to examine another bag of specimens, some individuals of which stubbornly warped the alignment. He chose not to label these the "exceptions that proved the rule" but rather went on patiently observing and comparing both tektites and their environments in the hope of finally discovering an alignment which the stubborn facts would fit.

Dr. Beyer was laboring under handicaps, living alone in a shabby upper-floor room to which his collection was moved prior to occupation of Manila by the Japanese. The lack of facilities and equipment would defeat most men, but he never uttered complaints. His work room and library had no electric lights and his windows were constructed of thin blocks of dingy oyster shell. On even the brightest days, his room was so poorly lighted I could not have thought of working in it, yet he could point out structure and form in amazing detail. If we overstayed the five o'clock closing time even a few minutes on the December evenings we spent with him, we had to find our way out through a maze of halls by feel or by flashlight.

***

Next, we went to Saigon because we knew that tektites had been found in South Vietnam when it was a part of Indochina. They had been labeled "Indo-Chinites" by Professor Lacroix, the French tektite authority who first described them. Our only hopeful contact in that little country was Professor Edmond Saurin of the University of Saigon. Could he tell us where we might search?

"In the city of Dalat, about 150 miles northeast of here, between the Old Church and the Hospital Civil. We found a considerable number right in the road."

His classroom duties made it impossible for him to go with us to Dalat. Since it seemed hardly likely to us that we would find collecting to be as simple as the good professor seemed to expect, we sought out a conveyance and a driver who could not only transport us to Dalat, but serve also as guide and interpreter. We hired an English-speaking employee of our Saigon hotel who, with notebook and great eagerness, began to outline a lengthy plan for visits to all the shrines and places of interest of Dalat, and who was quite crestfallen when I explained we were seeking a peculiar kind of rock and had no time for sightseeing. He brightened as he began to list museums and mines where we could find rocks, but again I interrupted to show him a sample tektite. Quite unenthusiastically, he drove us along the road to which Professor Saurin had directed us.

Unfortunately for our purposes, the road had been paved, but we searched the narrow shoulders and canvassed the houses along the way, showing samples and asking for information, with no results. We seemed almost at a dead end when a young fellow told us that

488 IN SEARCH OF FALLING STARS

we "could get lots of that up at the university." I suspected that he was referring to a display of tektites, but we turned our car toward the institution at the edge of town. As we approached the entrance, our guide stopped the car on a blacktop area and pointed to the tar-coated pebbles.

"Ah, there are the tektites the boy saw," he declared.

I was afraid he had only made a good guess, but I got down from the car to look about anyway. In a cut bank, I noticed a zone of laterite pebbles (laterite is a brick-red, gravelly formation characteristic of most tropical countries) and I recalled that Dr. Beyer had found Philippine tektites associated with such a deposit. I scanned the bank and there, protruding from among the laterite pebbles, was the end of what appeared to be a tektite. I quickly removed the inch-long, grooved fragment of black glass. It had the characteristic appearance of Indo-Chinites of seeming to have been stretched while in a molten condition.

The guide and Addie quickly joined me, and the three of us carefully scanned the cut banks along the road, each of us finding one or two specimens. Then the laterite disappeared. We looked along a crossroad and saw in the distance quite a lot of new road construction traversing low, rolling hills in which the cut banks showed clearly the brick-red hue of laterite. We headed in that direction, and as we came to the low hills, tektites began showing up on the road shoulders where they had been washed free by recent rains, and in the banks and ditches. The three of us scurried about finding more tektites than we supposed ever had been found in one locality. By the end of three hours, we had gathered more than 300, and with our guide enthusiastically producing his share, we gathered teardrops, elongate teardrops, dumbbells and disks, ovals and hollow tubular forms, mostly fragments but some complete individuals.

The next day, we worked another four hours and took back to Saigon a total of 825 pieces. Our guide was a complete convert to the hunt, and he continued for years afterward to find and ship tektites to us, thus making a better living than his job had provided.

***

The great island continent "down under" was the big goal of our trip. We had scheduled a two-month stay in Australia for a long trek into the outback to visit meteorite craters and to search for the Australian tektites, Australites. With Allan Kelly, a geologist friend from Carlsbad, California, we purchased a Volkswagen Combi-Van and set out on a seventeen-day, 3,000-mile drive into the interior, largely a barren land with few roads, most of them poor. Scarcity of water plagues the greater part of the continent, and accommodations in the outback were minimal. We carried camping and cooking equipment with us.

From Perth, we traveled north through 300 miles of wheat country somewhat like that of our own western wheat states, though the "stations" or farms are much larger, and the towns smaller and farther apart. The harvest of 1959 had been a good one (wheat being harvested in January this far south of the equator), and every little town had bulging granaries and great open bins filled to overflowing. Large flocks of beautiful rose-breasted, gray parrots were feeding in the open bins, almost completely covering the mounds of grain. When disturbed, they arose in great gray clouds to alight on the telephone wires nearby—such large, heavy birds that balancing on the wires was so difficult that many hung by their bills or feet.

There was practically no traffic. In 355 miles, we saw only two cars and one motorcycle.

We stopped in the villages and towns, inquiring as to where tektites might be found. We talked to science teachers and other local "identities." No one seemed to have any tektites, but nearly everyone had heard of them. "Yes, you find them in the bush." "They can be found on the Nullarbor." "You find them on the dry salt lakes." "A lot of them have been found around Kalgoorlie."

On the outskirts of nearly every village were the Aborigines' camps, clusters of makeshift shelters made of brush, or a blanket or canvas spread over a low bush. Occasionally a camp was set up out in the bush, some distance from any village or town. In each camp, we showed our tektite samples to the natives. Often they were shy and wary, but occasionally, after persistent questioning, someone would produce a tektite from somewhere on his person and be persuaded to part with it for a few shillings. All seemed to be familiar with tektites but showed a "What of it?" attitude. Years ago, tektite collecting and trading by the Aborigines had been a general practice, but as the government began to look after their needs, the natives saw no further advantage in searching for the bits of glass. Occasionally, we were directed to a large cattle or sheep station. Usually we got pretty much the same answers to our inquiries. "I used to have a lot of them but they were no good for anything and I gave them to the kids. They're all lost." "I had a whole cigar box full but I threw them out. Nobody wanted them." "Years ago there were lots of them."

An exception was the 640,000-acre station of Mr. Barton Jones near Kalgoorlie. Here the family had made a practice of preserving all of the tektites found on their ranch, and over a period of forty years had amassed a collection of 1,166 specimens. Mr. Jones admitted that he had seen less than half of his ranch and that most of it was still "bush." In view of the fact that they had simply collected those tektites which they came across accidentally, and those were recovered at a rate of slightly more than two per square mile, they must represent a rather small fraction of the total number present in the soil. We purchased the marvelous Jones collection entire, but still we made no finds in the field.

Countless times we stopped in likely looking areas of barren soil, blown fields, sand dunes, openings in the bush, where we walked about scanning the ground. We were camping out every night, and as we made camp and gathered firewood, we scrutinized the ground carefully. We never walked anywhere without looking at the ground, but never a tektite did we see.

From Kalgoorlie, our way lay across the great Nullarbor Plain, which most Australians know only as an area to be flown over or crossed by train. We were told that we were foolish to try it by car. The road was rough and sandy, the terrain barren and treeless.

We loaded our van with supplies, including gas and water sufficient to carry us the entire 1,500 miles to Adelaide on the eastern edge of the Nullarbor, and started off. The trek across the Nullarbor, a region resembling the Navajo reservation country of Northern Arizona, took five days. All the time, we continued our individual searching of the ground whenever we stopped. Still, we arrived at the end of our journey without having found a single tektite in the field.

Later I discovered in talking with scientists in Melbourne and Adelaide that very few Australian geologists had ever found a tektite. An exception was Dr. George Baker of the University of Melbourne, whose curiosity twenty-five years ago led him to explore an almost inaccessible, rapidly eroding slope resting precariously on a vertical cliff that plunged into the sea. Here, where other men had seen no reason to venture, the normal deposit of tektites had remained, a few being set free by erosion each year. Dr. Baker made regular excursions to this "private" hunting ground, gleaning a few choice specimens each trip. In twenty-five years, he thus gathered more than 800 Australites.

Dr. Charles Fenner, who spent many years studying Australites, and who by gift and purchase amassed a great collection for the Museum of South Australia, estimated that from one to ten million tektites had fallen in Australia. This seems a large number until one reflects that the portion of Australia within which specimens have been found comprises some 2,000,000 square miles. An average of two or three little tektites on each square mile is spreading them out pretty thin. It is no great wonder that our brief expedition made no finds.

At Adelaide, we disposed of our car and camping equipment, and flew to Alice Springs, the kickoff point for side trips to the meteorite craters. The Alice Springs country was hot, insect-ridden, and suffering from prolonged drought. One station had lost 5,000 head of cattle. There was absolutely no forage left except leaves on the trees. It was hard to visualize this country green and lush as we were told it would be in the "wet."

The Australian Bureau of Mining furnished cars and drivers for a one-day excursion from Alice Springs to the thirteen Henbury craters. I found spheroids, a number of small meteorite fragments, and some impactite. We purchased additional meteorites from the widow of the man who had supplied us with Henbury specimens many years before.

At the small Boxhole crater, a pit some 500 to 600 feet in diameter, 150 miles from Alice Springs, we found no meteorites.

To reach the Wolf Creek crater, Australia's largest, two-thirds the diameter of Arizona's, it was necessary to fly to Hall's Creek and from there drive a hundred miles by jeep. Because the trip would be so rough, Addie was strongly advised not to go and finally consented to remain behind.

The jeep trip required seven hours. The last nine miles and four hours were through a kind of spiny grass called spinifix, whose tight sod hummocks grew six inches high and two to twenty inches apart. The jeep bucked along like a galloping mule with a broken leg. Wolf Creek, as expected, yielded many fragments of oxidized meteorite.

We had visited the Dalgaranga crater on the third day out in our Combi, which we chose to call the "Kangaroo," and had found some small meteorites. I was surprised at the scarcity of material, since this beautifully shaped, well-preserved crater was little-known and little-visited. I was determined to return at a later date with a detecting device and a man and equipment to excavate. After our return to Sedona, we were surprised, on polishing one of the little Dalgaranga specimens, to find it was not a medium octahedrite, as described, but a mesosiderite. Other specimens were like it, and some others were true siderites, but none fit the published description.

The Museum of Western Australia endorsed our plan to extend the surface search and excavate to some extent within the Dalgaranga crater, and my son-in-law Glenn Huss and I returned the following October. We collected 207 specimens, all quite small, a third of them averaging about one-ninth of an ounce and the largest weighing two ounces. All of the larger ones were completely oxidized. When later we cut and polished more than half of the specimens, they showed a variety of composition—most were siderites, some were mesosiderites, and others a combination of siderite and mesosiderite. Two were aerolitic. Many showed deformation by pressure, and some showed heat alteration. The material from within the crater itself was badly disintegrated and oxidized and bore no resemblance to the meteorites found on the surrounding plain. We theorized that the crater was made by a mass of ten to twenty tons, predominantly stony, which was thoroughly shattered and which for the most part has decomposed.

2

When Addie and I left Australia, we had been away from the United States five months. Only visits to New Zealand and Hawaii and the long sea voyage lay between us and Los Angeles harbor, where we docked May 14, 1959.

We had arranged for purchase of a hilltop site in Sedona while we were away and had approved plans for building a home. Until it should be finished, we went back to trailer life. Margaret and Glenn Huss with their family were living in and operating the museum.

We found the museum just barely carrying itself. We would have to make a final decision as to the museum's and our future. Once more, I ruminated with myself for long hours over the alternatives of closing and remaining open. The sale of the first portion of the collection had given us a breather, but it was plain our problem was not solved. I wrote memoranda to myself: "Why the Meteorite Museum Should Not be Closed," "The Future of the Meteorite Museum," "The Program of the American Meteorite Museum."

Even with the removal of the great number of specimens acquired by the British Museum, the worth of our exhibit as an educational instrument had not suffered appreciably.

No visitor ever went through our little museum without opportunity to ask questions. Because our whole effort was devoted solely to meteorites, it was possible to concentrate and absorb in a fashion that is impossible in the ordinary museum, where hundreds of different subjects are presented.

We brought to the general public the opportunity to see and handle meteorites in their natural condition, just as they are found in the field. We furnished verbal instruction with special emphasis on those facts that would enable agricultural and other outdoor people to recognize meteorites. We carried on continuing field work, thus not only adding to the material available to view but enlarging the fund of knowledge concerning distribution of meteorites. We exhibited the widest possible variety of meteorites both in their natural form, displaying such surface features as fusion crust, pitting, orientation, and flight markings, and in cut, polished, and etched slices, showing interior structure and components.

What would be the source of materials in the future? The majority of all the meteorite finds of the past generation had directly or indirectly resulted from the educational program carried on by our laboratory, first, and our museum, later.

I did some multiplying. By my figures, the museum had served a half-million visitors during more than 4,600 days of operation since October 19, 1946, and the verbal instruction given to groups of two to fifty would equal nearly 7,000 one-hour lectures.

Our last hope had been that the museum could pay its way and support Margaret and Glenn, who would operate it, leaving us free of its burdens of attendance and care. But it was apparent it could not. We inventoried our remaining collection, this time using the data of our sale to the British Museum as a guide to value. Our resulting invoice was a lesson on the importance of taking time for inventory, for it showed, on that basis, a value of more than $530,000.

By Thanksgiving Day, 1959, we had moved into our new house atop one of Sedona's red hills, our first real home since leaving our rented house in Denver.

I helped the workmen select lichen-covered stones for both the interior and exterior rock work of our bright, contemporary home. Addie shopped for furnishings. The delight of our house was its hungry fireplace, for which we sawed and hauled firewood from the hills.

In the fall of 1959, Arizona State University again indicated a desire to acquire the collection. New and sudden interest in space, missiles, rockets, and satellites had made meteorites the obvious and most probable source of information vital to progress in all of these fields. Addie wrote to one of the children that "scientists are *hounding us to death* for material we don't want to part with." As the requests for material and information kept multiplying, we felt it would be almost tragic if we should drop out after all of our struggle, just as the long years of effort seemed to be fruiting. On the other hand, the time was late for us, so we concluded that we really had

496 IN SEARCH OF FALLING STARS

no choice.

We indicated to Arizona State that we would be willing to dispose of the collection at about half its catalog value, but that we might move to offer the collection out of state if definite interest was not shown soon.

Glenn and I returned from the second Dalgaranga trip in October. At the end of November, I made a quick flight east for lectures at the Massachusetts Institute of Technology and Harvard.

In accepting an invitation from the Departments of Metallurgy and Geology at MIT to speak before "interested persons," I wrote that my chief interest always had been and still was "to see a genuine, broadly conceived and well-financed program of meteoritics." I added, "At this stage of the game, I recognize that meteoritics has outgrown me in many of its aspects, for which I am greatly delighted. However, I am constantly being reminded that there are at least a few aspects of the subject wherein my years of critical observation can be of considerable use to the technological and the specialized researcher." My topic was, "A National Program of Meteoritics."

At Harvard, I addressed a group of scientists representing eastern universities and governmental agencies who were concerned with building an active program of meteoritics. Here, my subject was "A Plan for a Nation-Wide Coordinating Center for Meteoritical Research."

Most of the proposals I made at Harvard on December 2, 1959, were not markedly different from plans suggested in the past or from actual methods of field work of our own program, only updated.

In 1933, I had outlined a suggested program of research on meteorites.[36] In 1935, I published a plan for a "Proposed National Institute

---

[36] "A Suggested Program of Research on Meteorites," *Popular Astronomy*, Vol. XLI, No. 9, Nov., 1933, p. 521.

of Meteoritical Research."[37] Then, twice more, in connection with the effort of Denver businessmen to establish an American Foundation for Meteorite Research and the plan presented to the Penrose Foundation, I drew up detailed suggestions for a coordinating research program. My book, *Out of the Sky*, carried similar recommendations in 1952.

With the coming of the space age, it seemed there were more listeners and perhaps a greater willingness to undertake such a program. One suggestion in particular, almost identical to a suggestion of 1935, attracted interest—a plan for thirty batteries of four automatic, wide-angle cameras, to be set up at distances of 300 miles over the entire nation and, if feasible, a cooperating sister network in Canada. The cameras of each battery would face in four directions in order that any fireball of important magnitude could be promptly and accurately triangulated.

I suggested also a parallel plan for a lookout network of fireball observers, to consist of correspondents in principal universities and colleges, the US Forest Service, airport control towers, the US Weather Bureau, the Air Force, and other similarly situated individuals who would report fireball sightings to a central control center. I presented the same "Plan" before the "Space Science Board" of the National Academy of Sciences at the California Institute of Technology in mid-December 1959. Following the Caltech presentation, Dr. Harold C. Urey requested and received mimeographed copies. Others were sent to Fred Whipple, who requested further information on methods of implementing the program. Written appreciations for the filling of both these requests rested in my files.

Published accounts the next year, 1960, of the initiation of a

---

[37] "Proposed National Institute of Meteoritical Research," *The Pan-American Geologist*, Vol. LXIV, Sept., 1935, pp. 107-124.

nationwide search for meteorites with the Smithsonian Astrophysical Observatory at Harvard as the organizational center, credited the "launching" of the program to a meeting of the Meteoritical Society the previous September. No mention of any previous such discussion at the Meteoritical Society meeting was made to me during the brisk question-answer period that followed my December talk at Harvard.

The "Prairie Network" for observation and recovery of fireballs covers the plains area of the nation, with camera batteries set at intervals of 150 miles.[38] It is the sort of thing that I longed for back in the 1930s, when I was exhausting all of my resources in trying to plot the course of meteors from the testimony of witnesses who had only accidentally seen the phenomena, and none of whom ever had opportunity to make an instrumental record of what they had seen.

Even with the functioning of the Prairie Network of cameras, with advance preparation for accurate plotting of the courses of meteors, the recovery of meteorites from fireballs still demands that field workers have a very thorough training in the most elementary aspects of meteoritics, namely a complete familiarity with all of the known varieties of meteorites and a good understanding of the methods of carrying out an effective search.

Early in 1960, interest in purchase of the collection heightened at Arizona State University again, and then again it slowed. In the spring, finally, we sent to a small number of major institutions a brief announcement that the Nininger Collection of Meteorites was available for purchase. This stirred up a good deal of activity and brought an inquiry from abroad, of which we notified the National Science Foundation in May, since Arizona State pinned its hopes for financial help on the Foundation. The National Science Foundation assured

---

[38] The Prairie Network operated from 1964 to 1975. Other fireball networks throughout the world are operational today. *Ed.*

us of definite plans to buy, but in June, they still were moving very slowly and had imposed restrictions which we feared could take the collection out of Arizona. By then, it had become my firm wish that the meteorites remain in the state, but I threw up my hands, ready to start all over again.

Letters from Frederick Leonard during this period reflected that he and I were sharing some of the same emotions as life was advancing and, with it, at a pace a little fast to keep up with, the science of meteorites in which we had taken such a keen and sometimes lonely interest.

There seem to be a great many people these days who are working on certain phases of meteoritics. I cannot help wondering, however, how seriously interested they are in meteoritics as a whole or in meteorites per se.

I wish that it were possible for me to sit down with you and have a long conversation on a number of matters meteoritical. ... In spite of all the activity that seems to be going on at present, I must confess that I have mixed feelings in regard to some of it, and that at times, I am downright disturbed about the future of meteoritics in this country. I wonder whether you have similar reactions or whether you are more optimistic. Unfortunately, the old-time meteoritics (of 25 or 30 years ago) seems to be disappearing; who for instance, is going to carry on the field work that you have done—and who is interested in meteorites for their own sake and not simply in the answers to certain restricted problems that may be obtained from studying them?

Under date of March 24, 1960, Frederick offered some suggestions of additional institutions and individuals whom I might notify of the availability for sale of the remainder of the collection. He expressed a hope that the collection might remain in the United States, preferably the west, and added,

Although I have nothing against the British Museum, I cannot help regretting, as a loyal American, that it acquired 21% of your collection—but you obviously were not to be blamed for selling it to them if they were the only prospective purchaser!

Frederick was looking forward with mixed feelings to retirement in another three or six years. He remarked that were I to become financially independent, I should be able—and he quoted from a letter of mine to him—to "'get many of the things done that have lain unfinished through the years.'"

He added, "I wonder whether we do not all feel that way . . ."

The following month, April 1960, Frederick brought his two young sons and a class of students for a visit to the museum and the crater. He and I had made many field trips over the years, including our trip into the California wilds when he ecstatically hugged the great Goose Lake iron, but this was the last. He hoped to return for a week in August, but his final illness interfered. He died in June of that year.

Addie and I wished to attend an International Congress on Meteoritics in Copenhagen in August; our son Bob planned to attend a Geological Congress in the same city at the same time. We decided to close the museum immediately and to begin packing the collection. If a sale came through, it would be that much nearer ready for shipment; if not, we would store the meteorites and sell or rent the building. We went right ahead with plans to sail from New York on July 21.

Suddenly, the slow-grinding institutional mills completed their operation, and the remaining majority of the Nininger Collection was assured a lasting place in the halls of Arizona State University, at what we considered a fair price, $275,000, far below the listed value.

Four days later, Margaret, Glenn, and their family followed their belongings to Denver for a new life, and to carry on the name and work of the American Meteorite Laboratory. That evening, Addie and I went to Phoenix to catch a flight to New York for our sailing date. An era was over. The museum was gone.

***

In four months, we toured fourteen European countries, combining usual sightseeing goals with visits to universities and museums, examining meteorite and tektite collections. We listened with pride as Bob discussed the genesis of uranium deposits before the international gathering of geologists in Copenhagen.

The Munich collection of meteorites had been destroyed in the war, but Vienna had a marvelous collection which we photographed extensively. Prague, too, had meteoritical information, and the university there provided us with a guide who drove us 150 miles to visit the tektite fields. Near Tübingen, Germany, we were escorted by a university faculty member to inspect the nearly circular valley of Nordlingen Ries (Ries Kessel), about twenty-seven miles across, which has been since proven to be a meteorite crater by finding coesite in its walls.

While we were tracing information about an important meteorite which fell in Zweibrüchen, Germany, some ninety years ago, we stumbled into what seemed to be the ancient abode of the Niningers. The name is prominent in the village of Bad Dürkheim and appears several times in the Manheim telephone directory. It seemed consistent that a meteorite hunt should end at a dwelling place of Niningers. From Strasbourg, we made a prized excursion to the little Alsatian village of Ensisheim, where a large meteorite had fallen about the time Columbus landed at San Salvador. The meteorite had been pronounced a "miracle" by the clergy and other wise men of the place

502 IN SEARCH OF FALLING STARS

and time and is preserved to this day. The stone is no longer kept in the church, but in the *rathaus* (town hall), in a little upstairs "museum" that holds a few old records, an implement or two, and a few relics. We were led to the meteorite by a plump little lady about four-feet-six, who scurried about through several rooms of the small city hall until finally she emerged with a huge key. Waving this at us, she led us down the stairs, across the street, and upstairs again in an older and more ornate building, where we followed her up worn granite steps to a room which opened to her huge key. Here was her museum, with the meteorite displayed in the center of the room, lying atop a rickety old pedestal covered by a glass top. Nearly all the fusion crust was gone from the stone, and all of its prominences had been knocked off, leaving it about half its original size and distorting the meteorite to a mound-like shape.

In Heidelberg, we were guided by Dr. P. Ramdohr, who after showing us the small but important collection of true meteorites opened a small cabinet drawer labeled *pseudometeorits*. Looking over this collection, I noted that European scientists seemed to have been plagued by about the same array of mistaken identities as us Americans who had encouraged the untrained public to send in objects suspected to be of extra-terrestrial origin. There were nodules of iron sulfide, magnetite, iron concretions, basalt, and other minerals and rocks commonly mistaken for meteorites.

One small specimen caught my eye at first glance, but I waited until all the others had been examined before taking it up for inspection. (This is a habit of mine which I cannot explain. When I am in the field searching for a certain kind of specimen, and I spot one, usually I look all around it for similar ones before picking up the important find.) As I picked up that most important of the specimens contained in the collection of *pseudometeorits*, I said to the good Doctor Ramdohr: "Do you think this is a *pseudometeorit?*"

He looked at it, noted its number, then read the label in its small tray which bore on an attached pink tab the same number (96) as on the specimen, and answered "Yes," reading aloud the label: *"Pseudometeorit 108.47 q. Fund: 1909 in der Nahe de Lahnufers bei Marburg."*

"Dr. Ramdohr," I replied, "I think this is a genuine meteorite of the pallasite variety."

Seeing that he showed no interest in questioning further the accuracy of the classification which had been assigned, I asked if he would allow me to take the specimen, cut it, and return one portion of it to him with my own classification. We agreed that as a *pseudometeorite*, the stone really had no value, but should my judgment prove correct, it would constitute an important addition to the thirty-five falls and finds that had been recorded in Germany during the past century and a half.

I gave Dr. Ramdohr a receipt for the little specimen, and upon our return to Arizona, I had it bisected. It was revealed to be one of the most beautiful little pallasites on record.

3

March of 1961 found us back in Sedona, still settling into our new home, laying a hearthstone that boasted a fine series of tiny fossil tracks, and finding places for the small furnishings we had brought from all over the world to add a cosmopolitan flavor to the Indian and Mexican rugs and baskets and western paintings we had always favored. We spent a good deal of time unpacking, sorting, grading, and repacking for storage quantities of tektites that our Vietnamese friend had continued to send for our purchase.

We planted shrubs, geraniums, mounds of chrysanthemums, gilia, century plants, yucca, cactus, and pyracantha for the wintering birds. Our entry and patio and the garden descending a steep hill behind the house began to flow with color against the red ground rock. At the side of the house, wild white primroses congregated in a gleaming mass. Watering chores were never-ending. Few things other than the natural flora could grow in that soil and climate without the attentions of hose and spray. Even inside the house, my precious lichened rocks of fireplace and planter had to be kept moist. Care of living things is of itself rewarding, and I would rise early and have my chores well started and a walk completed by a respectable breakfast hour.

Harvey Nininger in front of his home in Sedona, Arizona, 1961.

The frequent retirement problem of what to do with one's time bothered neither of us. We merely extended our old interests and found new ones. Addie discovered a new hobby. She had saved stamps over many years from our foreign correspondence, so she bought reference books and stamp books, dragged out all her old envelopes and packages of stamps and all those she had collected on our Far East and European trips and earnestly set to work to build a creditable collection. I dragged out my old bundles and piles of manuscripts, some written as long ago as college, saved against the day when I might wish to set down an autobiography. I easily grew restless under the harness of writing chores, finding it difficult to work for more than two or three hours at a time. I had given up trying to learn the use of a typewriter years before, because my mind always raced ahead of my fingers so persistently that I was kept in a state of constant nervousness.

I doubt that the world holds another woman who would put up with such demands as my program had placed upon Addie. When she had accompanied me on field trips, it had not been merely for the outing. She had helped with the driving, with field notes, and whenever we were using a trailer, she had kept house and meals ready just as she had at home, while I had dashed about after all sources of information, scurried off on strenuous field work, or delivered a lecture or several lectures in a day. As the children grew, she had expanded her activities. Over the years, besides our own three children, from time to time, we had living with us several nieces, several nephews, and children of friends, who all shared our home for weeks or months at a time.

There was an almost constant stream of visitors—relatives, former students, old friends, scientists—to our home on the hill. When there were no visitors, we were often away. In the summer of 1961, we drove to Edmonton, Alberta, Canada, to make a magnetic rake survey of the field adjacent to the Bruderheim shower of March 4, 1960, for the University of Alberta.

This fall had occurred not far from Edmonton, and very near the village of Bruderheim, a Ukrainian settlement. At the time of the fall, there was a six-inch snow cover over frozen ground. All sizable stones had bounced, splashing black soil in all directions that made their finding an easy matter. About 700 pounds were recovered, the largest fragment weighing seventy pounds. The morning after the fall, farmers had noticed black dust or gray snow covering some of the ice of the broad Saskatchewan River and the fields just east of it. Searchers had gathered the meteorites, including several hundreds of small pea-sized stones.

Our purpose was to try to collect the meteorite dust, though it was now sixteen months after the fall. We spent many hours pulling our specially constructed little rake over the fields, but our efforts yielded only one tiny meteorite. Bruderheim was a stony meteorite; most of the dust simply was not magnetic.

Great areas, apparently hundreds of acres of snow, had been gray with small particles, which had sunk to the bottom on the first sunny day after the fall. Had anyone considered this dust important at the time, it could have been easily and meaningfully collected. These small particles were the most important feature of the fall, in my opinion.

There suddenly seemed to be demands for my time and services from individuals and groups heretofore uninterested in the Nininger name and work. A request came for a list of projects to be submitted to NASA under the auspices of the Rocky Mountain Association of Universities. I was being pressed for a field manual on discovering and collecting meteorites. These requests, together with my autobiography, a proposed book of meteorites of the world, the cataloging of about a thousand tektites a month, and a few other things seemed to be almost enough to keep a retired meteoriticist busy.

Two scientific meetings held in November 1961—one in Washington, D.C., and the other at Pennsylvania State College—appeared to herald the acceptance of theories I had advanced as far back as 1943, but which at that time had made no noticeable stir.

The Washington meeting was concerned with tektites and drew

a large crowd of interested people—mostly from the United States Geological Survey. Only a few years before, that organization was still failing to officially recognize the existence of meteorite craters. Practically the entire program on this occasion was devoted to the question as to whether tektites are produced on the moon or on the earth *by impact*. Not impacts of the size that produced the Arizona crater, but huge asteroid impacts such as those that produced the lunar maria.

At the Penn State meeting two days later, I lectured on "Cosmic Blitz," presenting the matter of asteroid impacts as the logical culmination of a fair and factual consideration of observed rates of infall of meteorites of various size. My talk was so well-received, the audience reaction indicated that the idea had finally arrived.

Several of the speakers on the Washington program credited me for their interest in tektites, and two went so far as to say that I had fathered the whole study of impact as the origin of tektites.

***

In February 1962, when a symposium on meteorites was held at Arizona State University, it was our pleasure to announce the establishment at Arizona State of a $25,000 trust fund from which an annual $1,000 prize would be given to a student in any college in the country for the best research paper on meteorites. At the same time, Herbert Fales made a grant for education and research on meteorites. We enjoyed our continuing association with the university people. Dr. Carleton B. Moore had become director of the Nininger Meteorite Collection (later the Center for Meteorite Studies) at Arizona State. Gerard Kuiper had left Yerkes Observatory in Chicago to join the University of Arizona at Tucson.

Our greatest satisfaction came from seeing the awakening

508   IN SEARCH OF FALLING STARS

realization among scientists that the general field of meteoritics was basically important and that certain aspects of this young science had vital bearing on problems that have long plagued geologists and astronomers. It seemed the years of battles and scoffing were at an end. Elusive recognition and honors for our work began to flow. The City of Denver honored me among other former citizens on the occasion of its centennial birthday celebration in 1958; McPherson College bestowed a special citation during commencement ceremonies that same year. I was granted a life membership in the American Association for the Advancement of Science. In June 1963, Arizona State University awarded me an Honorary Doctor of Laws degree.[39]

4

How many hundreds or thousands of specimens, some of them unusual in type and spectacular in size, must there be in odd nooks and crannies of our world? How many of them could we retrieve were there a more general interest in and knowledge about meteorites?

One of the largest, finest, and most unusual stony meteorites yet discovered came from the Bondoc peninsula of Luzon in the Philippine Islands only because of coincidental events and the persistence of an ingenious American. Conditions for finding meteorites in the Philippines are very poor. Most of the islands consist of rugged, heavily forested volcanic mountains. The combined 7,000 separate islands constitute only 1/570th of the earth's land surface, and very little of this is under cultivation.

The Philippines had recorded only one meteorite besides the witnessed fall of 1938 at Pantar, Lanao. The man who had given most

---

[39]Nininger received an Honorary Doctor of Science degree from Pomona College in 1976. *Ed.*

attention to meteorites in the islands had been Father Miguel Selga, a priest whose collection had been lost during World War II. Father Selga had since died, and I was positive that many Filipinos must have encountered meteorites. We had been in the Philippines ten or twelve days in 1959 when I visited the office of the National Bureau of Mines, sure that among the thousands of samples that inevitably reach such an office, there must be an occasional meteorite. The only question in my mind was: Would a meteorite have been recognized and saved?

"Well, yes, once in a while such a sample comes in," I was told, "but who cares about meteorites?"

I had the answer. Might not the Bureau have such specimens on hand? Well, the director thought, there should be one. It had come in not so long ago. He sent a girl to find it. She failed, so another young lady was dispatched. The second messenger returned shortly and handed over the specimen, a rounded, rusty lump of what appeared to be nickel-iron, very badly weathered. Despite its unusual shape— meteorites almost never are round or even approach that form—it was obviously meteoritic. The lump had some stony matter adhering to one side. This fact, together with its rotundity and the statement by the Bureau official that the lump had been found to contain nickel and that it had been detached from a large mass, set my mind racing.

The official assured me the Bureau had no interest in the specimen and that I was free to pursue the matter.

The sample had been brought in by a prospector as evidence that he had found an iron deposit, and the case had been duly referred to Senator Tanada of the Philippine Congress, whom I thereupon visited. But this gentleman had transferred the matter to his son, a lawyer, whom I then called upon only to learn that he had invited his friend Mendoza, another young lawyer, to share the new prospect with him, for Mendoza lived very much nearer to the location of the deposit, described as in a remote jungle far down the Bondoc Peninsula.

The pair of lawyers had contacted two Japanese geologists who were seeking an iron mine, but when these two were escorted to inspect the prospect, they were disappointed and disgusted. This was no iron outcrop, the Japanese geologists had said in frustration, but only something that had fallen from the heavens, and they turned away.

Perhaps the state of mind and muscles of these two geologists should be considered against the background of the ordeal required to reach the remote jungle location. After several hours aboard a slow train, they had waited for a bus which, when the weather and roads were not too bad, could cover the forty miles or so to a coastal village in a day. On bad days, the bus didn't run at all. From the village, a small boat carried them to and into the mouth of a river. After passage by water as far into the jungle as possible—the boat passage was also dependent on favorable weather—they had tramped for ten hours through crocodile and serpent-infested jungle to find not a vein of ore, as they had expected, but a lone metallic lump which they judged to be of low-grade quality.

After this failure of hopes for a rich iron ore prospect, the sample had been given to the Bureau of Mines. Had we not visited Manila, no doubt it would be yet lying there gathering dust, its identity perhaps lost for all time among the multitude of "no value" samples with which it shared storage space.

After I heard young Tanada's hard-luck story, I asked if he and his partner would be willing to relinquish their filing rights. They would be glad to do so, they told me, and for a small fee would see that I was guided to the site. Mendoza had relatives who lived in Gumaco, much nearer to the location of the meteorite, and arrangements were made by telephone for a family member to guide me to the meteorite. Accordingly, I purchased camp clothes and equipment, and we boarded the train for Gumaco the next morning.

We arrived at Gumaco at 1 p.m. As I stepped from the train to help Addie down, I crumpled with a badly turned ankle. Our host was most solicitous, installed us in cheerful quarters, and provided a vessel and hot water for my injured foot.

Toward evening, I felt so much better that we set out to shop for a pair of khaki trousers. There were four or five stores in the village, and each handled just about everything from clothing to hardware. The purchase of a pair of trousers seemed an easy chore. But though each store carried khaki clothing, for the first time in my life, I found myself inconvenienced, at five feet five inches, by being too large. It was only then that I noticed how small the people of the Bondoc Peninsula were. I ordered a pair of tailored trousers—and within two hours, I was ready.

However, a doctor who was called to examine me at our host's home ordered me to do no walking, thus ending the excursion. Sadly disappointed, we returned to Manila by train.

If the profession of meteorite hunting deals out some cruel blows, it also springs marvelous surprises. Ten years earlier, a visitor to our museum on Route 66 from Manila had shown a keen interest in meteorites. He was John A. Lednicky, a University of Kansas graduate who had lived in Manila most of his life. After his return to the Islands, he had sent us three fine tektites, and we continued to correspond. I had telephoned him while we were in Manila, and an invitation to dinner at his home awaited us on our return to that city. During the evening, I told him about the sample that the Bureau of Mines had turned over to me, my plans for investigating it, and my disappointment. The report given to me by Señor Tanada indicated a mass of a ton or more, and perhaps several tons. I mentioned that I might make a return trip.

Mr. Lednicky said that if and when I needed any assistance, he would be glad to help. He had access to all kinds of equipment, and

512   IN SEARCH OF FALLING STARS

he liked such work.

In the following months, as we traveled in Vietnam, Thailand, Australia, and New Zealand, and during our leisurely ship journey home, I pondered the Bondoc story from time to time. Everything pointed to an unusually large meteorite, there were indications that it was a new variety, and its geographical location was especially important since it was only the third find in a land area that should have yielded several times that number.

I decided to write to John Lednicky to request the help he had offered. His reply was prompt. He had given the matter careful thought and would go to the site as soon as weather would permit, Bondoc being a narrow peninsula that suffers very severe monsoon rains and windstorms and is subject to typhoons. Lednicky added that while awaiting good weather, he would look carefully into all of the legal aspects of the project.

I sent letters to Tanada and Mendoza introducing Lednicky as my representative. On September 15, 1959, Lednicky wrote that he was forced to wait until after the national election because bandits were operating on the peninsula. There had been considerable shooting. But after the election, there followed more rains and more typhoons, and then an illness kept Lednicky in the hospital for some time.

February 13, 1961, he reported that he had visited the site after great difficulty, having had to walk nine hours after following the crocodile-ridden river as far as navigable by boat. But he wrote that the stone looked like ordinary iron ore, although he could find no vein connected with it. On March 24, he wrote sad news. "The so-called Bondoc meteorite is just a hunk of low-grade hematite and not a meteorite at all." He had shown samples to his father, an experienced mining man who prided himself on knowing meteorites, and to two geologists, who had just "laughed it off" as "some of that low-grade iron ore" that had been "coming out of the Bondoc peninsula for the past fifty years."

I then knew that Lednicky either must have reached the wrong rock, or this was a new type of meteorite. The sample submitted to the Bureau of Mines could not be mistaken for low-grade iron ore because it was metallic, consisting of bright, tough metal inside its surface rust. If it had come from the same rock as that which my friend had visited, then here was a stony meteorite which bore large lumps of metal. I managed to reach Lednicky by phone and asked him to send his specimens to me at once via air mail. I was sure they were meteoritic. This evidently was a new type of meteorite, and I wanted it more than ever.

Lednicky had to await legal clearance before he could send the samples, but on June 25, they arrived. Tests proved them to be just as I had suspected—typical stony meteorite in which most of the small metallic grains had oxidized. I urged him to make every effort to recover this great meteorite, whatever the cost.

A reply from Lednicky dated July 6 expressed surprise that I still considered the specimen to be a meteorite, because among local geologists, the belief was unanimous that such "ore" had been coming from the Bondoc since time immemorial. However, he was taking steps to have the mass removed and shipped to me.

On November 23, Lednicky wrote that the stone was much larger than had been estimated. The field crew thought now that it would weigh eight tons—such a size would necessitate the use of a large bulldozer at a cost of $100 a day, a crane at $250 a day, with the probable total cost, $10,000. He advised caution. I telephoned him that the measurements that his men had reported would mean far less weight than the estimated eight tons, that my specific gravity tests indicated not over 3,000 pounds, and I asked him again to get it out at whatever cost was necessary.

As I learned later, his men reported on January 9, 1962, that they had been able to load the meteorite on a wooden sled, but that three carabaos had been unable to move it. Lednicky advised them to try a small bulldozer. When that was not sufficient, he himself went down with a larger bulldozer, and the meteorite was moved to the mouth of the river.

On February 21, Lednicky wrote me again. He had gone personally to the site for the third time. By making careful measurements, he found the meteorite to be much smaller than his men had reported. Also, he had dug around and under the original resting place and recovered many small pieces. He found, as I had predicted, that these were not of the same specific gravity as the first metallic sample.

Bondoc meteorite, Luzon, Philippines, 1962. John Lednicky, who coordinated the recovery of the meteorite, is on the left.

A raft was being built on which it was proposed to tow the meteorite to Manila. Recovery costs, Lednicky estimated, would not run over $3,700.

SLOWING DOWN   515

On May 29, he wrote again. The meteorite at last was in Manila:

The last phase of the recovery was rather risky and gave us some worried moments. The day we got it on the bamboo raft for the trip down the river and across the stretch of sea, a typhoon showed up. The water got so turbulent that we had to hire two motorized fishing boats to stabilize the raft. Night came, and one of the boats started to founder, so we steered close to shore as I was afraid to lose about $3,000 worth of recovery gear that I had borrowed from the office, besides the meteorite. As we got near the coast of Mulanay, one boat sank, and we almost lost four men trying to keep the raft from collapsing as the waves were unusually big. It was a nightmare all the way—so near and yet so far.

Once it was ashore, I found it hard to hire a truck for the run to Manila. All our office trucks happened to be out of town, and none of the trucking firms in Manila wanted to rent us a truck either because they were advised that the roads were lousy or due to the uncertainty of the typhoon. Luckily, I ran into a friend who would let me use one of his large trucks provided I had it back within twenty-four hours. It was sent after the meteorite and everything seemed fine until the truck wasn't showing up when due to return.

It gave me several anxious hours waiting as I worried that they got held up as they were lugging $2,000 with them to pay for the recovery team in the field. It finally showed up just before the deadline with the precious cargo. I was able to sleep soundly after that for a while.

John Lednicky put three and a half years of effort and frustration into the "favor" he had offered in late 1958. Without him, the Bondoc meteorite never would have been recovered.

Even after the great meteorite finally reached Manila, there were more delays and red tape before it could be carefully crated and shipped to the United States. It weighed 1,955 pounds—a shade under one ton and the second largest stony meteorite ever recovered. At the Flagstaff station, in August of 1962, a wrecker hoisted the massive stone onto a special small trailer, and we hauled it down Oak Creek Canyon to Sedona. It was installed just inside a window of my little studio off the main part of the house. Here I shared working quarters with it while I studied its exterior. Surprisingly, I found that it held a complex system of magnetic fields—the first meteorite found with such magnetic properties. The Bondoc meteorite has magnetic poles, negative and positive, miscellaneously scattered every few inches over its surface.

Anxious to see the inner structure, I solicited the help of C. H. Brandmeyer, friend and neighbor, to build a reciprocating saw. Then I set to the task of cutting the meteorite. I estimated the time that would be required, and Addie and I sent out invitations for the "opening." After 162 hours of saw time, a twenty-eight-inch slab, weighing 120 pounds, was removed. The inside was as interesting as the outside, bearing large metallic inclusions such as are known to occur in only one other stony meteorite.

The end piece was carried as our gift back to the laboratory at Arizona State University by Carleton Moore, who was among the sixty guests at the Bondoc "opening party."

Across the world, the real hero, John Lednicky, accepted only expenses and the satisfaction of success for the rather large fraction of his life's energies that Bondoc had cost him. As partial thanks, I sent John a meteorite suitable for a wall decoration.

5

When I was in Mexico in 1929, a veteran geologist, Dr. Aguilera, told me of a meteorite he had seen in a yard in Loreto, Baja California Sur.

SLOWING DOWN

Harvey Nininger, the Bondoc meteorite, and the homemade saw used to cut it, Sedona, Arizona, 1962. The first cut required 162 hours of saw time to remove a 28-inch slab weighing 120 pounds.

Then, in about 1950, a Los Angeles engineer and oil man, John B. Quinn, wrote to me with a similar story. He was virtually sure a meteorite was lying in a man's yard in the village of Loreto. He had tilted it and estimated that it weighed about 150 pounds. There was little chance of my being able to go into Baja in the near future, so I had filed the report for future investigation.

In 1952, Addie and I decided we could rake together enough for a drive down to Guaymas on the Mexican mainland, and from there I could fly across to Santa Rosalia in Baja California. Mr. Peter Mahieux, superintendent of the large copper mine there, a friend of Quinn's, had offered to assist me in reaching Loreto. I boarded a cargo plane used by the mining company to commute across the gulf. Between Guaymas and Santa Rosalia, the door to the cockpit opened and Mr. Mahieux, whom I had not met and whom I had no idea was aboard the plane, came back and introduced himself. He had noticed my name among the dozen passengers listed. He told me that he had at his home a small meteorite which he would give to me.

Mr. Mahieux took me to his quarters in Santa Rosalia, filled with interesting relics from various parts of the world. He brought forth from his mineral cabinet a beautiful little meteorite of the pallasite variety, saying that it had been brought to him by a native from somewhere near the village of Ignacio. This little meteorite was an exceptionally fine example of a rare and beautiful type, and also it represented a large geographical area never before known to yield a meteorite of any kind.

The following day, Mr. Mahieux drove me by truck some twenty miles south to where a young Mexican maintained a machine shop, explaining that this young fellow had previously been employed by his mining company to pilot the trans-gulf plane we had ridden from Guaymas, but they had been compelled to discharge him because of a very disturbing trick that he frequently played on passengers. He would take off with a group, then when he got out near the middle of the 100-mile-wide gulf, he would topple over and pretend he had passed out, remaining "unconscious" until the passengers were all in near hysterics. He seemed to derive great fun from this performance, but, said Mahieux, after one passenger had suffered a stroke, the firm had simply fired him.

It was this man whom Mahieux would ask to fly me to Loreto. I must have shown that the prospect scared me because Mahieux hastened to inform me that the young man was "the best pilot" he'd ever known.

"He built the plane you will fly in, made it out of the parts of various wrecked planes and automobile parts, and it's a better plane than a new one straight from the factory."

His words were only partially reassuring, but there was no other course of action, nor could I back out when my host was going to such lengths to accommodate me. So I said nothing, and arrangements were completed for the flight.

A retired navy captain and a young geologist were the other two passengers. We flew down the coast. The weather was fine. Below us, Concepción Bay was so glass-smooth and clear that I could see schools of fish in the water beneath us. We made a perfect landing in Loreto.

I called upon the padre at the venerable old church. He could speak some English, and he guided me to one of Mahieux's former employees who also spoke fair English and by whom I was escorted to the house of Señor Davis, in whose yard lay the meteorite I had come so far to see. It was a true meteorite, and instead of being only half or a third as large as I had been told, it was even larger than Mr. Quinn had estimated. Instead of 150 pounds, it weighed 209. We agreed upon a price, and arrangements were made for shipping it home.

I returned to the flying field, where my two co-passengers were ready for the return flight. But where was our pilot? He had been seen in a bar about an hour ago.

"Oh, my god!" exclaimed the captain.

"That S.O.B.!" cried the geologist.

Our pilot was found. He seemed to be in fair shape. We took him to lunch at once and saw that he drank only coffee, then went promptly to the plane. He took off without a hitch, and when we had reached his cruising altitude, he reached for his tobacco and cigarette papers and with a word or two and several motions directed the geologist (whom I shall call "G.") to take the wheel. Mr. G. protested that he knew nothing about flying, whereupon the pilot simply shifted the wheel over to him and applied himself to the rolling of his smoke. Mr. G., the back of his neck appearing as red as a beet to us behind him, refused to touch the wheel. The navy captain and I nudged each other and braced ourselves—to what purpose I don't know—and the plane began to bank and turn. The pilot reached over, righted it, then by insistent gestures instructed Mr. G. to look after it while he, the pilot, relaxed and slouched back for a rest.

Never have I wanted so strongly to hit a man over the head in all my life, but as I said to the captain beside me, just what good could that do? None of the three of us had ever received a flying lesson. The landing field in Santa Rosalia was a short one, ending at the very edge of a vertical cliff below which was the ocean.

Finally, G. took hold of the wheel, but when the plane wavered, he only protested again, for he didn't know what else to do. Meanwhile, our pilot leaned back and enjoyed his cigarette, occasionally reaching over to right the plane. It was only a thirty- or forty-minute ride, but during it, his three passengers lived just about that many days.

When the landing field came into view, our pilot took the wheel and made a perfect landing, but none of us felt inclined to overwhelm him with thanks.

Señor Alexandro Davis, from whom I had purchased the Loreto meteorite, related to me that his late father had told the family the meteorite had come from a site in the mountains about a six-hour ride by mule from the nearest ranch on the gulf shore, Luegi. He had also told the family that a much larger iron lay very near the one he brought down, but that he was unable to move it. I could not learn of anyone then living who knew the exact location.

The 209-pound iron I had bought bore on one side an area that showed plainly that it had been torn from a larger mass. This break appeared to have occurred either at the time of impact or very shortly before. Thus, the meteorite itself seemed to bear witness to the truth of the old man's story of another iron left in the hills.

***

In 1964, Addie and I decided to see the Baja Peninsula and to search again for the Loreto meteorite. We fitted out our International Harvester Scout with lockers to carry camp equipment and food supplies sufficient for a six-week drive in primitive country, and persuaded four friends, in Volkswagen and pickup truck, to make the trip with us. We left Sedona on January 17, 1964—my seventy-seventh birthday—and drove to Mazatlán.

After a four-day wait, we boarded the ferry to Baja, a very small, unkempt freighter with just enough room at the rear for our Scout and the pickup, parked crosswise. Bob Ayres, Glenn Keller, and the Volkswagen were up front with the freight, where Bob and Glenn had space to spread their sleeping bags and enjoy the trip. For the rest of us, the crossing, which took forty-eight hours instead of the thirty anticipated, seemed a never-ending misery. Floyd and Lois Hinsley sat in their pickup cab. The fit was so tight, the doors could scarcely be opened. We started out with the Scout's tailgate down to make up our bed, but during the first night, waves washed the entire stern and forced us to draw in our heads, close the tailgate, and stay put. Fumes from the engines nearby made us all seasick.

La Paz harbor is a beautiful one, and once we had retrieved our cars, we headed down the Baja side of the gulf to a beautiful little beach at the very tip of the peninsula. Marine life was colorful, vegetation was dense and mostly new to us, roads were rough but passable. When we headed back up the peninsula and drove into the village of Commondu, school was dismissed so that the children could see the Americanos. We sought out a man named Smith to inquire about the Loreto meteorite. He told us of seeing small meteorites in the mountains, and he had also heard that an American had bought a large one in Loreto about ten years ago—obviously my purchase of 1952.

From Commondu, we climbed a terribly steep, narrow, one-way road paved with rocks, hampered by sharp curves, and with no possible way to pass opposing traffic, of which fortunately there was none. The grade, the rocks, the curves and the mountains were repeated again and again, the hundred miles to Loreto, requiring a journey of two days.

I found Señor Davis, whose father had brought down the 209-pound meteorite, and learned that an elderly "brother" (probably a cousin) who had been present when Señor Davis had found the meteorite sixty years before, still lived in the mountains nearby. We organized a trip to meet this old man, whom I shall call "Señor Lejo." Very striking with his handle-bar mustache and patched and re-patched clothes, he impressed all of our party as a man of integrity and intelligence, quiet and serious, and of the old Spanish manner. Through our interpreter, Señor Bill Benziger, manager of the Oasis Hotel in Loreto, we conferred with this patriarch at some length. The interview added a few facts to those given me years before. Señor Lejo told me that he and Señor Davis had found the two irons on the edge of a high mesa, surrounded by a circle of large stones that had been set up by "ancient people," for what reason was anybody's guess. When he and his "brother" found that they could not move the larger of the two irons, they covered it up and took the smaller one by mule down to the town.

The old man made a drawing in the sand to show us how the mass lay within the circle of stones which he indicated was approximately ten to twenty meters in diameter. When asked if he could lead us to the place, he said without hesitation that he could, but it would require six hours by mule and three hours more on foot, for the last part of the trip would be too rough for animals.

We decided that we could not make such a journey on that occasion, so I gave the old man some money to go to the spot and bring back a sample of the iron. Señor Lejo said then that he understood there was a kind of detector that would enable him to locate the exact spot, that they had covered it over, and that he might have to dig up the whole area within the circle of stones in order to find the iron. I promised to send a metal detector in care of Bill Benziger and decided tentatively to make a return trip in July.

Traveling northward up the gulf coast, we indulged in beach lunches of fresh-dug clams, camped in lonely and beautiful surroundings, and received car service and civilized accommodations in occasional resort towns along the way. At San Ignacio, we checked out a meteorite report that turned out to be "just another story."

As soon as we returned to Sedona, I made arrangements to ship the detector to Loreto. We waited a bit impatiently to hear from Señor Lejo. About midsummer, Bill wrote us that he had learned from one of the young men of the Lejo family that they planned to obtain the meteorite and sell it to someone else. "We can get a lot of money for it," the young man told him. Bill refused to hand over the detecting instrument. I wrote to Bill that he should remind the Lejo group that the meteorite belonged to the Mexican government, and that I held a permit to obtain the specimen with the requirement that half must go to the University of Mexico, and I was ready to pay for help in recovering the meteorite.

After seven months, another letter came from Bill saying that a younger Lejo brother had come to him, deploring the action of his elder brother and offering to take me to the site of the meteorite. This Lejo youth suggested that I pay him by the day and then, when the meteorite was found, pay whatever bonus seemed right. He convinced Bill that he could take me to the location. It would require a six-hour mule ride and three hours of walking.

Bob Ayres and Glenn Keller, who had been along on the previous trip, agreed to go with me again to Baja. Carleton Moore recommended a student at Arizona State who was good with the Spanish language and would like to go. The fifth member of our party was a retired physician, Dr. Quade.

We arranged to fly to Loreto in a four-passenger plane owned by Wes Roberts, one of Sedona's more adventurous retired citizens, piloted by Ray Steele. He delivered the five of us in two flights. At almost exactly a year from the time of our previous visit, we were ready to begin an actual search for the meteorite with Pancho, our young guide, and his grandfather, Señor Lejo.

The morning of January 30, 1965, we arrived at the outlying ranch that was to be our starting point. There was Pancho, not with six mules, as planned, but eight. He had not three burros, but eight. He also had four men besides himself and his grandfather. This seemed strange, and I wasn't sure that I liked it, but I told myself, "Perhaps this is good. If we find that the meteorite is not heavier than a ton, perhaps we can manage with all this manpower to carry it down on a hammock by men and mule power." I had a nylon rope in my pack sufficient to bear 4,000 pounds. My earlier expectation had been that a helicopter would be the only feasible way to remove the iron, if it was as large as it was made out to be and rested where it was reported to be, on a formidable, high mountain ridge that looked to be 4,000 to 5,000 feet in altitude despite the fact that it rose only ten miles or less from the gulf coast. With these thoughts hurrying through my mind, I made no objection to the extra man and animal power, though I was quite aware I would have to pay for it.

At least a dozen times I had been told the trip would require six hours by mule and three hours on foot, but now, reviewing the plans, Pancho said it would require twelve hours by mule and six on foot. Our prospects were looking no better pretty fast. We asked about water, having come prepared to carry only five quarts for each man. We were assured the water was good.

Pancho said that we would camp the first night about midway to the mesa where we were to find the meteorite, and the next day, we would travel on to the foot of the mesa and set up a permanent camp, leaving the mules with a boy while we went on foot the next day to the top to locate the iron. We were told also that a much larger meteorite lay on the mesa about an hour's walk from the one which was buried.

The truck which brought us from town was instructed to wait at the ranch for our return, and off we went. We rode four hours, then stopped for the night by a waterhole so rich with animal and plant life and covered by such ugly scum that only the animals and the grandfather drank from it. The mules were fed branches of palo verde, and we ate and bedded down.

Our Mexican friends rose about midnight, built a big fire, and huddled around it to keep warm. I got up a little later and stood by the fire an hour or so, then returned to my cold sleeping bag, but slept little.

After breakfast the second day, we rode three and a half hours, stopping at noon by a better waterhole in a relatively flat valley from which a rugged mountain rose to the north. And we were told that on the slope of that mountain was the spot where the meteorite lay. Asked to point out the place, Pancho pointed to a general area just over a rugged hump a few hundred feet above the plain. A little later, Grandfather pointed in a direction about twenty degrees farther west. I sketched in my notebook an outline of the mountain and asked Pancho to mark the spot. He marked a spot in the direction the old man had pointed. But when his grandfather looked at the sketch, he said, "No!" Instead he seemed to point to an area on the gentle slope below the base of the mountain. I drew a line to represent the base of the mountain, and he made his mark *below* it on the edge of the valley floor.

By then, I was completely unsold on the quality of our guidance. Where was the three- or six-hour climb that was too steep for the

526   IN SEARCH OF FALLING STARS

burros? What of the prospect, painted in my mind's eye from the descriptions and explanations I had received from several parties, that had led me to inquire by long-distance telephone from Sedona to Phoenix, Yuma, San Diego, and Los Angeles as to the availability of helicopters to lift a meteorite from a terrain that was inaccessible to burros? The area we were in could be negotiated easily even by Clydesdale or Percheron draft horses.

Ten minutes farther along the way, we were shown a large rock, more or less on edge. This, we were told, was the marker Señors Lejo and Davis had set up. A glance at the surrounding terrain showed plainly that the ground had not been disturbed for centuries. Nevertheless, said Grandfather, the rock marked the spot where the meteorite had been buried. Our instrument showed nothing. We were led to a nearby arroyo. Here, said Grandfather, was the spot where they had toppled the iron in to bury it. Again, no signal from the detector.

I followed the arroyo up and down for perhaps 200 yards, testing every foot of the way. Nothing. By then, the old man had found another marker—a big stone with chipped edges. He told us that he had thus marked it. We tested again; no signal.

At no time before this day had Señor Lejo mentioned a marker. Indeed, he had requested the instrument for the very reason that he would not be able to spot the location within the circle of stones where he said the meteorite had been buried. Nothing was to be seen of any circle of stones.

Our next tour was along the base of the mountain, somewhat above the plain, where we were led from place to place with no indication by either the instrument or the nature of the terrain that a meteorite ever had been buried, and seeing no spot that even remotely resembled the clear picture that the old man had described repeatedly and consistently in 1964.

When the search along the base of the mountain failed, our

guides asked us to test a spot on the plain. I pointed out that Señor Lejo had never indicated that the meteorite was in a place so accessible. Now Pancho's elder brother—he of the proposed double-cross—explained that Grandfather and Señor Davis had by means of a rope hung the meteorite on a pole and thus had carried it 100, 200, or 300 meters before giving up and burying it.

I said nothing, but my thoughts were burdened. "Our first account was that the two men could not move the iron, then the story was that it was 'toppled' into a ditch, and now we are told that they carried it a considerable distance." I decided that I was finished, and so was our search.

Richard, our student interpreter, and I went to Pancho, telling him we were to go back to the ranch at once. He insisted again that they could find the meteorite. I directed Richard to tell him that when they had found it and brought it down to the ranch, I would give them 12,000 pesos (about $1,000) for it.

It was clear that our guides were not happy about returning to the ranch so soon. We were convinced by this time that the plan had been to keep the search going for several days to draw as much in wages as possible. I had tipped my hand; it might be assumed that I carried with me not only money for wages but also for a bonus. I wondered whether my hasty revelation might inspire them to collect on the spot whatever money I carried with me. I knew that Richard had been asked more than once whether we carried guns, and he had said that he did not know, but as a matter of fact, we were not carrying arms.

It was Richard who soothed ruffled spirits and quieted any unreasonable ambitions. He had been promising all along the way to cook tortillas for our quintet, so that night, he did so. During Richard's preparations, Pancho's elder brother came to watch and was given one of the first products of our interpreter's labors. Grandfather came, too, and enjoyed a sample. Bob had made chili to go with Richard's

528 IN SEARCH OF FALLING STARS

tortillas. It was our best meal of the trip. That night, our party slept near to each other, just across a natural hedge of palo verde, mesquite, cat's claw, and cactus that separated us from the guide crew. Most of our group talked far into the night to keep awake, but our guides were quiet. The next morning, all was pleasantness on both sides of our desert hedge. In five hours, we were back at the ranch, but the vehicle we had left there had departed. We were saved from a further mule trip by a one-and-a-half-ton truck which happened by, its high rack bed loaded with dry wood of all descriptions and with several passengers.

Richard negotiated for us to ride also. The deal was made, and all of our impedimenta, the five of us Americans, and Grandfather Lejo, were crowded aboard.

The following morning, Ray Steele appeared in response to our wire and flew us back to Sedona. I pronounced it a good trip, though failing its principal objective. There was nothing unusual about that in meteorite hunting. One day, perhaps, the second—perhaps even a third—Loreto meteorite will be uncovered by man or erosion, if the stories are more than legend, and will be brought to civilized view.

6

In the spring of 1965, Addie and I traded in our Scout for a camper mounted on a pickup truck and spent the three summer months touring Canada and Alaska. On our way north, we heard reports of a great evening meteor seen by many persons in British Columbia, Alberta, the Yukon, and the state of Washington the late evening of March 31, 1965.

Scientists of the Dominion Astrophysical Observatory near Penticton, B.C., the University of Alberta at Edmonton, and the University of British Columbia at Vancouver were impressed with the magnitude of the phenomenon and promptly collaborated in energetic efforts to locate the fall. On the basis of interviews with witnesses, they established by triangulation the approximate terminal point of the fireball.

The radio stations of British Columbia and adjoining provinces and states, moments after the fireball, had been besieged by calls from their listeners saying the fiery object had come to earth in their respective communities. There were reports of violent repercussions over a great area.

Calculations by an eminent physicist on the basis of seismic disturbances of different magnitudes and at different distances from the terminal point led him to an inescapable conclusion that the mass of the meteorite responsible for the fireball must have been on the order of a million tons. This conclusion was considered proof-positive that the meteorite must have produced a sizable crater, and a prompt aerial search was inaugurated by plane to be followed by helicopter.

The investigators at Penticton told me they had no encouraging results from their two months of effort, but they had not given up. On the very day of our visit, June 6, the aerial reconnaissance of the rugged wilderness area still was being carried on. The cooperating scientists had gathered what seemed to be a very satisfactory lot of data. They had determined with apparent accuracy that the fireball had approached from a westerly direction and had disappeared at a lower-than-normal altitude over a remote, uninhabited, and rather rugged terrain in southeastern British Columbia.

Patches of "black snow" were reported in a region in harmony with this determination. Requests for samples of the "black snow" brought a container of black mud to the observatory. This material was regarded skeptically and was forwarded to Ottawa for analysis. The reports returned indicated the presence of nickel but were far from positive.

I inquired if I might see a sample of the "black snow mud." All but a very small remnant had been sent to Ottawa, but they were glad to show me, somewhat apologetically, what remained. The sample

530 IN SEARCH OF FALLING STARS

was wrapped in white paper and stuffed into a glass jar. The rather crumpled paper showed smears of what appeared to be finger smudges from the black mud. In the folds were a few small, black, solid bits of matter, one of which was placed under a compound microscope for my examination.

When I brought the particle into focus, I could scarcely refrain from shouting with excitement. I was looking at as beautiful an example of fusion crust as I had ever seen, and it exactly resembled the crust that I had examined on several type I carbonaceous chondrites.[40] Nearby, as a portion of one of the black finger marks, I could discern a small chondrule, perhaps two-tenths of a millimeter in diameter.

After expressing my conviction that they had a carbonaceous chondrite to deal with and that any search for a crater would almost certainly be futile, I was told that none of the group had visited the site where the black snow was collected, either while the good snow cover of March 31 still lay on the ground or after it had gone.

Evidently none of the men concerned with this survey was sufficiently familiar with the various types of meteorites to recognize the spoonful of "mud" as the product of disintegrated carbonaceous chondritic material.

On being unexpectedly faced with the overwhelming confusion of reports from excited and terrified witnesses, the men to whose lot it fell to deal with this very outstanding event had thought first of that feature which had been the central theme of ninety percent of meteoritic literature—the great Arizona crater. Actually, such a disorder of reports suggests first the smash-up of a stony mass of uncertain magnitude and the showering of several square miles with large and small

---

[40] Type I, now called CI, carbonaceous chondrites are a very rare group with bulk chemical compositions that, of all types of meteorites, most closely match that of the Sun. *Ed.*

fragments and more miles with sand, gravel, and dust.

The great good fortune of this situation should have been the snow cover on the terrain at the time of fall and the reports received of the black snow. There should have been no time lost in reaching the area by helicopter and snow shoes. From the site of the blackened snow, the party should have radiated in a search for small disturbances in the snow—not for a crater, but for small bits of black rock or masses of the deliquesced remains of such.

A brief laboratory examination of the mud that had been black snow should have told the investigators at once that the chances were a hundred to one that no mass larger than a few pounds should be expected to have survived, and that most of the material would be in the form of particles too small to cause any noticeable disturbance of the snow cover. From such a large mass, one could expect thousands of small, fully encrusted individual meteorites ranging in size from BB shot to a few ounces, with perhaps a few weighing several pounds, besides tons of dust spread over several square miles. The only evidence of such a deposit that could have been seen by plane was, of course, the black snow where heavy concentrations of dust had fallen. Masses of a few pounds would in most cases have been crushed as they struck the frozen ground beneath the snow and would not have rebounded as did stones of the firm-textured Bruderheim fall in the same province five years before.

The essential ingredient was swift action. The snow cover might hold small fragments briefly or show the pockmarks of their fall, but as the sun's rays reached them, they would have sunk from sight faster than the surrounding snow due to their capacity for heat absorption. By the time of snow melt a few weeks later, they might have been badly disintegrated, and in any case would present a target formidably difficult to search for.

It was at a point somewhere between advanced technology and

laboratory methodology that the investigation had bogged down. Lack of familiarity with meteorites had lost what was probably the most important meteorite fall in a century in North America for these highly trained and capable men. Carbonaceous meteorites are exceedingly fragile, and the evidence from field work is that the great bulk of such falls is reduced to dust long before they reach the ground. Therefore, they are rare, at least in collections. The actual frequency of their occurrence is one of the unknowns I hope a modern and thorough approach to fireball recovery will validate.

Carbonaceous meteorites of type I tend to alter rapidly. In a matter of two or three days, if subjected to moisture, they may simply change from a more or less firm stone to a mass of muddy consistency. Such alteration, of course, must mean chemical changes, as well. Apparently, in the wilds of British Columbia, there had been delivered to the soil of our planet a large mass of the most sought-after variety of meteorite which now may have to be recorded as a mere inconsequential trace.

Addie and Harvey Nininger in their home in Sedona, Arizona, c. 1965.

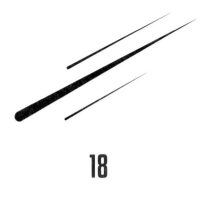

18

New Paths and Byways

Truth Is eternal, but her effluence,
With endless change, is fitted to the hour;
Her mirror is turned forward to reflect
The promise of the future, not the past.

—James Russell Lowell, *A Glance Behind the Curtain*

Man is not yet ready to write a chemical formula for the word meteorite. Twice within a half-century, crater-forming meteorites have punctured the land surface of the earth. Within the same period, mountain-size asteroids have encroached upon the orbital precincts of our planet no less than five times—in one case with what might be described as a lucky near-miss. Looking then at the pock-marked face of the moon, one is forced to conclude that meteorite impacts constitute a geological problem of considerable magnitude. Furthermore, we cannot expect that meteorites collected so far represent all of the varieties that fall to earth.

534 IN SEARCH OF FALLING STARS

Some varieties have been encountered only once, others twice or three times, and some quite frequently. It is only reasonable to suppose that some varieties arrive on earth only once in 1,000 or perhaps 10,000 years.

Among the more than 1,800 falls that have been cataloged, some varieties are represented by only one, two, or three falls. Such rare forms probably have arrived more often than our records show. It is equally reasonable to conjecture that there are other varieties orbiting our sun, examples of which have not reached our planet since scientists began recognizing meteorites. Beyond the problem of varieties that may go unrecognized is the uncertainty regarding the existence or prevalence of combustible, icy, or otherwise perishable varieties of meteorites which may exist in space, but which by their nature never succeed in depositing any surviving remnants on our planet.

We do not yet have an adequate sample of the overall increment of meteoritical matter. Until we do, all of the efforts at analysis and measurement of the earth's cosmic relationships are not very meaningful.

One of the fascinations which meteorites hold for man is that man simply can't know, by his present knowledge, when or what to expect in the way of a fall of great size or new variety.

The ancient civilizations of Europe and Asia, through superstition or out of curiosity, disposed of nearly all the meteorites with which they had come into contact long before their meteoritic origin was understood. On the North American continent, where civilization is relatively young, man has had access to the accumulated increment of many thousands of years, most of the survivors being the hardier irons.

In the century and a half since the existence of meteorites has been recognized, man has recorded an average of about three or four falls a year from which specimens have been recovered.

As I first began to peruse the literature on meteorites in the 1930s, I was impressed with several dominant inconsistencies: The matter of the number of meteorites that reach the earth, the discrepancy between the dominance of irons among meteorites found by chance and of stones among witnessed falls, and the assumption that nearly all stony meteorites were chondrites, a type rather rich in nickel-iron and carrying as a prominent constituent little rounded bodies known as chondrules.

It seemed that quantitatively meteorites were considered to have no meaning. I asked the late Dr. F. R. Moulton why this attitude existed, telling him I had evidence that far more meteorites were falling on our earth than was consistent with the extant literature on the subject. Dr. Moulton replied that the question would be worth investigation if I could demonstrate that the meteoritic increment was several thousand times as great as had been assumed. He indicated that he was sure this could never be achieved.

Although I believed this to be true, I did not have enough evidence at that time. Dr. Moulton died December 7, 1952. Had he lived ten years longer, perhaps I could have convinced him. I was amazed to find that the mass of material represented by each of the recorded falls seemed to have been considered as approximately equal to the amount recovered.

An illustration of the magnitude of this error is the Plainview fall. My first experience with this great shower came by way of a Mr. Rightmire, who informed me that he had two meteorites he would like to sell. He said that Dr. Merrill had told him he had all he needed from that fall. Naturally, Merrill's response to the farmer in this case simply put an end to the search for stones. I chanced to learn later that at the very time Rightmire offered him those two stones, some of the neighbors had been throwing meteorites out of the fields into fence rows and into the water holes of the neighborhood. Merrill described Plainview as a fall of total weight of 68.2 pounds; ultimately, known stones totaled some 1,500 pounds.

As early as 1933, in *Our Stone-Pelted Planet*, I began to dispute the estimates of the amount of increment to the earth. At the rate of infall estimated by one respected writer, it would take fifty billion years for a layer one foot thick to accumulate on the seabed. Since man occupies only a very small part of the planet, and since experience shows that only in a very small portion of cases were falls discovered even though they were in inhabited areas, it is evident that the number of known falls is a small percentage of those that have actually landed on the earth. Work in the field led me to conclude that perhaps several thousand falls occur upon the planet each year, the majority of which, of course, land in the great seas and oceans, wide expanses of arctic snows, jungles, deserts, and other areas where man is not present to observe. In the course of a thousand years, there is an enormous amount of meteoritic matter sprinkled across our planet. Some of these meteorites are of such nature that they would last for many millennia before complete disintegration. Others would survive only a few years. A few would become nearly unrecognizable in a matter of a few days, or in wet weather might be reduced to mud in a few hours, but the majority of meteorites are of such character as to survive the action of the elements for at least a century.[41] It is not surprising that during this century and a half of man's recognition of meteorites, there have been comparatively few specimens found that were not actually seen to have fallen to earth.

I believed that inadequate field work in cases of witnessed falls left important small constituents of those falls uncollected, and also that the meteoritic dust and gravel carried in the cloud trails actually constituted the principal bulk of such falls. Setting out by field work to gather facts to replace assumptions, I came up with an estimate of meteoritic increment of 50,000 tons a day, which would amount to a

---

[41] Here I refer to the well-known kinds of meteorites.

layer nineteen feet thick in 60 million years. It is yet too early to say how nearly my estimate was justified, but every year, the "experts" come nearer.

My field work proved, as I had contended, that long-accepted theories as to the ratios of occurrence between metallic and stony, and between chondritic and non-chondritic specimens, were false because they were made without the existence of any meaningful or extensive field program on which to base conclusions.

I reasoned that there are unbelievable numbers of meteorites remaining unfound because people are unable to identify them as meteorites. Iron meteorites differed most from terrestrial rocks, and chondrites differed more than did achondritic stones. Our work demonstrated definitely that stony meteorites are some hundreds of times as abundant in the soil as had been supposed, and that the ratio between them and metal meteorites was not different among witnessed and unwitnessed falls. In the course of chasing down as many fireballs as possible, half of all the finds we made proved to be of the achondritic class.

In an area of 450,000 square miles, we recovered representatives from 200 previously unknown falls of stony meteorites in thirty-five years, yet probably not more than ten percent of the population of the area was ever contacted by our group either by word-of-mouth or by printed matter.

My program of instruction through schools, lecturing, and field activities was initiated in 1923, and by 1950, the State of Kansas, which had yielded only fifteen falls in a period of seventy-five years, had given us representatives of forty additional falls.

The large number of meteorites found in Kansas can be attributed to two factors—the land's relative freedom from rocks and the "interest factor"—the acquaintance of the people with meteorites.

It was our experience that when a barren area was properly

538 IN SEARCH OF FALLING STARS

investigated, it no longer was barren but might turn out to be a most productive area.

In the Texas panhandle, only one meteorite had been recorded prior to 1933. In that area, twenty different falls were recovered over the next seventeen years. At the same time, the area around Plainview, adjacent to the panhandle on the south, where we really concentrated our search, yielded a dozen more falls.

It is impossible to separate the influence of the museum from our earlier field work, the lecturing in schools, writing for small town newspapers, and other ways and means of attracting the public's attention to meteorites. It is certain that our whole educational effort did much to arouse an active interest in meteorites among scientists. Many of the papers that reach my desk from all over the world seem to reflect discussions held over our museum cases or at the crater. I kept no record of such discussions. There were hundreds of them. Many ideas that were presented and discussed have become widely accepted since.

Our museum and activities provided a forum and a kind of experimental laboratory for ideas. But besides exerting indirect influences, our program directly accomplished and proved certain facts and erased several erroneous notions that had been held by scientists for many, many years. Most of these findings were the outgrowth of our field work, by which our very small force of workers brought to light more meteorites than were being recorded by all institutions in the United States and the entire world.

Showers of stony meteorites were previously considered to be the exception, but we proved them to be the rule. We brought to light at least 222 previously unknown falls, aggregating more than 2,000 individual meteorites. A number of additional falls probably came to light as a result of this effort and the publicity which resulted, but these were not traced to us as the original alerting agency.

In the year 1937, thirty-one new discoveries were tabulated within a twelve-month period. This one year's harvest was a larger number of meteorite discoveries than had ever been recorded in any three-year period for the entire world. In his book, *Between the Planets*, published in 1941, Dr. Fletcher Watson of Harvard wrote that "At present, Nininger is accounting for half of all the discoveries in the world."[42]

We examined some forty-odd pseudo-meteorite craters that we proved to have been formed by causes other than impact. I published more than 150 papers, four books[43], and three booklets, and distributed at least 200,000 free leaflets. We were sent some 35,000 specimens thought to be meteorites. These were examined and free reports were provided to the senders.

A revered elderly astronomer once criticized my work by saying that I had not added greatly to meteoritical theory. My reply was that I had spent my time gathering facts, and the more facts I gathered, the less certain I was that we were yet ready for more theories. I have been asked about programs to be followed in studies of meteorites in the seventies, as we began to edge our way farther into space beyond the moon. In reply, I mentioned first the suggestions I have been making for a generation. Some should continue precisely the activities we pursued over the years and that remain unfinished. Other activities, we were never able to undertake adequately, and there were yet others that could not be foreseen until space and the atom were tamed.

We should continue to obtain useful quantitative data through recovery of fireballs and recognizable meteorites, by applying new techniques and an intensified effort to the matter of terrestrialization to discover reliable clues by which to identify even the completely altered remains of ancient meteorites, and by the study of all of the

---

[42] Watson, Fletcher G., *Between the Planets*, Philadelphia: The Blakiston Company, 1941, p. 144.

[43] The final count was seven books, including *Find a Falling Star*. Ed.

grist from the friction mill of the atmosphere—the ash from millions of "shooting stars" per hour, ablated particles from surviving meteorites, and terminal dust clouds and showers of sand and gravel discharged from stratospheric smash-ups, including particularly the non-magnetic portions. We must inquire further into the nature of crater-forming falls, the forces and effects attendant upon the instantaneous displacement of hundreds of millions of tons of rock by impact and the products of explosion and impact. Further investigation should be made of the effects of cosmic rays on meteorites, an aspect attracting probably more research than any other single question relating to these travelers from space.

To date, there has been little progress in the introduction of information on meteorites into the curricula of elementary and secondary schools and colleges, which I have urged for many, many years as essential to the truly substantial recovery of meteoritic materials.

2

Meteorites are in a class by themselves. All other material belongs to this earth, and so far as we know has been on the planet through all of geologic time, but meteorites are invaders from without. They represent other parts of the universe about which we as yet have no way of learning save the study of light that arrives from luminous sources, and through the cataloging and interpretation of photographs taken by the satellites that are the new miracles of our age. These lumps of stone and iron are still our most tangible source of information concerning the vast expanse of space stretching immeasurably in all directions.

A small boy once asked me, "When do meteors become meteorites?" Rather than to consider whether the change takes place in burning flight, or at the instant that destructive friction ceases, or upon striking the ground, it seemed simple and plausible to explain that an object is a *meteorite* invisible in space until it flares up during its struggle with the atmosphere, producing a light that is known as a *meteor*, and that what remains to reach the ground is still a *meteorite*. Friction has reduced its size, and has changed its surface texture and contour, but for the most part, the materials in the stone are not different from what they were in space.

Our solar system holds a great expansive multitude of meteorites, with the sun as their central controlling force around which each member revolves in an orbit. The number of meteorites in this great aggregation is too great to estimate in the present limited state of man's knowledge, but astronomers tell us that in the course of a year, our own planet making its annual journey around the sun is bumped into by more than eight billion meteorites, each of which is large enough to produce a "shooting star" or visible meteor, and that a thousand times more that are too small to see also plunge into our atmosphere. This would give us a figure of eight million million each year that are swept up by our own little planet.

Although we have pushed out the horizons of knowledge a long way since the time when the earth was considered the central feature of the universe, we still have only superficial knowledge of even our own insignificant planet, this speck of the solar system we call our world.

If a huge map the size of an ordinary city block were made of the known universe, the earth could not be represented at all. An ordinary punctuation mark, a period like those used on this page, would be more than fifty thousand times too large to represent even the outermost limits of our solar system.

Scientific investigations of our earth have been limited to a very thin surface layer. Using a large apple to depict our planet, cutting it crosswise between the blossom and stem end, the skin would represent a layer about twenty-five miles thick on the outside of the earth.

Most of our first-hand knowledge is limited to a depth of only about a mile, though in a few places, the earth exposes its own interior to a depth of a few miles. Our mines dip only a little more than two miles into the earth, and these penetrations occur only in scattered places. Even the planned Mohole project would reach only to a depth comparable to a quarter of the thickness of the apple skin.

We know very little aside from what we can see on the surface of the earth. Going upward, we have until very recently been limited to the stratosphere, and our space explorations have as yet been confined to a few vehicles, a few men, for small spaces of time and circumscribed areas in space, and under such restricted conditions as to afford only limited information.

Until we receive more and better information about the moon, the planets, and boundless space, we are limited to an extremely thin film on the exterior of one planet of one of the smaller systems of the universe. As we try in every way possible to find out more about this vast enormity of space that we live in, certainly we owe interest to the stones that come to us from the mysterious reaches of space.

The subject of meteorites was for so long relegated to the ranks of least important, with so little attention paid to it that even now there are states in our union that have provided not a single meteorite discovery, yet hundreds of meteorites are mingled with the soil of even the smallest states.

"Weigh yourself daily," say the scales at the corner drugstore. Our doctors echo the admonishment.

"Check your waistline every million years," we say to our planet.

Only as we reached the half-way mark of the twentieth century did science begin an intensive diagnosis of the effects of wanderers from space on the state of the earth. The most annoying occurrences of my museum days were those occasions when some visiting scientist would peruse the collection and then murmur, "Interesting, but of what use are these meteorites?"

As the space age dawned, scientists at leading research centers began to purchase bits of meteorites for analysis. These growing orders for research material more than once staved off impending decisions to give up our museum and our program.

The Dominion of Canada is supporting a project of exploration for and into craters and other impact scars of ancient vintage—"fossil scars" dating back half a billion years. The US Geological Survey has established a Laboratory of Astrogeology. The Astrophysical Observatory of the Smithsonian Institution has a nationwide program of meteoritical research under way, including the Prairie Network of photographic batteries for tracing incoming meteorites that is the very reality of the dream I advanced in the middle 1930s.[44]

In the United States, there are now scores of young astronomers and geologists, geophysicists, geochemists, and mathematicians devoting their talents to the problems of extra-terrestrial forces acting upon our planet. Many of their research projects are directed to refined chemical studies of various kinds, the mathematical formulation of hypotheses which explain the effects on our earth of great meteoritic impacts, and to nuclear means of measuring the ages of meteorites in space and so contributing to the understanding of the origins of our solar system.

Naturally, I did not foresee the extensive possibilities that have developed since the tool of atomic fission has come into such general use. But I did point out, as early as 1934, that there must be some relationship between cosmic rays and the meteorites immersed in them during their long sojourn in space.

Meteoritics truly is on the march. Many young scientists are now

---

[44] Although the Prairie Network ceased operation in 1975, other camera networks operate today. *Ed.*

engaged in astronomical, geological, and nuclear chemical studies and isotopic analyses that were not thought of nor possible a decade or two ago. In the complex world of the seventies, these young researchers make great contributions, yet most of them, by the nature of our changed world, are confined to in-depth studies in a few critical but specialized areas. Their brilliant minds are exceedingly well-trained, and their contributions are great, but I would not wish to trade the opportunity I had to explore a broad and relatively untouched field during a time when one had the ability to make his own path and to examine whatever he chose along the way.

The new researchers are yielding extremely valuable information regarding the evolution of our solar system and the greater universe. Underlying all of the specialized investigations should be an equally vigorous campaign for the recovery of material for these researchers to work on.

Some geologists have come to realize that our planet bears proof of having been bombarded by large as well as small chunks of matter and are beginning to look for evidence that such bombardment has had its effect on the processes of mountain-building. Eventually, the geologists will change their views and their textbooks in the manner of the astronomers, whose acceptance that the moon's face bears scars of impact, rather than volcanism, forced their fellow scientists to search for evidence of a similar bombardment of the earth.

I cannot say whether my paper of 1942, "Cataclysm and Evolution," ever made any impression or not, but certainly it pointed the direction in which the sciences of geology and astronomy must turn, and when the textbooks of geology are rewritten, as they must be, the ideas set forth in that short paper will fall neatly into place.[45]

---

[45] "Cataclysm and Evolution," *Popular Astronomy*, Vol. L, No. 5, May, 1942, pp. 270-272.

I have the satisfaction of having been a member of a rather small group who have marked a turn in the road that science must inevitably follow. The space program has revealed that Mars, like the moon, is pockmarked with meteorite craters. Eventually, methods will be developed to determine definitively that the earth, whose surface is continually subjected to the scrub brush of erosion, has been similarly bombarded by giant missiles during hundreds of millions of years. It must follow that millions upon millions of small masses have arrived and have been terrestrialized beyond recognition.

The earth's atmosphere constitutes a very effective armor against small meteorites, but not against large ones. For example, a meteorite one foot in diameter would be virtually stopped by the atmosphere and would strike the ground at a few hundred feet per second. One ten feet in diameter would weigh a thousand times as much, but it would have only one hundred times as much surface for the atmosphere to press against and thereby check its speed. For a mass a mile in diameter, there would be almost no reduction of speed.

Every now and then, for more than a generation, an introduction to an audience has described me as "the man who has discovered more meteorites than any other man who has ever lived." The record is one of which I am rather proud, but in my opinion, it is not my chief contribution to the young science of meteoritics. I would rather believe that the finding of meteorites only served as a means to the more important end of helping man to better understand the environment in which he finds himself. I would like to think that my interest in these immigrants from space reached a little farther than the number of meteorites resting on museum shelves.

Specimens like Estherville and Bruno bring messages written on their stone or iron faces, and chemical and nuclear tests tell us more about their composition and their origin. Yet interpretation of this information is difficult enough, and the facts bared by each specimen are so comparatively few, that we need to recover every meteorite and every kind of meteorite that it is possible to recover, so that as each specimen adds to our fund of knowledge and each helps to make some single point clear, we will be able to understand from the aggregate some of the mysteries of space beyond our planet and more about our own planet's development.

Perhaps my greatest contribution to meteoritics has been the creation of a proper interest in the subject, and the suggestion of lines of investigation which, though at the time considered out-of-bounds, have since become worthy topics of discussion and study.

Dr. Gerard Kuiper, in about 1955 or 1956, invited me to contribute a chapter for Volume Four of the series on the solar system that he was editing: *The Moon, Meteorites and Comets*. When the 800-page volume was published in 1963, it included a dozen chapters on meteoritics, and Dr. Kuiper and his co-editor, Barbara M. Middlehurst, referred in the preface to "the almost explosive growth of this subject in recent years."[46] Several of these chapters concern aspects in which I attempted to arouse interest back in the 1930s and 1940s when publishers regarded them as too far out for print, thus many ideas I advocated for without much research to back them up came to be respectable topics for expert treatment.

A few years ago, a long-distance telephone call came from a researcher at Harvard University. He inquired about a statement I had made some three years earlier to the effect that a certain meteorite was a fragment and not a complete individual iron from space. I explained that my conclusion was based on what I had learned from studies over the years of the surface features of many meteorites, including the one in question. My caller then told me that nuclear

---

[46] Kuiper, Gerard P. and Barbara M. Middlehurst, eds., *The Moon, Meteorites and Comets*, University of Chicago Press, 1963.

studies of this specimen indicated I was correct, "But I wanted to know how you knew."

What I knew about meteorites, I had learned from observation and from applying as well as I could all the facts that I observed. For more than forty years, I directed most of my time and thought to one subject, one overwhelming interest—meteorites. It should not be surprising that some knowledge and new ideas resulted from such single-minded attention.

The recognition and occasional plaudits are now gratifying, but it is somehow ironic that it took a Sputnik and the sale of my life's collection to seal my verdict that meteorites *are* important.

In March 1965, Carleton Moore brought two scientists from the Ames Laboratory at Moffett Field to discuss meteoritical problems, to take with them some of the enigmatic specimens that we had been accumulating over the past forty years, and to learn if a newly developed instrument for detecting ancient exposures to cosmic rays might render definite conclusions as to whether or not these specimens were of meteoritic origin.

Ours is the age of mathematicians. Stimulated by the great discoveries of Einstein, men have sought to use mathematics almost to the exclusion of other equally important scientific methods, particularly the ways of observation. With the invention of the microscope, the biologists concentrated on ever smaller and smaller units, often to the neglect of the organism as a whole. With the perfecting of more and more powerful telescopes, the problems of deep space absorbed the attention of astronomers to the exclusion of work on members of our own solar system.

The old saying that one may become so concerned with the individual trees that he fails to see the forest illustrates quite well the behavior of man before the advent of the telescope. If the saying is reversed, it applies equally well today—constant advances in the power and efficiency of instruments with which to explore ever greater and greater reaches of space have led men to neglect some of the aspects of our surroundings that are nearby, but, like the trees, are essential to the forest.

By their very nature, many meteoritical problems must be solved by a combination of field work, laboratory tests, and mathematical analyses. There has been considerable printer's ink wasted in efforts to explain why meteorites are absent from all but the most recent geological formations. Much greater quantities of ink have been used to demonstrate by mathematical analyses that the meteorite did or did not explode to produce the great Arizona crater. The overall image of meteorite composition was for a century calculated on the assumption that collections in museums constituted an adequate sampling of the total increment of meteoritic material.

Portions of the beautiful pallasite of Springwater, Saskatchewan, now shine forth in various collections of the world only because a farmer was a better observer than was the provincial assayer who, as a student, never had been shown a sample of pallasite meteorite, and so judged the farmer's specimen to be merely leakage from a furnace. Because of the farmer's acuity and refusal to accept this mistaken judgment, the world acquired a specimen that bears a mineral hitherto unknown either from the earth or sky, discovered in 1960 and christened *Farringtonite* by E. R. DuFresne and S. K. Roy in the course of modern-day investigations into the nature of specimens gathered and preserved against the day when the meanings they hold for us should become clear.

The Canyon Diablo irons were first described in the early 1890s by men who observed, and who decided rightly that the surfaces of all of these irons were weathered surfaces, that any fusion crust they ever bore had long since flaked off due to oxidation. In my personal examination of many thousands of the irons, I have yet to see the least bit of fusion crust. They were probably well-covered by such crust when they fell 20,000 to 50,000 years ago, but even weather-resistant nickel-iron suffers from oxidation in that length of time. Yet there are men today who think that on those irons is still to be found the original fusion crust. These are men who have never developed powers of observation. They accepted without qualification the statement that fusion crust is a typical feature of meteorites, so rather than recognizing the ability of nature to obliterate natural features, they identify as fusion crust the brown film of oxidation that has replaced the original fused surface during Canyon Diablo's centuries of weathering.

It has not been many months since I was showing an eminent astronomer some small meteorites I had brought home from a recent field trip. He spoke of them as being nickel-iron and, when corrected, responded, "Well, *most* meteorites are nickel-iron, aren't they?" They are not.

Scientists from all parts of the United States and from many other countries walked around the Arizona crater environs over the years, seeing the big hole, viewing specimens of meteorites, reading past accounts of the crater's making. They walked without knowing it over impactite bomblets by the millions and over more millions of metallic spheroids—the two most meaningful classes of material, found in tremendous quantities, in the vicinity of the great crater. They did not observe that which they had not read about.

3

I have now lived long enough to be able to look at my early life somewhat objectively, and from this vantage point, I realize that certain characteristics, commendable or otherwise, have dominated my behavior. Nothing could have turned me from my chosen field of action. Stubborn? One of my friends put it more kindly. "You are one of the last rugged individualists," he told me.

I have had a tendency to try my hand at doing something someone said couldn't be done, and a tendency to question everything. These characteristics have affected my life all the way through, sometimes as afflictions, sometimes to advantage.

Men had studied meteorites many years before my time, but only as a special diversion from the professions of mineralogy, chemistry, or astronomy. The best of them tried to convince me that the subject was not worthy of a professional attack.

Mine was a strange, unique, and to many of my contemporaries, mysterious career, without precedent, without salary, without capital. There were no blazed trails to follow, there were no subsidies and no institutions to fall back on if my plan failed. The same goal was pursued steadily, sometimes painfully, through four decades. There were times when the financial fog got pretty thick, nearly obliterating the road ahead, but looking back down the zigzag trail, most of the rocky points seem to bear a mossy coat, many of the thorns share the stem with roses, and the shaky bridges by which quagmires were spanned appear steadier now.

Probably my "unique" career will have to remain unique. I was in the fortunate position of living in a time when it was possible for a man to make his way through some unusual pursuit, and I was privileged with the opportunity of studying all phases of meteorites—the means of their recovery, what they had meant to our earth, what they are made of, what tortuous descents through the atmosphere they had survived. In today's technological world, scientists are much more restricted to particular areas or kinds of research, and most will not experience the fun of the whole view, as I had it.

I insisted to myself that when the importance of my program became known, finances would be available for its completion. This has come to pass in time for me to be witness to it, but not in time for me to play as key a part as I had hoped.

At the last, by another of those strokes of good luck that so often salvaged something of my hopes and plans from destruction and disappointment, the disposition of the collection on which I had always built my plans itself provided the means to the end. At Arizona State University, where the bulk of the collection would reside, there would be a research program that would have some reflection in public education, despite the closed doors of my meteorite museum.

Each individual who spends his lifetime in natural science becomes wrapped up in that particular phase which claims his attention. There are no dull days for the man who devotes his eyes and mind to the lives of wild animals, fishes, reptiles, birds. His life is one continuous series of surprises as he finds and investigates new forms of life and uncovers new behaviors and characteristics.

Our new world offers more leisure. The natural sciences stand to play an important part in the occupation of that free time and in making it useful as well as enjoyable.

The field of fossils carries us back into the realm of infinite time; astronomy carries us out into a speculative pursuit of infinite space; but meteorites are concrete messengers out of both infinite time and infinite space.

Fossils occur in myriad forms, in almost numberless duplications in many instances, for a newly found species may run into hundreds or thousands recovered in a single year. Names of insects newly discovered in a year contribute substantially to the catalogs. New birds, new fishes, new reptiles are added regularly to the lists, although there are legions that are rare, even extremely rare, and many species and subspecies that will soon be extinct.

But every new meteorite brings not only its own substance to be described, but also a potential source of totally new information that may add new dimensions to our knowledge of earth and space.

What else is there that can give one such a sense of wonder as holding in one's hand an object that has come from space?

Perhaps some of the wonder is gone for this new generation that stands on the threshold of the sky, yet it seems to me that I have stood at the foot of the sky while man has propelled and thrust himself and his machines into its reaches, and my wonder has only increased. The sky stretches still unknown beyond where stars have led and beckons still into new fields with paths in all directions.

Glossary

achondrite: a stony meteorite without chondrules

aerolite: a meteorite in which metal is either absent or present in amounts of not more than twenty-five percent; also called stony meteorite

asteroid: a small planet-like body, sometimes called minor planet; several thousand of these move in a zone between Mars and Jupiter

astrobleme: ancient impact scar produced by a large meteorite, asteroid, or comet that has collided with the earth

chondrite: a stony meteorite in which chondrules are present

chondritic: an adjective descriptive of a meteorite or portion of a meteorite in which chondrules are present

chondrule: a small, rounded grain of mineral, very prominent in many stony meteorites

coesite: a high-pressure form of silica produced by meteorite impact

554 IN SEARCH OF FALLING STARS

fireball: a large meteor, appearing a quarter of the size of a full moon or larger; meteorites may fall from these

fusion crust: a thin layer of slag left on the surface of fresh-fallen meteorites by the frictional heat of boring through the atmosphere at speeds of miles per second

hexahedrite: a nickel-iron meteorite crystallized in a cubic, or hexahedral, structure

impactite: blobs of slag produced as a spatter of melted country rock mingled with bits of meteorite in connection with the production of meteorite craters

mesosiderite: a stony-iron meteorite, the metal of which is not in the form of a reticulum

metallic spheroids: small metallic droplets which have condensed from the vapor cloud which results from the explosion of a huge meteorite in the formation of a meteorite crater

meteor: the light phenomena caused by a meteorite's passage through the atmosphere

meteor shower: the appearance of many meteors in the sky over a brief period of time, often appearing to radiate from a single point in the sky, resulting from the earth's passage through the debris left behind by a comet

meteorite: a mass of solid matter, too small to be considered an asteroid; either traveling through space as an unattached unit, passing through the atmosphere, or having landed on the earth and still retaining its identity

meteorite crater: a rimmed depression produced by the explosion of an impacting meteorite

GLOSSARY   555

meteorite shower: the fall of numerous masses from the disintegration of a meteorite during its passage through the atmosphere

meteoritic dust: finely divided meteoritic substance resulting from atmospheric friction or impact

meteoritics: the systematic study of meteorites, meteors, and associated phenomena

octahedrite: a nickel-iron meteorite crystallized in an eight-sided, or octahedral, structure

olivine: an olive-green mineral found in pallasite meteorites; gem-quality crystals of olivine are known as peridot

oriented meteorite: a meteorite which has presented one face toward, and has been shaped by, the atmosphere on its earthward flight

pallasite: a stony-iron meteorite consisting of a reticulum of nickel-iron, the meshes of which are filled with olivine

shattercone: conical structures produced by shock in any of several varieties of hard and compact rock

siderite: a meteorite consisting mainly of nickel-iron alloys

stishovite: a high-pressure form of silica, produced by still greater pressure than that which produces coesite

stony-iron: a meteorite consisting of about equal parts of stone and metal; pallasite is one variety

tektites: small bits of natural glass thought to have formed during the impacts of large meteorites on earth

Widmanstätten figures: the pattern of crystallization which appears when the cut and polished face of an octahedrite is etched with acid

List of Photographs

1. Nininger farmstead near Clarkson, Oklahoma, c. 1906. Nininger Family Photo. Page 21.

2. The Niningers, c. 1907. Nininger Family Photo. Page 23.

3. Harvey Nininger and Harley-Davidson motorcycle, 1912. Nininger Family Photo. Page 40.

4. Harvey Nininger, Senior Photo, McPherson College, 1914. Nininger Family Photo. Page 47.

5. Harvey and Addie Nininger at camp in California. Nininger Family Photo. Page 52.

6. Harvey Nininger and Warren Knaus at Cedar Breaks, Utah, 1921. Nininger Family Photo. Page 72.

7. Rocky Mountain Summer School brochure. Nininger Family Photo. Page 78.

8. The Harvey Nininger Family, 1925. Nininger Family Photo. Page 101.

9. Custom-made house-car (the "Nininger Runabout"). Nininger Family Photo. Page 102.

LIST OF PHOTOGRAPHS 557

10. Making camp along the way on the house-car trip of 1925–1926. Nininger Family Photo. Page 104.

11. Celebrating Bob and Margaret's joint birthday on March 28, 1926, in Sabino Canyon, Arizona. Nininger Family Photo. Page 119.

12. Fossilized lion tracks recovered near Cornville, Arizona. Nininger Family Photo. Page 123.

13. Caravan of cars on the "Natural History Trek" of 1927–1928. Nininger Family Photo. Page 133.

14. A campsite during the Natural History Trek of 1927–1928. Nininger Family Photo. Page 136.

15. Group photo at Carlsbad Caverns, Natural History Trek, 1927. Nininger Family Photo. Page 140.

16. Caravan of cars at Okefenokee Swamp, Natural History Trek, 1928. Nininger Family Photo. Page 144.

17. At camp en route to Mexico City, 1929. Courtesy of American Meteorite Laboratory Photo Collection, Collections Research for Museums, Denver. Page 157.

18. In search of meteorites in the village of Xiquipilco, Mexico, 1929. Courtesy of American Meteorite Laboratory Photo Collection, Collections Research for Museums, Denver. Page 165.

19. The *barreta* (crowbar-like tool) made from a meteorite, purchased from a blacksmith in Xiquipilco, Mexico in 1929. Courtesy of American Meteorite Laboratory Photo Collection, Collections Research for Museums, Denver. Page 166.

20. Harvey Nininger and meteorite saw, c. 1930. Courtesy of Denver Museum of Nature & Science. Page 201.

21. Huizopa meteorite prior to its shipment to Denver, Chihuahua City, Mexico, 1931. Courtesy of Denver Museum of Nature & Science. Page 210.

558   IN SEARCH OF FALLING STARS

22. Draft horses and wagon used in 1950 search of pasturelands where the Springwater, Saskatchewan, Canada pallasite was found. Courtesy of American Meteorite Laboratory Photo Collection, Collections Research for Museums, Denver. Page 219.

23. Pasamonte, New Mexico, fireball of March 24, 1933. Courtesy of American Meteorite Laboratory Photo Collection, Collections Research for Museums, Denver. Page 242.

24. Charlie Brown, who captured the historic photo of the Pasamonte, New Mexico, fireball of March 24, 1933. Courtesy of Denver Museum of Nature & Science. Page 242.

25. Excavation of the Haviland, Kansas, crater, 1933. Courtesy of Denver Museum of Nature & Science. Page 271.

26. The excavation "crew" at the Haviland, Kansas, crater, 1933. Courtesy of Denver Museum of Nature & Science. Page 271.

27. Bob and Harvey Nininger standing beside "meteorodes" recovered from the Haviland, Kansas, crater, 1933. Courtesy of Denver Museum of Nature & Science. Page 272.

28. The Niningers at the Haviland, Kansas, crater, 1938. Courtesy of American Meteorite Laboratory Photo Collection, Collections Research for Museums, Denver. Page 273.

29. Bob Nininger using a magnetic balance at the Odessa Crater, Texas, 1935. Courtesy of Denver Museum of Nature & Science. Page 275.

30. Aerial view of the Arizona Meteorite Crater, looking south, c. 1937. Courtesy of American Meteorite Laboratory Photo Collection, Collections Research for Museums, Denver. Page 289.

31. Harvey Nininger boarding a train for a lecture tour, c. 1937. Nininger Family Photo. Page 297.

32. The Hugoton, Kansas, meteorite after excavation, 1935. Courtesy of Denver Museum of Nature & Science. Page 318.

LIST OF PHOTOGRAPHS 559

33. The Hugoton, Kansas, meteorite, wrapped in newspaper, burlap, and plaster, being loaded for transport, 1935. Courtesy of Denver Museum of Nature & Science. Page 319.

34. Harvey Nininger "exhuming" a "deadman" (weight used to anchor a fence), 1937. The Miami, Texas, meteorite. Courtesy of American Meteorite Laboratory Photo Collection, Collections Research for Museums, Denver. Page 322.

35. Harvey Nininger with the Goose Lake, California, meteorite in situ, 1939. Courtesy of Denver Museum of Nature & Science. Page 328.

36. The Goose Lake, California, meteorite ready for loading, 1939. Courtesy of Denver Museum of Nature & Science. Page 329.

37. The Goose Lake, California, meteorite on the wagon used to recover it, 1939. Courtesy of Denver Museum of Nature & Science. Page 330.

38. The Goose Lake, California, meteorite after being transferred to a pick-up truck, 1939. Courtesy of Denver Museum of Nature & Science. Page 330.

39. The Niningers' temporary quarters in a trailer below Mt. Elden, near Flagstaff, Arizona, during the summer of 1946. Courtesy of American Meteorite Laboratory Photo Collection, Collections Research for Museums, Denver. Page 376.

40. The American Meteorite Museum on Route 66 near the Arizona crater. Courtesy of American Meteorite Laboratory Photo Collection, Collections Research for Museums, Denver. Page 390.

41. Postcard showing the interior of the American Meteorite Museum on Route 66, c. 1947. Courtesy of American Meteorite Laboratory Photo Collection, Collections Research for Museums, Denver. Page 390.

42. Harvey Nininger at the American Meteorite Museum on Route 66, c. 1947. Courtesy of American Meteorite Laboratory Photo Collection, Collections Research for Museums, Denver. Page 392.

560   IN SEARCH OF FALLING STARS

43. Harvey Nininger lecturing to a school group at the American Meteorite
    Museum on Route 66, c. 1947. Courtesy of American Meteorite
    Laboratory Photo Collection, Collections Research for Museums,
    Denver. Page 394.

44. A 269-pound Arispe, Mexico, meteorite that was used as an anvil on a
    Mexican hacienda for thirty-five years. Courtesy of American
    Meteorite Laboratory Photo Collection, Collections Research for
    Museums, Denver. Page 401.

45. Harvey Nininger inspecting fragments of the Norton, Kansas,
    meteorite, 1948. Courtesy of American Meteorite Laboratory Photo
    Collection, Collections Research for Museums, Denver. Page 413.

46. Addie and Harvey Nininger in front of the American Meteorite
    Museum on Route 66, c. 1951. Courtesy of American Meteorite
    Laboratory Photo Collection, Collections Research for Museums,
    Denver. Page 427.

47. Addie Nininger at the sales counter of the American Meteorite Museum
    on Route 66, c. 1951. Courtesy of American Meteorite Laboratory
    Photo Collection, Collections Research for Museums, Denver.
    Page 428.

48. Bob and Harvey Nininger stand beside a magnetic rake constructed on
    a trailer and towed behind their Studebaker, 1939. Courtesy of
    American Meteorite Laboratory Photo Collection, Collections
    Research for Museums, Denver. Page 431.

49. Harvey Nininger collecting magnetics at the Arizona crater, c. 1951.
    Courtesy of American Meteorite Laboratory Photo Collection,
    Collections Research for Museums, Denver. Page 436.

50. Addie and Harvey Nininger with their dog, Blondie, at the American
    Meteorite Museum on Route 66, c. 1951. Courtesy of American
    Meteorite Laboratory Photo Collection, Collections Research for
    Museums, Denver. Page 451.

LIST OF PHOTOGRAPHS 561

51. The American Meteorite Museum in Sedona, Arizona. Courtesy of American Meteorite Laboratory Photo Collection, Collections Research for Museums, Denver. Page 452.

52. Interior of the American Meteorite Museum in Sedona, Arizona, c. 1954. Courtesy of American Meteorite Laboratory Photo Collection, Collections Research for Museums, Denver. Page 453.

53. Harvey Nininger in front of his home in Sedona, Arizona, 1961. Nininger Family Photo. Page 504.

54. Bondoc meteorite, Luzon, Philippines, 1962. Courtesy of American Meteorite Laboratory Photo Collection, Collections Research for Museums, Denver. Page 514.

55. Harvey Nininger, the Bondoc meteorite, and the homemade saw used to cut it, Sedona, Arizona, 1962. Courtesy of American Meteorite Laboratory Photo Collection, Collections Research for Museums, Denver. Page 517.

56. Addie and Harvey Nininger in their home in Sedona, Arizona, c. 1965. Nininger Family Photo. Page 532.

Original Dedication
Find a Falling Star

To the hundreds of school officials who, seeing the educational value of our program, gave encouragement to the new science of meteoritics.

To the farmers, ranchmen, and others who live close to the land and were willing to broaden the quest beyond measure.

To my long-suffering family for the support of their love, for their loyalty and their enthusiasm for the search that continues.

To my friends—hundreds of them—who have shared the talking and the seeking, and to a particular friend, Herbert G. Fales, who would not allow me to abandon this book.

And last but not least, to Mr. George Boyd, Coordinator of Research at Arizona State University, who grasped the full significance of our program and would not rest until he saw the fruits of our labors safely installed in that university.

Original Preface
Find a Falling Star

Everyone agrees that it is important that man understand something of the history of the earth and of the universe, particularly the Solar System. This book, and indeed the author's whole life, reflects in a small and modest way an attempt to help with that understanding.

It seems high time that man cease propagating the myth that the earth was created one afternoon about 6,000 years ago and that he gear his thinking to a recognition of the fact that he lives in the midst of a process the individual incidents of which are so widely separated in time and space that no one ever sees enough of it to adequately describe it; but by careful evaluation of recorded incidents, he may build a reliable concept of Nature, including himself, and his race, basking in the radiance of increasing knowledge and of the ingenuity behind the whole performance.

It is in this spirit that the investigations of the author and his wife, Addie, have been carried on for more than a half-century.

Although we were not able to get the necessary financial support from educational or research institutions for more than the equivalent

564   IN SEARCH OF FALLING STARS

of a few months of field work during our thirty most active years, I cannot refrain from expressing our gratitude for the moral support that came from many quarters. Letters and invitations to lecture in hundreds of colleges, universities, high schools, museums, etc. all provided an outlet for our message. Writers, who found in our program an exciting and rewarding prospect for the farming population, ranchers, hunters, fishermen and nature lovers. Hundreds of lucky finds kept the program alive.

Scientists who were especially helpful by way of encouragement were: Dr. Oliver C. Farrington, Head, Department of Geology, Field Museum; J. D. Figgins, Director, Colorado Museum of Natural History; Dr. Clyde Fisher, Curator of Meteorites, American Museum of Natural History; Prof. Curvin H. Gingrich, Professor of Astronomy, Carleton College; Dr. Frederick C. Leonard, Professor of Astronomy, UCLA; Dr. Forest R. Moulton, Permanent Secretary, A.A.A.S.; Prof. F. K. G. Mullerreid, University of Mexico; Prof. Harlow Shapley, Department of Astronomy, Harvard Univ.; Edward Steidle, Director, Penn. State College; Dr. Fletcher Watson, author of *Between the Planets*, Harvard Univ.; Dr. Fred L. Whipple, Department of Astronomy, Harvard Univ.; Dr. Alexander Wetmore, Director, US National Museum; Prof. E. J. Workman, Director, New Mexico School of Mines; and many others.

Writers who kept our activities before the reading public were: Neil M. Clark, *Saturday Evening Post*; F. C. Cross, magazine writer; Roscoe Fleming, magazine writer; Gene Lindberg, science writer, *The Denver Post*; Alexis McKinney, City Editor, *The Denver Post*; Ernie Pyle, roving reporter; Alan Swallow, publisher; and a host of newspaper editors.

Participants in field work were: H. G. Fales, Dean Gillespie, John Hilton, Irving Hoglin, Alfred Knight, Robert Nininger, Alex Richards, Jim Rothrock, Don Thompson, O. J. Walters, and the US

National Museum.

President D. W. Kurtz of McPherson College was a constant source of encouragement.

Very special credit goes to our daughter, Mrs. Doris Banks, for her weeks and months of editing the original manuscript.

H. H. Nininger
Sedona, Arizona
July 1972

Original Introduction
Find a Falling Star

H. H. Nininger and *meteorites* go together like word pairs such as *bread* and *butter*, at least for those who study these rare samples of cosmic debris. Meteorites were, indeed, bread and butter for the Nininger family, the first ever to survive by finding, collecting, trading, selling, and exhibiting meteorites. This book is a fascinating autobiographical account that clearly delineates the single-minded purposeful persistence of a man who refused to be diverted by seemingly overwhelming obstacles from pursuing an almost impossible dream. Nininger's early prophetic vision of meteorites as scientific Rosetta Stones has grown to reality in his later years, a development that he substantially furthered by his missionary effort.

I am delighted that Nininger's experiences are now preserved in this volume.

As one among the few with a personal active scientific interest in shooting stars or *meteors* for some four decades, I have been amazed that he could indeed survive while persisting in his unique profession. His secret, of course, is exposed in these pages—a rare ability to communicate his knowledge and his enthusiasm, with an engaging, rustic simplicity of approach. Applying this talent in the field, Nininger gains both the interest and the confidence of untrained rural people. If approached in a sophisticated or patronizing manner, they would become aloof. Furthermore, his basic honesty and integrity in his field activities stand out in his account, coupled, I must admit, with the shrewdness of a Yankee horse trader.

To illustrate his practical realism in the field, I recall his cooperation with me in a Harvard project at the White Sands Proving Grounds in New Mexico, December 1946. We were to photograph the first artificial meteors to be made by shaped charges in grenades exploded from a V-2 missile at high altitude. In his spare time while we were waiting, Nininger dug a "foxhole" as a personal safety precaution in case the unguided V-2 should land near our site, by no means an impossibility.

Nininger expresses his frustration at the lack of scientific interest in meteoritics before the middle of this century. I, too, shared that frustration, but in fact we both have profited by that lack of interest. It permitted us to work alone at our own pace in virgin scientific ground without much competition. Nininger wisely fails to recount some unhappy incidents where too vigorous competition actually thwarted him from receiving the just rewards of his efforts. Today, it is a pleasure and a satisfaction to see the tide turn as numerous young enthusiasts push forward the meteoritical research that we have long felt to be of vital importance.

It is not, however, just the search for solutions to practical problems of ballistics, missiles, satellites, and space probes—generated by the space age—that has produced the current impetus in meteoritical research, nor is it the availability of Moon samples. Advances in physics and chemistry have produced research techniques and methods of analysis so sophisticated and so subtle as to be almost miraculous when compared with the tools of research available early in this century. Today, the meteoriticist can analyze a sample the size of a pinhead to determine its composition and mineral content, its age since it was formed, since it cooled, and since it broke away from a larger parent body in space, its temperature and pressure at formation, and other details of its life history and, indeed, of the history of the solar system. Thus, meteorites are truly far more precious than diamonds, because they carry cryptic messages of happenings somewhere in the solar system more than four billion years ago. Oddly enough, we still cannot identify the parent body from which any meteorite was broken off.[47] Presumably, meteorites come from asteroids by collision, and certainly they are of solar system origin. But where? To answer even this simple question, we may have to send space probes to asteroids and possibly even to old comet nuclei. As we seek these answers, Nininger's life's work will serve as an increasingly important foundation. He has directed our attention both to the importance of meteorites and to the means of finding them. Now let us carry on from this accelerated start.

Thus, we all owe a debt of gratitude to H. H. Nininger for his successful educational program, both to the layman and to the professional, and for his material contribution of so many actual meteorites. In addition, we must thank him for telling us in such a vivid fashion the story of his life's work.

The life vividly exemplifies what he urged of his students, "Do something that needs doing."

Fred L. Whipple, Director
Smithsonian Astrophysical Observatory, Cambridge, Massachusetts

---

[47] This is no longer the case. Since the original publication of *Find a Falling Star*, the origins of certain meteorites have been traced back to the Moon, Mars, and Vesta (one of the largest objects in the asteroid belt). *Ed.*

569

Index

Names of meteorites are in **boldface** type.

Abernathy, 284, 369

Acme, 416

Admire, 180

Aguilera, Jose, 162, 516

Ake, Olin E., 325, 327

Alamosa, 196

Albareto, 190

Alberta, University of, 505, 528

Allen, H. Julian, 470

American Association for the Advancement of Science, 270, 348, 351, 508, 564

American Meteorite Laboratory, 204, 330, 501

American Meteorite Museum: on Route 66, 374, 378, 379-381, 385-398, 404-407, 415-428, 450-451; in Sedona, 451-454, 472-475, 494-495; value of, 421-423, 473-474, 494-495, 538 (see also: Nininger Collection)

American Museum of Natural History, 90, 122, 123, 564; meteorite collection, 224, 227, 371, 396, 564

American Philosophical Society, 375, 431, 434, 441

Ames Laboratory, 469-470, 547

Amherst College, meteorite collection, 154, 224, 227; tuition, 302-303

Anders, Edward, 455

Anderson, Pete, 243, 244, 245

Anvil (Arispe meteorite), 400-401

Archie, 231-235, 246, 335

Arispe, 139, 400-401

Arizona State University, 477, 495-496, 498, 500, 507, 508, 516, 551, 562

Arizona, University of, 118, 138, 139, 196, 374, 475

Artesia, New Mexico, 367

Asteroids (see: Collision geology)

Astroblemes, Ries Kessel, 501; Sudbury Basin, 287; Vredefort Structure, 361-362, 363

Ayres, Bob, 521, 523

Bachelay, Father, 190, 194

Bacubirito, 166

Bailey, Alfred M., 288, 372, 382

Baker, George, 491

Baldwin, Ralph B., 363

Ballinger, 128

Banks, Doris Nininger, 67, 76, 101, 119, 123, 136, 273, 303, 390, 565

Barnes, Virgil, 484

Barnett, G. L., 274, 276

Barreta - crowbar (Xiquipilco meteorite), 165-166

Barringer, 225, 375, 377-378, 418, 440; Brandon, 377; D. M., 265, 373-374, 429-430, 434, 446, 449; Lewin, 287

Beardsley, 220-222, 369

Benziger, Bill, 522, 523

Bethune, 369

Beyer, H. O., 485-487, 488

Bibbins, A. B., 269

Bible Institute at McPherson College, 22

Big Cypress Swamp, 144

Birthplace, 2

Blondie (dog), 391, 424-425, 451

Blunt nose (see: Missiles and space flight)

Bondoc, 508-516, 517

INDEX

Boogaldi, 465

Boy Scouts, 301

Boyd, George, 562

Boyer, Fred, 405

Brandmeyer, C. H., 516

Braunau, 295

Brenham, 92, 103, 172, 179-180, 259-264, 267-273 (see also: Haviland under Meteorite craters, Kansas Meteorite Farm, Meteorodes)

Briscoe County, 196

British Columbia, University of, 528

British Museum (Natural History), 209, 476-477, 480-482, 483, 500

Brown, A. M., 89, 96, 97

Brown, Barnum, 122

Brown, Charles M., 235-236, 240-243, 293

Brown, Harrison, 244

Bruderheim, 505-506, 531

Bruno, 217, 463, 465, 471, 545

California Academy of Sciences, meteorite collection, 224

California Institute of Technology, 497

California, University of (Berkeley), 51-52, 57

California, University of (Los Angeles), 226, 227, 326, 327, 499, 564

Cally, Sim, 257

Camp Verde, 196

Cananea, Mexico, 139

Candy business, 34-38

Canyon Diablo, 142-143, 179, 195, 196, 225, 296, 339, 369, 375-377, 385, 387, 402-404, 418-421, 422, 431, 442, 443-444, 548-549; diamonds in, 444, 455 (see also: Arizona under Meteorite craters)

Canyon Diablo #2, 444

Carleton College, 303, 564

Carlsbad Caverns, 99, 117, 140

Casas Grandes, 197

Casey, John C., 211

Casey, Lee, 286

Cedar Breaks, 71-73, 74

Center for Meteorite Studies, 507

Chabot Observatory, 326

Chamberlin, T. C., 350

Charlotte, 465

Chicago, University of, 475, 507

Chihuahua City, 166

Chilcoot, 197

Chladni, E. F., 194

Chupaderos, 166

Church of the Brethren, 6-8, 32-33, 48, 263

Claringbull, G. F., 480

Clark, Neil M., 564

Coes, Jr., Loring, 454

Coldwater (iron), 89, 95, 96, 97

Coldwater (stone), 89-90, 95, 96, 97

Collision geology, 266, 359-364, 441-442, 506-507, 533

Colorado Museum of Natural History, 199, 204, 212, 270, 285, 288, 350, 351, 382, 564 (see also: Denver Museum of Natural History)

Columbia University, meteorite collection, 224

Composition of meteorites, 444; chondrules, 176, 530, 535, 553; copper, 211, 212, 213, 215; diamonds, 444, 455; gold, 243, 244, 245; olivine, 264, 267, 268

Conly, Frank, 286

Corpus Christie (meteorite sales), 107-108

Cotesfield, 196

Cottonwood, 368

Covert, 170, 305

Cragen, F. W., 261, 262

Craik, E. L., 83, 84, 286

Cross, Frank Clay, 285, 286, 351, 564

Crow, H. E., 174, 177

Cullison, 306

Cumberland Falls, 215

Cummings, Byron, 138

Dachille, Frank, 364

Daly, Reginald, 361, 362

Danforth puzzle, 342-344, 405

Dashani, Dr., 472, 473

Davis, Alexandro, 519, 520, 522

Dawson, George E., 195, 196

Denver, University of, 347, 350

Denver Museum of Natural History, 353, 354, 371-372 (see also: Colorado Museum of Natural History)

Department of Mines and Special Surveys (Canada), 362

Detrick, Herbert J., 369-371

Detrick, Lula, 369-371

Diaz, Porfirio, 155

Dietz, David, 286

Dietz, Robert S., 287

Dollarhide, R. A., 314, 315

Dominion Astrophysical Observatory (Canada), 528-529

Doyleville, 223

Drake University, meteorite collection, 224

Draeger, Harold, 404

Du Fresne, E. R., 548

Dunkards (see: Church of the Brethren)

Durala, 197

Early life, Kansas, 2-4

Early life, Missouri, 4-9

Early life, Oklahoma, 9-23

Eaton, 211-216

Ebel, B. E., 74, 75, 76

Ebner, A. D., 217

Elkhart, 196

English, George L., 294-296

572 IN SEARCH OF FALLING STARS

Enon, 285

Entomology (master's degree), 50-51

Estacado, 228, 402

Estemere (Rocky Mountain Summer School), 76, 77, 78

Estherville, 213-214, 335, 461-463, 465, 469, 470, 471, 545

Evans, John, 230

Everley, I. C., 325, 327

Evolution (teaching), 47-50

Explosion products, 266, 272, 273, 430-437, 447; coesite, 454, 455, 501, 553; impactite, 439-440, 441, 445, 455, 549 554; metallic spheroids, 421, 436-437, 445, 549, 554; shattercones, 287, 555; stishovite, 455, 555 (*see also:* Arizona *under* Meteorite craters)

Fales, Herbert, vii, 287-289, 507, 562, 564

Farmington, 306

Farringtonite, 548

Farrington, O. C., 91, 128, 129, 172, 173, 215, 225, 226, 266, 304-305, 313, 580

Fenner, Charles, 491

Fermi Institute, 220, 455

Field Guide to the Birds of Central Kansas, 53

Field Museum of Natural History, 91, 172, 347, 564; meteorite collection, 81, 224-225, 227, 396

Figgins, J. D., 212, 241, 270, 371-372, 560

Fireballs, 83-84, 187-188, 190-191, 220, 233-234, 235-238, 251-258, 293-294, 300, 342, 349-350, 369-370, 394-395, 399, 455-459, 497-498, 554; dust clouds, 233, 236, 238, 241, 246, 255-256, 293-294, 324, 362-363, 370, 462; sound phenomena, 84, 87, 180, 187-188, 233-234, 245, 250, 253, 254, 257, 300, 324, 370, 459; surveys, 83-86, 180-183, 231, 233-235, 235-243, 246-251,

251-256, 283, 455-459, 528-532 (*see also:* Prairie Network)

Fisher, Clyde, 546

Fleming, Roscoe, 286, 564

Fletcher, Joe, 183

Foote, A. E., 162, 222, 422, 446

Foote, W. M., 335

Ford, William E., 225, 461, 463

Foreign travel, 479, 483-493, 500-503, 516-523, 528

Fort Hays State College (Kansas), 358

Fossil meteorites, 355-358

Fossil tracks, 120-123, 124, 125, 138-139, 140-141

Foster, W. H., 211-213, 214

French, Bevan, 287

Freed, J. K., 93

Friends University, 174

Gaines, W. Roy, 220-221; (Mrs.) 220

Gard, O. P., 239

Garnett, 213

Giacobinnid-Zinner meteor shower, 385-386

Gilbert, G. K., 265, 266, 446

Gilgoin, 228

Gillespie, Dean, 206-208, 225, 274, 292, 314, 351, 352, 431, 564; meteorite collection, 208

Gilmore, C. W., 124-125

Gingrich, C. H., 564

Gladstone, 320

Glen Rose (*see:* **Rosebud**)

Goddard, R. H., 459-461

Goose Lake, 325-330, 500

Grand Canyon, 124, 134

Grasshoppers, 4, 42-45

Grinnell College, 303

Groves, Guy, 243-244, 245

Hahn, Harold, 411, 412-413

Haight, Tom, 20, 23

Harnley, H. J., 43-44

Harrisonville, 231-232

Harris's hawk (Mexico), 157, 167

Harvard University, 292-293, 430, 476, 496, 498, 546, 564; meteorite collection, 154, 224, 226, 227

Haviland, 369 (*see also:* Haviland *under* Meteorite craters, Kansas Meteorite Farm, Meteorodes)

Hawley, F. G., 244, 269, 436

Henderson, E. P., 477-480

Henning, Lloyd, 335

Hey, Max, 476

Higgins, Bill, 458

Hilton, John, 399-400, 564

Hilton, William A., 50

Hinsley, Floyd, 521

Hinsley, Lois, 521

Hisey, Sam, 314-315

History of meteorites, 187-194

Hoagland, Peter, 331

Hodges, W. H., 183

Hoglin, Irving, 431, 564

Hokanson, Elmer, 385

Holbrook, 335-336, 369

Horse Creek, 196

House-car trip, 99-127

Howland, Frank, 212

Hubbard, Wendell 316-319

Huff, Dean, 282-283

Hughes, Ivan, 331-332

Hugoton, 317-319, 382, 384, 387, 481

Huizopa, 208-210

Hummiston, Lee, 213

Huss, Glenn, 453-454, 474, 493, 494, 495, 500-501

INDEX

573

Huss, Margaret Nininger, 67, 76, 101, 118, 119, 273, 303, 375-376, 391, 425, 453, 474, 494, 495, 500-501

Instituto Geologica (Mexico), 160, 209

International Nickel Co., 287

Iowa, University of, 226; meteorite collection, 224

Iron Creek, 197

Ivory-billed woodpecker, 144, 145

Jefferson, Thomas, 188-189

Jerome, 91

Jerome, Arizona, 118, 119, 120, 138

Johnstown, 324-325, 354

Jones, Barton, tektite collection, 490

Kaaba, 193

Kansas Meteorite Farm, 261-264, 270 (see also: **Brenham**)

Kansas State Agricultural College (Manhattan Kansas), 66

Kansas, University of, 89, 92-94

Keller, Glenn, 521, 523

Kelly, Allan O., 364, 489

Kentucky , University of, 80

Kimberly, Frank, 260, 261, 262, 263, 267, 268, 269, 270

Kimberly, Mary, 260, 261, 262, 263, 267, 268, 270 306

King Ranch, 109, 112, 141

Klopping, Edd, 65

Knaus, Warren, 67, 68, 69, 70, 71, 72, 73

Knight, Alfred, 564

Kuiper, G. P., 475, 507, 546

Kurtz, D. W., 66-67, 96, 128, 565

Lafayette, 313, 463-464

L'aigle, 190-191, 194

La Lande, de, Jerome, 190, 193

Lanfor, Fidel, 240

LaPaz, Lincoln, 408, 410, 412, 413

LaVerne College, 46, 47-50, 51, 53, 56

Lednicky, John, 511, 512-515, 516

Lee, Chester, 288

Leeds, 312-313

Lehman, John, 145

Leonard, F. C., 226, 326, 327-328, 330, 335, 336, 346-347, 412, 499-500, 564

Lindberg, Gene, 286, 564

Linsley, Earl G., 326, 327

Lipschutz, Michael, 455

"Little brown jugs," 15-18

Loomis, F. B., 302-303

Loreto, 516, 517, 519, 520, 521-528

Lost Lake, 196

Lynch, Jr., John D., 317-318, 319

Magnetic rake, 431, 432, 433, 505, 506

Mahieux, Peter, 517-518

Marburg, 503

Markman, H. C., 212

Marriage, 46

Marsland, 298

Martin, H. T., 89, 90-91

McKinney, 228

McKinney, Alexis, 286, 564

McPherson College, 22, 32, 33, 34, 35, 38, 41, 42, 46, 47, 48, 66, 67, 74, 80, 82-83, 131, 133, 199, 203, 269, 369, 399, 410, 508, 565

McPherson Scientific Expedition of 1921, 67-74

Meade, Judge J. A., 321

Means, Evans, 140-141

Melbourne, University of, 491

Melrose, 236, 243-245

Mendoza, Señor, 509, 510, 512

Merrill, G. P., 95, 172, 173, 174, 176,

177, 178, 226, 266, 278, 535

Meteor, defined, 82, 540-541, 554; shower, 385-386, 554 (see also: Fireballs)

Meteorite, defined, 82, 540-541, 554; shower, 555; varieties, 92-93, 215, 220, 240, 263, 295, 493, 503, 518, 530, 535, 548-549, 553-555

Meteorite collections, 154, 224, 225, 226, 227, 228, 289, 290, 397, 419, 421, 422, 423, 498, 499, 500, 501, 502 (see also: Nininger Collection)

Meteorite craters, 266, 267, 287, 288, 359-362, 441, 446, 448, 449, 532, 543, 544; Arizona, 125-126, 225, 265-266, 288, 289, 340, 341, 373-378, 379, 380, 402-404, 406, 407, 418, 419, 420, 425, 428-449, 450, 454-455, 477, 530, 548, 549; Boxhole, 492; Dalgaranga, 492-493; defined, 554; formation of, 276, 429-431, 434, 435, 436, 439, 440, 443-445; Haviland, 267-273, 277, 351, 355; Henbury, 270, 439, 492; lunar, 357, 360, 361, 363, 507, 533; on Mars, 545; Odessa, 269, 273-277, 288; Wabar, 439; Wolf Creek, 492 (see also: Astroblemes)

Meteorite Recovery Program, 96, 152, 154, 203-205, 228, 232, 233, 284-286, 291-299, 305-311, 338-342, 417-418, 424, 536-540, 563-564; interest factor, 305, 306-307, 537; instruments, 218, 264, 265, 274-276, 377, 402-403, 404, 416, 431-432; personal finds, 367-369 (see also: surveys under Fireballs, specific meteorites)

Meteoritical Society, 347, 498 (see also: Society for Research on Meteorites)

Meteoritic increment, 534-540; dust, 336-337, 345-346, 483-484, 506, 529-532, 536, 555; overlapping falls, 283-284; ratio of stones to irons, 129, 304-305, 354, 534, 537

Meteoritics, defined, 128, 172, 555; future of, 539-540; growth of, 226, 227; importance of, 178, 396-397, 461-472, 506-508, 540-552

574 IN SEARCH OF FALLING STARS

Meteorodes, 268, 269, 270, 271, 272, 355

Mexican National Railroad, 167-169

Mexico expedition (1929), 155-169

Mexico, University of, 523, 564

Miami, 321-322

Michigan, University of, 224

Middlehurst, Barbara, 546

Miller, A. M., 80, 81, 82, 94

Miller, C. F., 196

Miller, Loye Holmes, 52, 53, 54

Mills, Glen, 211

Minnesota, University of, 224, 331

Missiles and space flight, 461-472

Mizzell, Hamp, 144

Modoc (1905), 93, 300, 306

Mohole Project, 542

Monnig, O. E., 227, 277, 290, 408

Moore, Carleton B., 507, 516, 523, 547

Morland, 314-315, 382, 481, 482

Morris, Charles, 84, 85

Moulton, F. R., vii, 133, 175, 176, 178, 179, 181, 182, 237

Mount Elden, 375, 376, 432

Muller, Hermann J., 141-142

Mullerried, F. K. G.,160, 161, 162, 163, 164, 165, 564

Muroc, 196-197

Muroc Dry Lake, 196-197

Museo Historia Natural del Estado Chihuahua, 209

Museum of South Australia, tektite collection, 491

Museum of Western Australia, 493

National Academy of Science, 91, 351, 497

National Bureau of Mines (Philippines), 509, 510, 511, 513

National Museum (Mexico City), 161, 166

National Science Foundation, 476, 477, 498-499

Natural History Trek, 131-151

Navajo, 197

Nebraska, University of, 413; meteorite collection, 224

Nedagolla, 197

Ness County, 179, 180

Newsom, 196

New Mexico School of Mines, 564

New Mexico, University of, 408, 413, 414, 418

Nininger, Addie D., 46, 47, 50, 51, 52, 53, 57, 60, 66, 67, 76, 99-100, 101, 103, 104, 105, 106, 110, 111, 116, 120, 124, 127, 128, 130, 131, 133, 135, 136, 139, 141, 147, 148, 181, 184, 198, 205, 216, 218, 219, 230, 231, 246, 263, 273, 292, 309, 325, 326, 327, 335, 336, 338, 366-367, 371, 373, 375, 376, 388, 391, 392, 404, 405, 406, 407, 408, 409, 412, 413, 417, 418, 419, 423-424, 426, 427, 428, 432, 451, 452, 474, 480, 482, 488, 492, 493, 495, 500, 501, 504, 505, 516, 517, 521, 528, 532

Nininger Award, 507

Nininger Collection, 222, 227, 228, 352, 372-373, 374, 395, 406, 424, 472-473, 474, 507, 551; moving of, 382-385, 451, 481-482; sale of, 476-482, 495-496, 498-501 (*see also:* American Meteorite Museum)

Nininger, Edgar, 246, 247

Nininger, Jake, 2, 9, 21, 23, 25, 31, 34-35

Nininger, James (Harvey's father), 4, 9, 10, 21, 22, 33

Nininger, John, 2, 9, 10, 21, 23, 47, 278, 279, 280, 281, 283

Nininger Laboratory, 204

Nininger, Mary Ann (Harvey's mother), 2, 5, 6, 9, 10, 21, 22, 23, 53

Nininger, Naomi, 2, 9, 21, 23, 216

Nininger, Robert (Bob), 65, 76, 101,

106, 118, 119, 121, 122, 124, 135, 141, 145, 235, 236, 237, 270, 272-273, 274, 275, 276, 301-303, 309, 316, 317, 318, 322, 431, 452-453, 500, 501, 546

Nininger, Roy, 2, 3, 9, 23, 32-33, 34, 35

Nininger Runabout (*see also:* house-car trip), 100, 101, 102, 103, 104, 105

Niningerite, 125

Niswanger, Ray, 199

North American Air Defense Command, 456

Northwestern State Normal College at Alva, 23, 24-32, 39-41

Norton County, 215, 407-415

Notes from a Night's Camp, 290-291

Odessa, 274, 275, 276, 369 (*see also:* Odessa, *under* Meteorite craters)

Okefenokee Swamp, 144-147

Olivier, Charles P., 226

Oroville, 224

Ortiz, I. Andres, 210

Ownership of meteorites, 230-231, 331-332

Palache, Charles, 226

Palmer Lake, Colorado, 74, 76, 77, 78, 127, 132

Pantar, 369-371, 508

Paradise, Kansas (lectures), 103, 169, 170

Paragould, 180-185

Parkinson, Raymond, 180, 181, 182, 183, 184

Pasamonte, 235-243, 257-258, 283, 362-363, 462, 463, 465

Passenger pigeons, 54-55

Peabody Museum (*see:* Yale University)

Peary, Admiral, 197, 371

Peña Blanca Spring, 215

INDEX 575

Pennsylvania State College, 506, 564

Penrose, Spencer, 352

Perry, Stuart H., 181, 220, 227

Perseids, 192-193

Philippines, University of the, 485

Plainview, 278-284, 369, 535

Pojoaque, 196

Pomona College, 50, 51, 374, 508

Port Orford, 230-231

Potter, 323

Prairie Network, 498, 543

Proposed National Research Program, 293, 348-350, 351-353, 474, 496-497 (see also: Prairie Network)

Puente Del Zacate, 166

Puente-Ladron, 367-368

Pyle, Ernie, 286, 564

Quade, Dr., 524

Quinn, John B., 517, 519

Ramdohr, P., 502-503

Rancho De La Presa, 166

Red River, 197

Richards, Alex, 139, 155, 156, 157, 160, 161, 163, 167, 169, 170, 201, 273, 309, 316, 320, 333, 453, 564

Richardton, 179

Risinger, Charlie, 122

Roach, Otto, 288

Rochester, University of, meteorite collection, 227-228

Rock, Kenneth, 148-150

Rocky Mountain Summer School, 74-79, 99, 127, 132, 199, 200

Rosebud, 174-178, 313

Rothrock, J. H., 213, 564

Roy, S. K., 548

Sabetmahet, 197

St. Genevieve County, 296-297

St. Louis Academy of Science, meteorite collection, 225-226

Santa Rosalia, 518

Saonlod, 197

Saurin, Edmond, 487

Saw, meteorite, 172, 199-201, 207-208, 481, 516, 517

Schmidt, Clarence, 325, 326-327

Secco, Joseph, 325

Sedona, Arizona, 391, 451, 452, 475, 476; new home in, 493-494, 495, 503-504, 505, 532 (see also: American Meteorite Museum)

Selga, Father Miguel, 509

Shapley, Harlow, 293, 294, 361, 430-431, 564

Shepard, C. U., 224, 226

Shirk, Claude, 38-39

Siblings, 2

Sikhote-Alin, 361, 419, 448

Simonds, F. W., 175, 176, 177

Sioux County, 245-251

Smith, J. L., 226

Smithsonian Astrophysical Observatory, 498, 543

Smithsonian Institution (see: United States National Museum)

Society for Research on Meteorites, 347, 348, 407, 466 (see also: Meteoritical Society)

South Dakota, 58, 59, 60; Brookings, 58, 59, 60, 65-66; blizzard, 60-65

Southwestern College (Winfield, Kansas), 66

Spencer, L. J., 270, 484

Sprague, George, 342-343

Springfield, 196

Springwater, 217, 218-220, 548

State Museum of North Carolina, meteorite collection, 224

Steidle, Edward, 564

Sternberg, George, 358

Stevens, G. W., 20, 26, 28, 29, 30, 31, 32, 39-40, 41, 53, 54, 69

Stockwell, H. O., 264-265, 273, 276, 408, 413

Struve, Otto, 419

Surface features of meteorites, 212, 225, 461, 465-467, 494, 546-547; flight markings, 215, 462-464, 465, 466, 478; fusion crust, 240, 251, 294, 313, 336, 337, 502, 530, 548-549, 554; orientation, 217, 313, 462-472, 555; pitting, 212, 313, 466; velocity factor, 214, 463-465, 468, 469-470

Swallow, Alan, 564

Tanada, Senator, 509

Tanada, Señor, 509, 510, 511, 512

Tektites, 484-486, 506-507; Australites, 489-492; defined, 555; Indo-Chinites, 487-489, 503, 506; lunar origin, 485, 507

Thompson, Arthur, 285, 343, 344, 392

Thompson, George (Don), 392, 393, 404, 405, 406, 417, 418, 423, 564

Thompson, Miz, 392

Thompson, Ruth, 392, 393, 406, 417, 418, 423

Thornhill, W. W., 321-322

Tilghman, B. C., 434

Tlacotepec, 166

Toluca (see: **Xiquipilco**)

Tonganoxie, 92

Townsend, A. J., 196

Tremaine, Burton, 375, 377, 378, 418

Trenton, 264-265

Triavestigia Niningeri, 125

Troili, D., 190, 194

Tucson, 401

Tufts College, meteorite collection, 224

Tulia, 175

Udden, J. A., 175, 176

United States Bureau of Entomology, 58, 66

United States Geological Survey, 265, 446, 507; Laboratory of Astrogeology, 543

United States National Museum (Smithsonian Institution), 103, 148, 172, 173, 174, 176-177, 199, 202, 203, 213, 227, 230, 231, 245, 277-278, 326, 329-330, 397, 460, 477, 478, 479, 481, 564, 565

Urey, Harold C., 497

Venus (daylight star), 19-20, 135

Walters, O. J. (Monte), 402, 403, 564

War Production Board, 365

Ward, H. A., 90, 162, 222, 422

Ward's Natural Science Establishment, 198, 201, 202, 227, 228, 294, 295, 425

Wash stand (Estacado meteorite), 402

Washburn College, 261

Wasson, W. F., 285

Waters, Jack, 174-175

Watson, Fletcher, 539, 564

Weathering of meteorites, 353-359, 376, 545, 548-549; oxide, 376, 421, 432, 434, 435, 436, 493; (see also: Meteorodes)

Webb, Robert, 327

Western Reserve University, meteorite collection, 224

Weston, 187-189, 191, 227

Wetmore, Alexander, 230, 564

Whelan, Ab, 377, 416

Whipple, Fred, 476, 497, 564

White, Dale, 59, 60, 64, 65

Whitney, W. T., 374

Whooping cranes, 55-56, 112-115, 141

Wichita County, 197

Widmanstätten figures, 161, 165-166, 312, 334, 555

Willamette, 197, 331-332

Wilmot, 368-369

Winchell, Horace, 224, 331

Wisconsin, University of, meteorite collection, 224

Workman, E. J., 564

World War I, 57, 58, 59

World War II, 208, 365, 371, 468, 509

Wylie, C. C., 226

Xiquipilco, 161-166, 334, 369

Yale University, 188-189, 191; meteorite collection, 91, 154, 224, 227, 461, 463

Yellowstone National Park, 133-134

Yerkes Observatory, 475, 507

Yohe, Homer, 250

Zacatecas, 166

Zion Canyon, 135